# Space Technology Proceedings

Volume 7

The Space Technology Proceedings series publishes cutting-edge volumes across space science and the aerospace industry, along with their robust applications. Explored in these conference proceedings are the state-of-the-art technologies, designs, and techniques used in spacecraft, space stations, and satellites, as well as their practical capabilities in GPS systems, remote sensing, weather forecasting, communication networks, and more.

Interdisciplinary by nature, SPTP welcomes diverse contributions from experts across the hard and applied sciences, including engineering, aerospace and astronomy, earth sciences, physics, communication, and metrology.

All SPTP books are published in print and electronic format, ensuring easy accessibility and wide visibility to a global audience. The books are typeset and processed by Springer Nature, providing a seamless production process and a high quality publication.

To submit a proceedings proposal for this series, contact Hannah Kaufman (hannah.kaufman@springernature.com).

More information about this series at http://www.springer.com/series/6576

H. Paul Urbach • Qifeng Yu
Editors

# 6th International Symposium of Space Optical Instruments and Applications

Delft, the Netherlands, September 24–25, 2019



Springer

*Editors*
H. Paul Urbach
Optics Research Group
Delft University of Technology
DELFT, Zuid-Holland, The Netherlands

Qifeng Yu
College of Aerospace Science and Engineering
National University of Defense Technology
Changsha, Hunan, China

ISSN 1389-1766
Space Technology Proceedings
ISBN 978-3-030-56487-2 ISBN 978-3-030-56488-9 (eBook)
https://doi.org/10.1007/978-3-030-56488-9



This Springer imprint is published by the registered company Springer Nature Switzerland AG
The registered company address is: Gewerbestrasse 11, 6330 Cham, Switzerland

# Preface

In recent years, space optical payloads are advancing towards high spatial resolution, high temporal resolution, high radiometric resolution, and high spectral resolution and becoming more and more intelligent. Commercial remote sensing industry has made steady progress in terms of the scope of satellite systems and applications. Meanwhile, space optical remote sensing data has been extensively applied to monitoring of resources, meteorology, ocean, environment, disaster reduction, and many other fields.

On September 24–25, 2019, the Sixth International Symposium of Space Optical Instrument and Application, sponsored by the Sino-Dutch Joint-Laboratory of Space Optical Instruments, was held in Delft, the Netherlands. Like previous years, the joint-laboratory continuously put efforts into encouraging international communication and cooperation in space optics and promoting innovation, research, and engineering development.

The symposium focused on key innovations of space-based optical instruments and extensive exchanges on spectrometer design, calibration, optical manufacturing, as well as the demand for air quality and climate monitoring, and achievements made by China and Europe in space optical instrumentation, which have laid a foundation for follow-up technical cooperation.

There were more than 100 attendees at the conference. The speakers were mainly from Chinese, Dutch, and other European universities, space institutes, and companies.

The theme of this year's symposium is on utilizing remote sensing of atmospheric composition for climate change and air quality applications.

The main topics included:

- Space optical remote sensing system design;
- Advanced optical system design and manufacturing;
- Remote sensor calibration and measurement;
- Remote sensing data processing and information retrieval;
- Remote sensing data applications.

The joint-laboratory will continue to strengthen technical cooperation and personnel exchanges between the Chinese and Dutch space industry, promote cooperation through exchanges, and reach win–win outcomes through cooperation, so as to boost the development of space science and technology.

Delft, The Netherlands H. Paul Urbach
Changsha, China Qifeng Yu

# Organization

第六届空间光学仪器与应用国际研讨会 ISSOIA
6th International Symposium of Space Optical Instrument and Application

Hosted By:

SINO-DUTCH JOINT-LABORATORY OF SPACE OPTICAL INSTRUMENT

Organized By:

中国航天 北京空间机电研究所 Beijing Institute of Space Mechanics& Electricity

TNO

SRON

TU Delft

**Venue:**

Science Centre Delft
Mijnbouwstraat 120, 2628 RX Delft

**Date:**

Sep 24th and 25th, 2019

**Chairman:**

H. Paul Urbach, TU Delft
Qifeng YU, NUDT

**Executive Chairman:**

Kees Buijsrogge, TNO
Bin FAN, BISME

**Secretary-General:**

Andrew Court, TNO
Peng XU, BISME
Avri Selig, SRON

**Organizing Committee:**

Jing Zou, TNO
Sandra Baak, TNO
Yvonne van Aalst, TU Delft
Weigang WANG, BISME
Yue LI, BISME

# Contents

# On-Orbit Data Verification and Application of FY-4A Lightning Imager

**Tang Shaofan, Liang Hua, Bao Shulong, Cao Dongjie, and Lu Zhijun**

**Abstract** The FY-4A Lightning Imager (FY-4A LMI) is the optical payload of lightning detection in geostationary orbit developed by China (and also the first optical payload of lightning detection). It is one of the first two lightning imageries of geostationary meteorological satellites in the world and fills the gap of space-based lightning optical detection in China. FY-4A was launched on December 11, 2016. Other indicators are basically consistent with European MTG-LI and American GOES R-GLM. The LMI field of view covers China and its surrounding areas. The real-time observation of total lightning in China and its surrounding areas can solve the lack of observation ability in mountain areas and oceans. The real-time observation of severe convective weather phenomena can provide a basis for lightning forecast and global climate change research and also provide a basis for the strong convective weather in the mainland of China and its surrounding areas. Observing, forecasting and researching provide necessary scientific data and provide scientific basis for the research of global atmospheric circulation. The success of FY-4A LMI is a major technological breakthrough in lightning space-based detection in China. Lightning imager uses near-infrared spectral channel detection, spectral bandwidth of 1 nm, two lenses for field-of-view mosaic and high-speed (2 ms) imaging detection, with real-time lightning event processing capability on board, and thus the overall performance of the camera is excellent. The whole process of occurrence, development, movement and extinction of strong convective thunderstorm system in China has been observed many times during on-orbit test. On-orbit test shows that lightning imager can detect lightning correctly and realize real-time monitoring, tracking and early warning of severe convective weather system. This paper aims at validating LMI on-orbit algorithm and data application. The correctness of satellite lightning observation can be seen by comparing with ground load.

T. Shaofan (✉) · L. Hua · B. Shulong · L. Zhijun
Beijing Key Laboratory of Advanced Optical Remote Sensing Technology, Beijing Institute of Space Mechanics and Electricity, Beijing, People's Republic of China

C. Dongjie
National Satellite Meteorological Centre, Beijing, China



H. P. Urbach, Q. Yu (eds.), *6th International Symposium of Space Optical Instruments and Applications*, Space Technology Proceedings 7,
https://doi.org/10.1007/978-3-030-56488-9_1

**Keywords** Geostationary orbit · Lightning · On-orbit algorithm verification · Meteorological

# 1 Introduction

FY-4A meteorological satellite is the first satellite of the second-generation geostationary orbit meteorological series independently developed by China. It was successfully launched at the Xichang Satellite Launch Center on December 11, 2016, and fixed at 99.5° east longitude equator on December 17, 2016. Four main loads of FY-4A are: multi-channel scanning imaging radiometer, interferometric atmospheric vertical detector, lightning imager, space environment monitoring instrument package, etc. Among them, lightning imager is mainly used for real-time observation of total lightning in China and its surrounding areas to solve the lack of observation ability in mountain areas and oceans. Lightning imager is the first satellite lightning imager developed in China and one of the first two geostationary satellite lightning imageries in the world. The success of FY-4A LMI is a major technological breakthrough in China's lightning space-based detection. Lightning imager uses near-infrared spectral channel detection, spectral bandwidth of 1 nm, two lenses for field-of-view mosaic and high-speed (2 ms) imaging detection, with real-time lightning event processing capability on board, and thus the overall performance of the camera is excellent. Lightning imager requires a lot of testing and calibration before and after launch, including focusing, stray light, crosstalk, solar flashing, dark level drift, gain, noise, linearity, dynamic range, cluster filtering and false alarm removal related to lightning characteristics. During the on-orbit test of FY-4A LMI, the whole process of occurrence, development, movement and extinction of the strong convective thunderstorm system in China has been observed many times. The on-orbit test shows that the lightning imager can detect lightning correctly, realize real-time monitoring [1], tracking and early warning of the severe convective weather system and compare the lightning observation data between satellite and ground. The correctness of satellite lightning observation is verified.

# 2 Brief Introduction of Lightning Imager

## 2.1 Principle of Lightning Detection

The emergence of lightning is accompanied by the sudden release of electric energy, which is converted into the rapid heating of the gas around the lightning channel, which generates shock wave and electromagnetic radiation. One of the strongest radiation regions is in the optical wavelength. This optical radiation originates from the dissociation, excitation and close combination of the elements of the

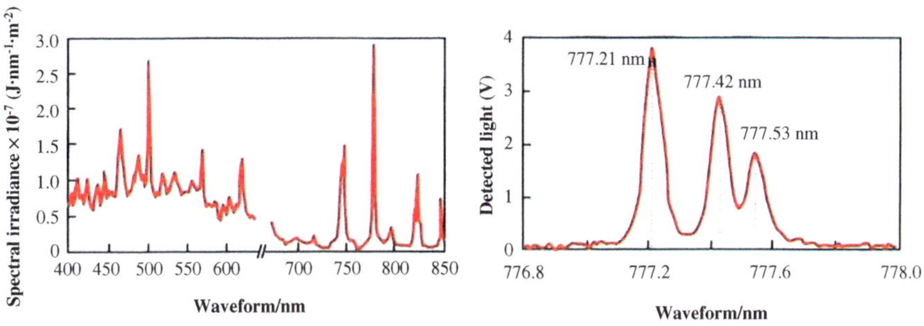


**Fig. 1** O3's strong absorption spectrum at 777.4 nm

atmosphere. The measurements from NASA U-2 aircraft show that the strongest radiation characteristics of cloud top optical spectra are generated from the spectral lines of neutral oxygen and neutral nitrogen in the near red band. Oxygen triplet (O3) at 777.4 nm is always a strong characteristic spectral line [2]. The detection of space-based lightning can be realized by observing the strong absorption spectral lines of O3 (Fig. 1).

## 2.2 *World Lightning Imager*

Beginning in the 1950s, humans began to observe the lightning at the top of lightning. The satellite is used as the ideal platform of lightning observation.

Geostationary meteorological satellite GOES-O, which NASA formally planned to launch in 2003, equipped with Lightning Map Sensor (LMS), to achieve high-resolution and high-detection rate of the lightning observation, however, due to the development plan and the development of the change of LMS, adjusted for the next generation of geostationary orbit satellite GLM and has been launched in November 2016, currently working in orbit.

During the same period, Europe is developing the LI of the third generation geostationary earth observation satellite. LI is the first optical payload of the geostationary orbit lightning survey in Europe and is not launched so far in the second half of 2017.

The technical indicators of MTG-LI and GOESR-GLM are as follows: 8 km spatial resolution, 2 ms frame and global coverage with lightning detection rate larger than 70% (ideally larger than 90%) [3].

## 2.3 *LMI*

LMI is the optical load of lightning detection in geostationary orbit developed in China (and also the first optical load of lightning detection) [4]. It is one of the first two lightning imageries of geostationary meteorological satellites in the world and

Fig. 2 LMI

fills the gap of space-based lightning optical detection in China [5, 6]. The Wind Four satellite was launched on December 11, 2016. The lightning imager is basically consistent with the European MTG-LI and American GOES R-GLM except that the field of view covers China and its surrounding areas (Fig. 2).

The main functions of LMI are as follows:

1. The convection movement information about total precipitation rainfall, storm of torrid zone and temperate zone,and core zone of storm system;
2. The relationship between lightning distribution and storm microphysics together with movement dynamics, the relationship between territorial weather circumstance and its relevant change and the relationship between rainfall and cloud type will be integrated in the diagnosis and forecast model of global rainfall cycle and water cycle;
3. To develop the global lightning meteorology, to study the distribution and variability of lightning frequency and to evaluate the influence of SST and road surface temperature variations on thunderstorm distribution and its intensity;
4. Observational modelling of the global electrical system and the factors that cause it;
5. A study of the generation, distribution and transmission of lightning track gases, determining the contribution and variability sources of global track gas consumption.

The main performance indicators are shown in Table 1.

## 3 LMI On-Orbit Data Verification and Application

After the launch of FY-4A and on-orbit testing and evaluation, the test load is directly transferred to the operational load, which is incorporated into the national meteorological forecast operational system, realizing the real-time detection of lightning and short-term lightning disaster prediction and early warning in China and its surrounding areas. It has been applied to the launch base, aviation, lightning disaster prediction and other fields.

On-orbit test shows that LMI can detect lightning correctly, realize real-time monitoring, tracking and early warning of severe convective weather system, effectively serve real-time lightning disaster monitoring, short-term real-time lightning

**Table 1** LMI performance indicators

| | |
|---|---|
| Weight | 65 kg |
| Size | 528 mm (long) × 346 mm (wide) × 1032 mm (high) |
| CCD array | 400 × 600 |
| Imaging rate | 500 frame/s |
| Central wavelength | 777.4 nm |
| Bandwidth | 1 nm |
| Spatial resolution | 7.8 km@nadir |
| Total field of view angle | 4.98° (North and South) × 7.47° (East and West) |
| Lightning detection rate | 90% |
| False alarm rate [7] | 10% |

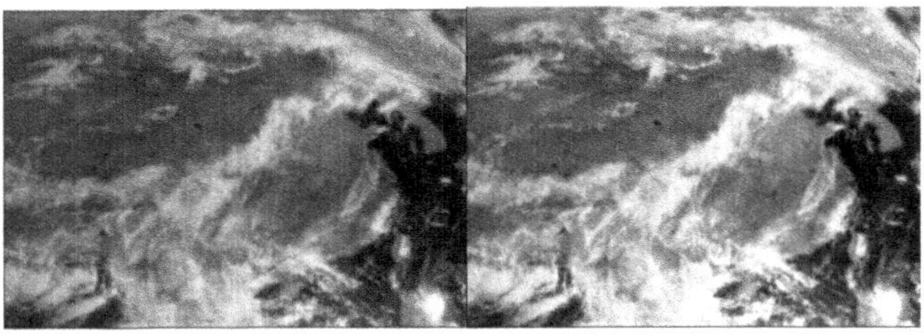

**Fig. 3** LMI 2- and 4-ms integral time imaging results

disaster early warning and numerical weather prediction in China. During the on-orbit test of LMI, the occurrence, development, movement and extinction of severe convective thunderstorms in China were observed many times. The preliminary comparison between satellite and ground lightning observation data verifies the correctness of satellite lightning observation.

## *3.1 Location of LMI*

LMI can obtain various imaging pictures with different integration time in landmark observation mode. Figure 3 is the image of the lightning imager under the integration time of 2 and 4 ms.

From the results of landmark imaging, the image of terrain and cloud in daytime is clear, the level of cloud is clear, the dynamic range is large, and the coastline and other characteristic targets can be used for positioning [8]. In the daytime, the precise positioning is achieved by landmark navigation, and the positioning accuracy reaches 1 pixel.

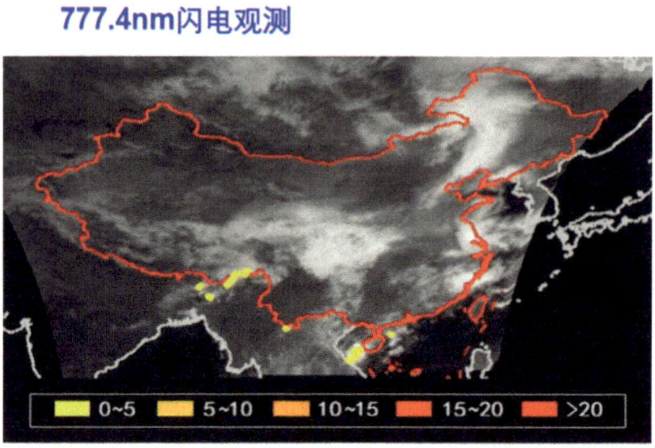


**Fig. 4** LMI real-time lightning monitoring

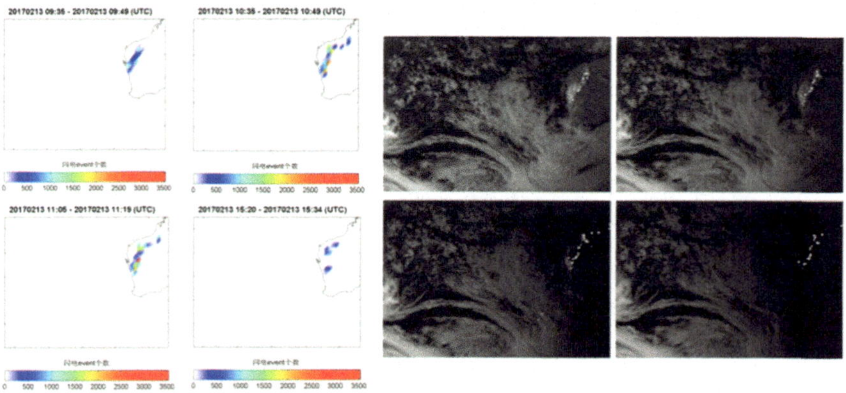


**Fig. 5** LMI strong thunderstorm process in Swan Valley, Western Australia, February 13, 2017

## *3.2 Lightning Monitoring of LMI*

From the results of lightning detection, lightning detection mode can realize real-time detection of lightning events with different intensity. Lightning imager has the ability of complete monitoring and tracking of a strong convection process (Figs. 4 and 5).

## *3.3 Verification of Detection Results of LMI*

The basic detection unit of satellite lightning imager is lightning event. On the basis of lightning event, data such as lightning group, flash and lightning area can be generated by clustering. Flash has temporal and spatial extensibility and can be regarded as a lightning in traditional sense (Table 2).

**Table 2** LMI performance indicators

| LMI | Product | Product element | Spatial resolution (km) |
|---|---|---|---|
| 1 | L2 product | LMI event | 8 |
| | | LMI group | 8 |
| | | LMI flash | 8 |
| 2 | L3 product | Lightning density distribution | |

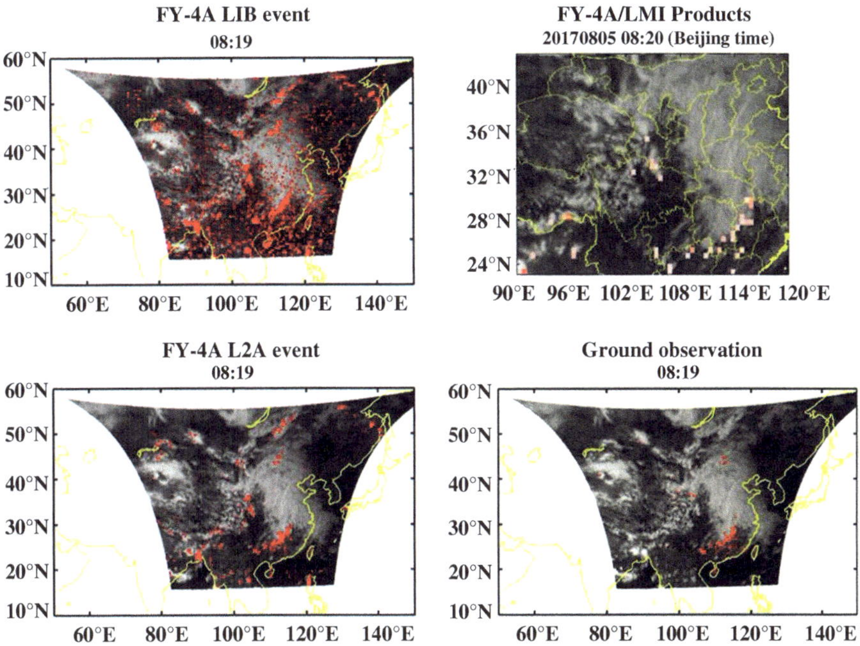


**Fig. 6** Efficient elimination of false lightning noise signals

LMI product generation algorithm consists of two parts: false signal filtering and clustering analysis. First, according to different types of noise sources, the false signal (non-lightning event) in lightning events is filtered by using the false signal filtering algorithm. Second, the lightning detection products such as "group" and "lightning" are obtained by clustering analysis of lightning "events" filtered by false signals.

Figure 6 shows that the lightning products formed by filtering false noise signals and lightning clustering are basically consistent with the ground observation, and the quality of observation is good.

On July 1, 2017, LMI successfully observed thunderstorms in Jianghuai, Jianghan and Northeast China, which was consistent with the trend of ground observation (Fig. 7).

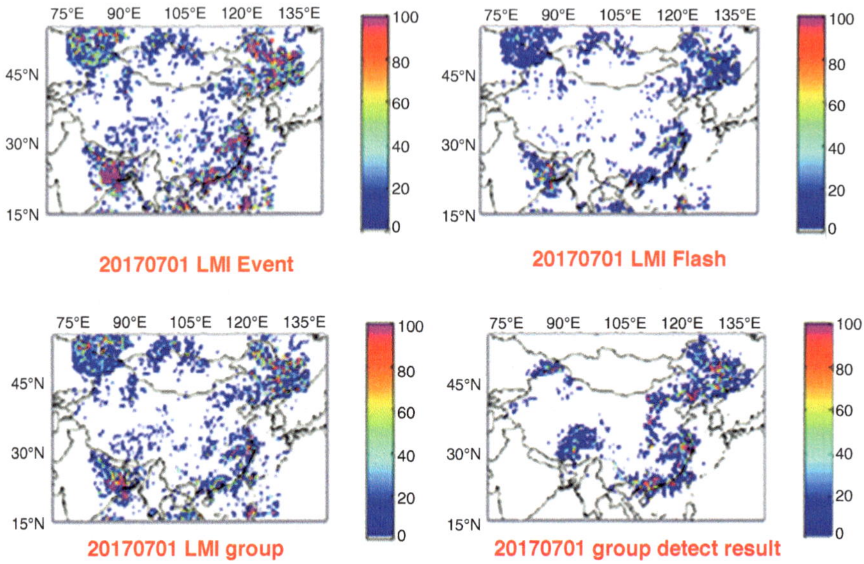


**Fig. 7** LMI observed thunderstorms in Jianghuai, Jianghan and Northeast China on July 1, 2017

In the afternoon of March 20, 2019, the lightning imager successfully captured the lightning phenomena occurring in the thunderstorm process in Beijing area from west to east. From 14:00 to 15:00, the lightning "event" density distribution maps of lightning imager (+ and o represent positive and negative ground flashes observed by lightning ground detection network, respectively) are shown in Fig. 8.

The ground lightning data in the above figures are derived from the National Lightning Monitoring Network of China. The national lightning monitoring network is a relatively complete ground lightning detection system covering China. The ADTD lightning location system developed by the Center for Space Science and Applied Research of the Chinese Academy of Sciences is used to record ground lightning return data. The measurement parameters include the occurrence time, location, polarity, intensity and steepness of ground lightning return. Because of the difference of the principle of satellite-ground lightning observation, the time–space distribution characteristics of the two products are different, so they can be used as qualitative comparison. Quantitative analysis and comparison need to be further studied. After many comparisons between the lightning detection data of the lightning detector in the mainland and the lightning ground network, the lightning imager detection results basically coincide with the ground network data.

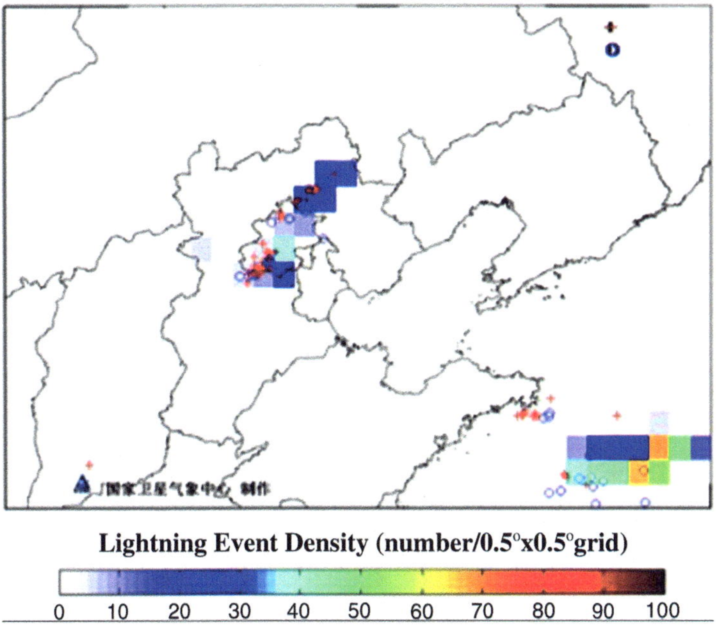


**Fig. 8** Thunderstorm process observed by LMI in Beijing area on March 20, 2019

## *3.4 Contribution of Lightning Observation Data to Weather Forecast*

An important function of the Wind Four Lightning Imager is to provide short-term and imminent prediction by assimilating the numerical weather prediction (NWP) system (Fig. 9).

Chinese meteorologists are conducting research on the evaluation and assimilation of lightning imager observation data, establishing lightning observation operators and linking up the relationship between observation and model variables. By assimilating the lightning imager observation data into the numerical weather prediction system, the preliminary conclusions are as follows: after assimilating the FY4 lightning imager observation, the ability of local precipitation prediction is enhanced; to a certain extent, the problem of missing report of precipitation is reduced.

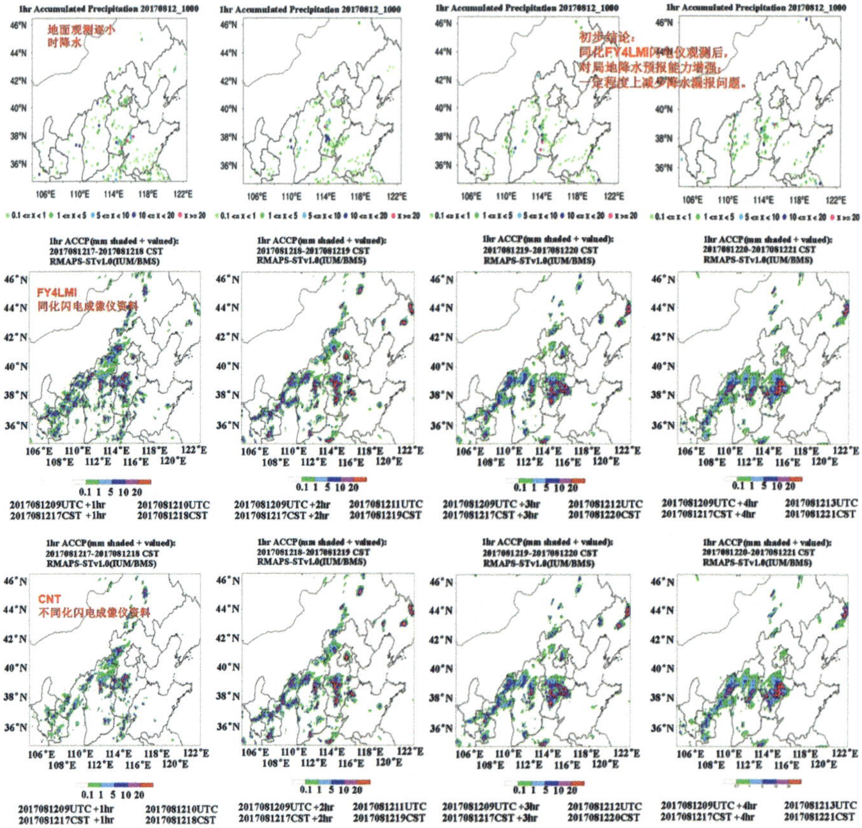


**Fig. 9** Lightning assimilation process of lightning imager

## 4 Conclusion

FY-4A LMI can detect lightning events of different intensity in real time. The lightning imager has the ability to monitor and track a strong convection process, to forecast local precipitation, and plays an important role in disaster prevention, mitigation and numerical weather prediction.

**Acknowledgements** Thank Zhang Zhiqing, Lu Feng, Guo Qiang, Huang Fuxiang, Han Wei, Yang Hanzhe and Wang Jing of National Satellite Meteorological Centre, for their work in on-orbit data processing, product generation and product application

## References

1. Bao, S., Tang, S., Li, Y., et al.: Real-time detection technology of instantaneous point-source multi-target lightning signal on the geostationary orbit. Infrared Laser Eng. **41**(9), 2390–2395 (2012) (in Chinese)
2. Finke, U., Hannover, FH.: Lightning observations from space: Time and space characteristic of optical events. 5th MMT Meeting, Ischia, October 2007 (2007)
3. Durand, Y., Hallibert, P., Wilson, M., et al.: The flexible combined imager onboard mtg: from design to calibration. SPIE Remote Sens. **9639**, 108–122 (2015)
4. Tang, S., Liu, Z., Bao, S.: Laboratory calibration of the lightning imaging sensor. In: The 17th Academy Meeting of Space Exploration Professional Committee of China Space Science Society, pp. 38–40 (2004). (in Chinese)
5. Liang, H., Bao, S., Chen, Q., et al.: Design and implementation of FY-4 geostationary lightning imager. Aerosp. Shanghai. **34**(4), 43–51 (2017) (in Chinese)
6. Hui, W., Huang, F., Zhu, J.: Technology of optical detection of lightning from space. Opt. Precis. Eng. **24**(10), 361–369 (2016) (in Chinese)
7. Huang, F., Guo, J., Feng, X.: Simulating calculation of lightning detection efficiency and false alarm rate for lightning imagery on geo-satellite. Acta Photon. Sin. **38**(12), 3116–3120 (2009) (in Chinese)
8. Bao, S.L., Tang, S.F., Li, Y.F., Liang, H., Zhao, Y.H.: Real-time detection technology of instantaneous point-source multi-target lightning signal on the geostationary orbit. Infrared Laser Eng. **9**, 114 (2012)

# On-Orbit Performance Analysis of AIUS/GF-5 Instrument

**Jiang Cheng, Xu Peng-Mei, He Hong-Yan, Cao Shi-Xiang, Ma Zhong-Qi, Zhang Yu-Gui, and Hou Li-Zhou**

**Abstract** GF-5 is a hyperspectral remote sensing satellite in the GF series of satellite of China, which was successfully launched on 9 May 2018. The key instrument, the Atmospheric Infrared Ultra-spectral Sounder (AIUS), is the first space-borne occultation sensor with ultra-spectral resolution in China, as well as the space-borne Fourier transform spectrometer (FTS) with finest spectral resolution. The instrument covers wavelengths from 2.4 to 13.3 μm, providing the ultra-spectral resolution with 0.03 $cm^{-1}$. The AIUS instrument will measure trace gases in the atmosphere and analyse the chemical and dynamical processes by solar occultation from GF-5 satellite. This paper describes the on-orbit test activities and performances of key characteristics, including Signal-to-Noise Ratio (SNR), accuracy of spectral calibration and spectral resolution based on the on-orbit data obtained. The preliminary results from the on-orbit verification, calibration and validation after launch have shown that AIUS has good performance with good quality spectrum. The results show that the SNR will meet the requirement, the full width at half maximum (FWHM) will be smaller than 0.03 $cm^{-1}$ and the accuracy of spectral calibration will be greater than 0.008 $cm^{-1}$. The AIUS instrument on-board GF5 satellite will provide fine spectral transmittances of the atmosphere and offer the best atmospheric constituents observing performance in solar occultation.

**Keywords** Atmospheric sounding · Fourier Transform Spectrometer · Occultation On-orbit performance · GF-5

---

J. Cheng (✉) · X. Peng-Mei · H. Hong-Yan · C. Shi-Xiang · M. Zhong-Qi · Z. Yu-Gui
H. Li-Zhou
Beijing Institute of Space Mechanics & Electricity, Beijing Key Laboratory of Advanced Optical Remote Sensing Technology, Beijing, China

 
H. P. Urbach, Q. Yu (eds.), *6th International Symposium of Space Optical Instruments and Applications*, Space Technology Proceedings 7,
https://doi.org/10.1007/978-3-030-56488-9_2

## 1 Introduction

Remote sensing of atmospheric constituents using infrared solar spectrum by spectrometer has been developed over the past 40–50 years. The recorded solar spectrum from the ground and air-borne spectrometers is suffered from interference caused by absorption lines. Solar occultation from satellite-borne spectrometers can measure trace gas vertical profiles with high vertical resolution over the entire infrared spectral range [1, 2].

In the past years, there are several satellite-borne missions for measuring the concentrations of atmospheric constituents by solar absorption infrared spectrum. The Atmospheric Trace Molecule Spectroscopy (ATMOS) spectrometer, flown on four space shuttle missions in the 1980s and 1990s, provided good occultation. As the inheritance of ATMOS, the Atmospheric Chemistry Experiment (ACE), also known as a Canadian mission SCISAT-1, is a small satellite for sounding the atmosphere launched in 2003. The ACE includes three instruments: an infrared Fourier transform spectrometer (FTS), a UV/visible/near-IR spectrograph and a two-channel solar imager. During sunrise and sunset, FTS measures infrared spectrum with 0.02 $cm^{-1}$ spectral resolution and 3–4 km vertical resolution covering spectral range from 2 to 13 μm [3, 4].

In order to understand the dynamical processes in the stratosphere and upper troposphere, particularly in the Antarctic, the Atmospheric Infrared Ultra-spectral Sounder (AIUS) was added to the payload of GF-5 satellite, which is a hyperspectral remote sensing satellite in the GF series. AIUS is a solar occultation FTS with several scientific objectives. During an occultation event, AIUS records the infrared spectrum that contains trace gases information on different atmospheric layers, which provides the vertical concentration profiles of trace gases and other atmospheric components [5–7].

In order to achieve all the science goals, the AIUS instrument needs good performance on SNR, and fine spectral resolution for resolving the absorption lines of atmosphere. After the successful launch of GF-5 satellite in May 2018, the AIUS instrument has been conducted successfully, too. With nearly 20 days for cooling the infrared detectors, the first occultation data were acquired and transferred to AIUS team. The atmospheric absorption spectrum produced by occultation data was found to present splendid spectral features with good SNR over the whole spectral range. The AIUS instrument shows no degradation of performance or functionality after being observed since launch.

## 2 AIUS Instrument Description

AIUS is an infrared FTS covering from 2.4 to 13.3 μm, and it provides the solar absorption spectrum by occultation with ultra-spectral resolution (0.03 $cm^{-1}$) at different altitudes in the atmosphere. The instrument is a Michelson interferometer with an optimized optical layout. Inside AIUS, there is a folding mirror to increase the optical path difference. Similar with ACE of Canada, AIUS also includes a sun-tracker that can keep instrument pointing toward the radiometric centre of the sun with high accuracy. AIUS's field-of-view (FOV) is 1.25 mrad. In order to cover the spectral range, two detector elements InSb and MCT are used to obtain the desired spectrum. InSb detector element covers the 2.4–5.4 μm range, and MCT detector element covers the 5.4–13.3 μm range. The obtained atmospheric spectrum can help us to get detailed information on vertical concentration profiles of atmospheric constituents and temperature and pressure (Fig. 1).

The relevant parameters of the AIUS instrument are summarized in Table 1.

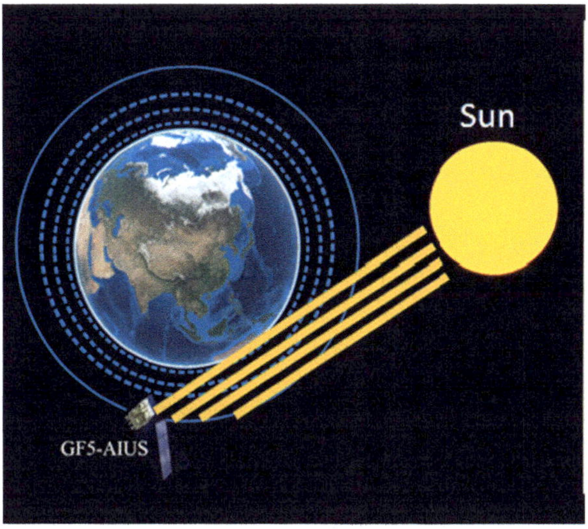


**Fig. 1** Geometry of solar occultation measurements of AIUS

**Table 1** Parameters of the AIUS instrument

| Parameter | Value |
| --- | --- |
| Spectral range | 2.4–13.3 μm |
| Spectral resolution | 0.03 $cm^{-1}$ |
| FOV | 1.25 mrad |
| Bit | 18 bit |
| SNR | >100 (@5800 K) |
| Detector | InSb: 2.4–5.4 μm<br>Mct: 5.4–13.3 μm |

# 3 On-Orbit Performance and Analysis

## 3.1 Spectral Resolution

Several inherent factors of the AIUS instrument can contribute to the spectral lines of a spectrum. Focusing on an elementary spectral line, spectral resolution was affected by the emission from a single gas molecule, interactions between the molecules. These factors expressed primarily by temperature and pressure will broaden the spectral line [8].

The expression including temperature and pressure is usually performed in the interferogram domain by multiplication or in the spectral domain by convolution. The AIUS instrument was designed to have the maximum optical path difference of 25 cm and a sampling window of 50 cm, which limited the FWHM of the instrument line shape (ILS) function as 0.025 $cm^{-1}$. As mentioned in section II, both the 1.25 mrad FOV and sampling window determined the 0.0259 $cm^{-1}$ spectral line width at 4100 $cm^{-1}$.

Figure 2 shows multiple absorption lines of $CH_4$ (Fig. 2a), $NH_3$ (Fig. 2b), NO (Fig. 2c) and $O_3$ (Fig. 2d) at different tangent height lines taken from a typical occultation sequence. Different colour lines in Fig. 2 represent atmospheric absorption spectrum obtained from different atmospheric layers. An AIUS spectrum for $CH_4$ absorption band is shown in Fig. 2a, taken from 5 altitudes of 10 km, 15 km, 20 km, 25 km and 30 km. It is found that the absorption positions are well approximately fitting.

Table 2 shows the results of spectral resolution on-orbit test for MCT and InSb detector, separately. All the spectral resolutions were estimated from atmospheric absorption spectrum at different tangent heights obtained by AIUS. According to the data from index 1 to index 15 in Table 2, all the spectral resolutions both in MCT channel and InSb channel met the pre-flight spectral resolution requirements. We averaged the 15 absorption spectra covering 2.4–5.4 μm and 5.4–13.3 μm spectral range. For the MCT detector element, the spectral resolution is approximately 0.0265 $cm^{-1}$. For the InSb detector element, the spectral resolution is approximately 0.0261 $cm^{-1}$.

## 3.2 Spectral Calibration Accuracy

In order to acquire the wavelength drift data caused by temperature variation at post-launch phase, a laser diode was used in the AIUS instrument. Unfortunately, accurate temperature measuring on-orbit on absolute scale could not be accomplished for AIUS. For each occultation event, we take the wavelength position of the solar spectrum as the reference wavelength and correct the wavelength position acquired by laser diode [9]. During the calibration process, the absolute spectral drift was measured.

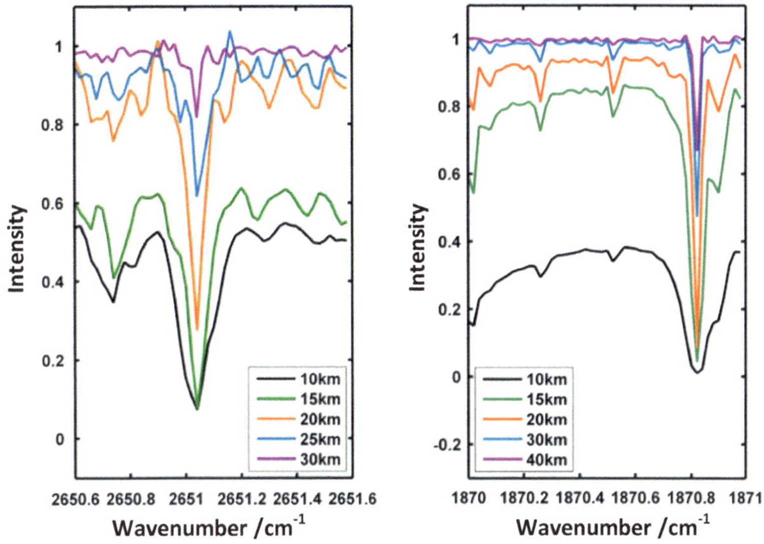


(a) $CH_4$ (b) $NH_3$

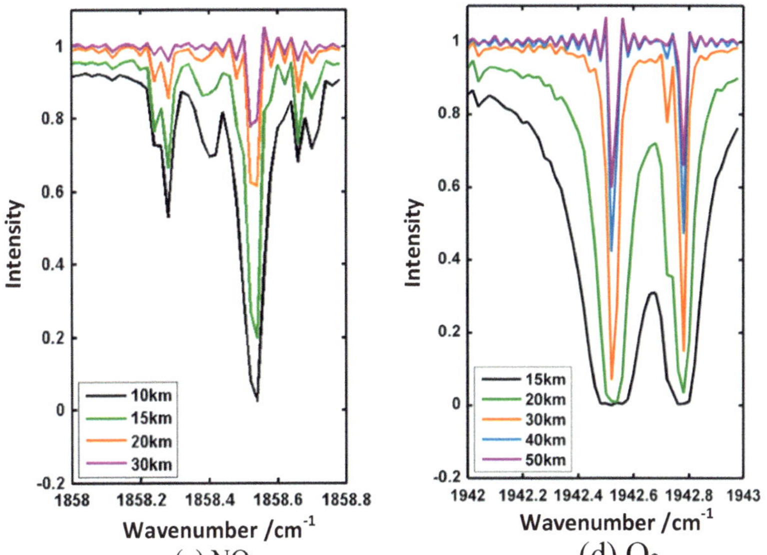


(c) NO (d) $O_3$

**Fig. 2** Absorption spectrum of AIUS at different tangent heights. (**a**) $CH_4$, (**b**) $NH_3$, (**c**) NO, (**d**) $O_3$

A full occultation sequence contains atmospheric spectrum at different layers and exo-atmospheric spectrum. We use the ratio of the occultation spectrum by the exo-atmospheric spectrum as the transmittance of the atmosphere. Table 3 shows the spectral calibration results estimated from absorption lines of $HNO_3$ and $CO_2$. The bias of reference spectrum and measurement spectrum indicates that the spectral calibration is greater than 0.008 $cm^{-1}$.

**Table 2** Results of spectral resolution

| Channel | Index | Spectral resolution ($cm^{-1}$) | Frequency ($cm^{-1}$) |
|---|---|---|---|
| MCT | 1 | 0.0273 | 1418.1441 |
| | 2 | 0.0251 | 1472.0887 |
| | 3 | 0.0233 | 1463.4921 |
| | 4 | 0.0268 | 1521.2226 |
| | 5 | 0.0261 | 1532.4325 |
| | 6 | 0.0284 | 1734.7721 |
| | 7 | 0.0288 | 1805.4322 |
| | 8 | 0.0289 | 1930.6739 |
| | 9 | 0.0253 | 1543.2481 |
| | 10 | 0.0280 | 1546.4936 |
| | 11 | 0.0241 | 1547.7659 |
| | 12 | 0.0242 | 1403.6201 |
| | 13 | 0.0297 | 1386.1736 |
| | 14 | 0.0251 | 922.7802 |
| | 15 | 0.0266 | 1026.1074 |
| InSb | 1 | 0.0275 | 2529.1812 |
| | 2 | 0.0268 | 3008.9614 |
| | 3 | 0.0271 | 3207.5645 |
| | 4 | 0.0271 | 3400.0018 |
| | 5 | 0.0272 | 2517.1339 |
| | 6 | 0.0265 | 2004.0546 |
| | 7 | 0.0277 | 2101.2215 |
| | 8 | 0.0293 | 2252.7869 |
| | 9 | 0.0261 | 2374.5217 |
| | 10 | 0.0289 | 2517.1303 |
| | 11 | 0.0231 | 2835.0752 |
| | 12 | 0.0228 | 3087.8274 |
| | 13 | 0.0205 | 3087.7524 |
| | 14 | 0.0270 | 3936.2664 |
| | 15 | 0.0234 | 3894.9623 |

**Table 3** Results of spectral calibration

| Index | Gas | Reference ($cm^{-1}$) | Measurement ($cm^{-1}$) | Bias ($cm^{-1}$) |
|---|---|---|---|---|
| 1 | $HNO_3$ | 868.104 | 868.107 | 0.0030 |
| 2 | $CO_2$ | 942.3833 | 942.391 | 0.0077 |

## *3.3 SNR Performance*

Each infrared spectrometer consisted of noises to be dealt with in the data process phase. Typical noises that exist in the AIUS instrument design chain include the shot noise, the detector and electronics noise, the quantification noise, sampling jitters

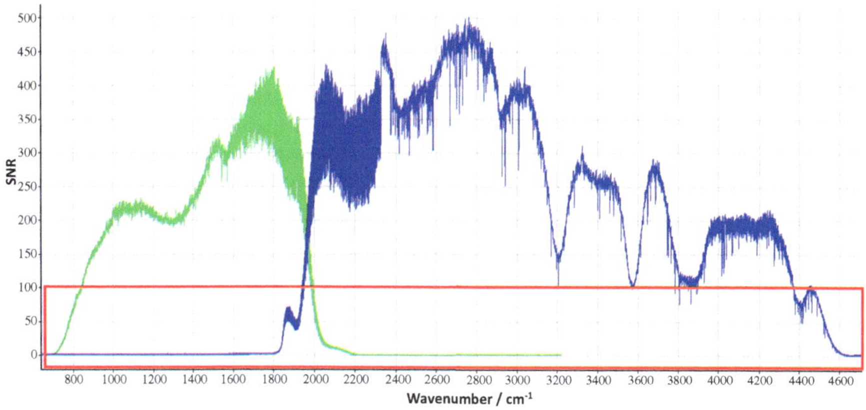


**Fig. 3** On-orbit SNR performance on 19 October 2018

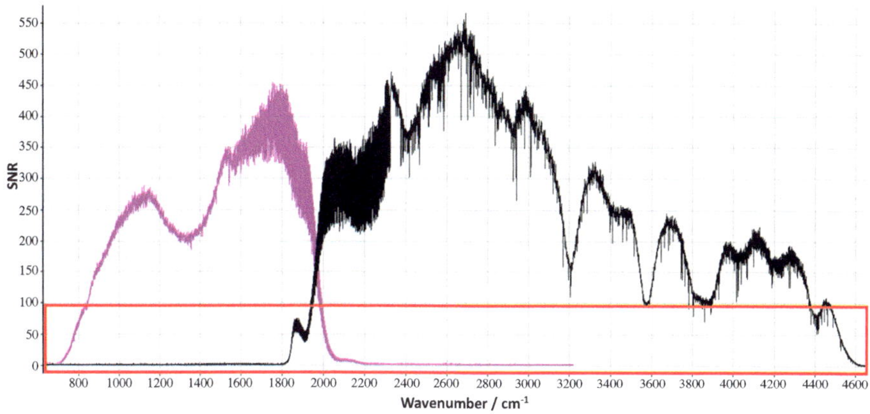


**Fig. 4** On-orbit SNR performance on 10 December 2018

noise and non-linearity noise. In order to evaluate the pre-launch SNR performance of the AIUS instrument, the instrument is verified in thermal vacuum chamber at the Beijing Institute of Space Mechanics & Electricity (BISME) in early 2017.

The post-launch SNR has been evaluated from exo-atmospheric measurements. Figures 3 and 4 show the post-launch SNR performance based on two occultation measurements on October and December 2018. In the low frequency (750–850 $cm^{-1}$) of the spectrum, the signal slightly exceeds the noise. This phenomenon is due to the inappropriate combination of signals of exo-atmospheric. Note that, since there is an overlap near 1900 $cm^{-1}$ between the InSb detector and MCT detector, this represents lower SNR performance in part of this spectral range.

## 4 Summary

AIUS represents a new era of ultra-spectral resolution sounding capabilities for atmospheric gases in China. The AIUS instrument will measure trace gases in the atmosphere and analyse the chemical and dynamical processes by solar occultation from GF-5 satellite. This paper described the on-orbit test activities and performances of key characteristics, including SNR, accuracy of spectral calibration and spectral resolution based on the on-orbit data obtained. The preliminary results from the on-orbit calibration (Cal) and verification (Val) after launch have shown that AIUS has good performance with good quality spectrum. The results showed that the SNR will meet the requirement, FWHM will be smaller than 0.03 $cm^{-1}$ and the accuracy of spectral calibration will be greater than 0.008 $cm^{-1}$. The AIUS instrument on-board GF5 satellite will provide fine spectral transmittances of the atmosphere and offer the best atmospheric constituents observing performance in solar occultation. The experiences obtained in the Cal/Val process of AIUS/GF-5 will be valuable for the next generation of atmospheric sounding satellite. The Cal/Val team of BISME will continue the calibration and verification work and long-term monitoring of the instrument on-orbit status to ensure the data quality of AIUS/GF-5 for applications.

**Acknowledgment** The authors would like to thank all of the Centre for Resources Satellite Data and Application (CRSDA) employees and the Institute of Remote Sensing and Digital Earth of Chinese Academy of Sciences (CAS) employees who provided great support and collaboration in Cal/Val process of AIUS/GF-5. The authors would like to thank all the engineers who worked hard to design and manufacture the AIUS instrument in BISME. This work was supported by the National Natural Science Foundation of China (61675012) and the Beijing Science and Technology Project (Z171100000717010).

## References

1. Rozanov, V.V., Rozanov, A.V., Kokhanovsky, A.A.: Radiative transfer through terrestrial atmosphere and ocean: Software package SCIATRAN. J. Quant. Spectrosc. Radiat. Transf. **133**, 13–71 (2014)
2. Blum, M., Rozanov, V.V., Burrows, J.P.: Coupled ocean-atmosphere radiative transfer model in the framework of software package SCIATRAN: Selected comparisons to model and satellite data. Adv. Space Res. **49**, 1728–1742 (2012)
3. Soucy, M.-A., Châteauneuf, F., Deutsch, C., Étienne, N.: ACE-FTS instrument detailed design. Proc. SPIE. **4841**, 12 (2002)
4. Boone, C., Nassar, R., McLeod, S., Walker, K., Bernath, P.: SciSat-1 retrieval results. Proc. SPIE. **5542**, 184–194 (2004)
5. Bin, F.A.N., Xu, C.H.E.N., Bicen, L.I.: Technical innovation of optical remote sensing payloads onboard gf-5 satellite. Infrared Laser Eng. **46**(1), 0102002 (2017)
6. Walker, A., Bernath, F.: Validation measurement program for the atmospheric chemistry experiment. Proc. SPIE. **4814**, 91–101 (2002)

7. Cheng, J., Tao, D., Hongyan, H.E.: Digital modeling and simulation of GF-5 instrument AIUS. Spacecr. Recovery Remote Sens. **39**(3), 94–103 (2018)
8. Fortin, S., Soucy, M.-A., Deutsch, C.: On-orbit commissioning of the ACE-FTS instrument. Proc. SPIE. **4814**, 3 (2002)
9. Saarinen, P., Kauppinen, J.: Spectral line-shape distortions in Michelson interferometers due to off-focus radiation source. Appl. Opt. **31**(13), 2353–2359 (1992)

# Research on Image Motion Measurement of the Space Camera Based on Optical Correlator Method

**Yao Dalei, Xue Jianru, Qiu Yuehong, Bu Fan, Cao Weicheng, and Wang Li**

**Abstract** During the imaging process of the space camera, the image motion is caused by the relative motion between the motion of the captured image and the transfer of the photogenerated charge packet. The image motion causes image degradation and distortion. Therefore, the specific analysis of image motion and the study of measuring the image motion of remote sensing camera have practical significance for improving the image quality of high-resolution remote sensing camera. This paper proposes a new method based on automatic threshold and subpixel correlation power spectrum correlation peak processing using spatial light modulator to measure image motion, correlation peaks of images obtained by optical joint transform optical modulator correlation, adaptive global threshold segmentation method and subpixel correlation method to accurately identify the position of the correlation peak. Finally, the simulation and actual measurement experiments are used, respectively. The simulation results show that the method is insensitive to the key factors affecting image motion measurement such as input image and image SNR (signal-to-noise ratio). It has strong robustness and uses spatial light modulator to build image motion. In the measurement test platform, the measured results show that the image motion measurement error is less than 0.25 pixels.

**Keywords** Space camera · Image motion measurement · Optical correlator · Image motion error

Y. Dalei (✉) · Q. Yuehong · B. Fan · C. Weicheng · W. Li
Xi'an Institute of Optics and Precision Mechanics of CAS, Xi'an, China

X. Jianru
Xi'an Jiaotong University, Xi'an, China



H. P. Urbach, Q. Yu (eds.), *6th International Symposium of Space Optical Instruments and Applications*, Space Technology Proceedings 7,
https://doi.org/10.1007/978-3-030-56488-9_3

## 1 Introduction

Using space remote sensing camera in the process of shooting the ground object, we hope to obtain two kinds of accurate information: the first is the shape and size of the object; the second is the brightness of the object. However, due to various reasons, the camera has a relative motion between the motion of the image of the illuminated object on the focal plane and the transfer of the photogenerated charge packet, causing image motion. Image motion causes image degradation, degeneration of the image causes two kinds of distortion: distortion of the shape and size of the image; and distortion of the brightness of the image. The distortion of the image causes the resolution of the image to decrease, the blur of the target edge, and the grayscale distortion of the object. As the resolution of the camera continues to increase, the effect of image motion on the quality of space camera imaging is particularly prominent. Therefore, the specific analysis of image motion and the study of measuring the image motion of space camera have practical significance for improving the image quality of high-resolution space cameras.

In order to overcome the image motion caused by satellite orbit perturbation, attitude drift, and on-board component chatter, it is necessary to adopt a certain image motion compensation method, where the precondition for compensation is to obtain the accurate image motion of the satellite in the focal plane. ESA K. Janschek et al. propose a method for measuring image motion using a joint transform optical correlator [1, 2]; the basic idea of the method is to send the current image and the reference image to a spatial light modulator (SLM), convert the digital signal into an optical signal, and through the Fourier transform lens, use the CCD (charge-coupled device) to detect the joint power spectrum of the two images. The joint power spectrum image is sent to the spatial light modulator, and the relevant image is detected by the CCD through the Fourier transform lens. Then, the image motion amount can be obtained by measuring the position of the related image by digital image processing.

Applying the measurement principle of optical correlator [3, 4], this paper proposes a new method of automatic threshold and subpixel-related power spectrum correlation peak processing to measure image motion. Through the simulation experiment, this method is insensitive to the key factors affecting image motion measurement, such as input image and image SNR, and has strong robustness. The relationship between the image motion and the measurement error under the actual motion of the space camera and the maximum measurement of image motion corresponding to the input image signal-to-noise ratio are simulated. Finally, the method is verified in the actual experiment.

## 2 Principle of Image Motion Measurement by Optical Correlation Method

First, the image acquired by the image motion measurement detector at time $T = T_0$ is used as a reference image, denoted by $f_{i-1}(x,y)$. The image motion is caused by satellite motion and disturbance, after the $\varnothing t$ time interval, and the current image acquired by the image motion detector is represented by $f_i(x,y)$, which has an image motion amount $\varnothing x$ compared with the $T_0$ time image. The reference image and the current image are simultaneously sent to the correlator for correlation operation, and as shown in Fig. 1, the two images subjected to the correlation operation are simultaneously read into the input surface of the spatial light modulator. For the convenience of expression, the triangular symbol is used to represent the common scene in the two images. It is assumed that the image motion in the $x$ and $y$ directions between the two images is $\delta x$ and $\delta y$ due to motion, and the input E of the correlator can be expressed as

$$i(x,y) = f_{i-1}(x, y-a) + f_i(x+\delta x, y+a+\delta y). \tag{1}$$

The input image $i(x,y)$ is subjected to a Fourier transform (SLM implementation) to obtain its spectral function $T(u,v)$:

$$T(u,v) = F_{i-1}(u,v)\exp(-2i\pi av) + F_i \exp\{2i\pi[u\delta x + (a+\delta y)v]\}. \tag{2}$$

Then, the power spectrum function of $T(u,v)$ is

**Fig. 1** Correlator input image arrangement

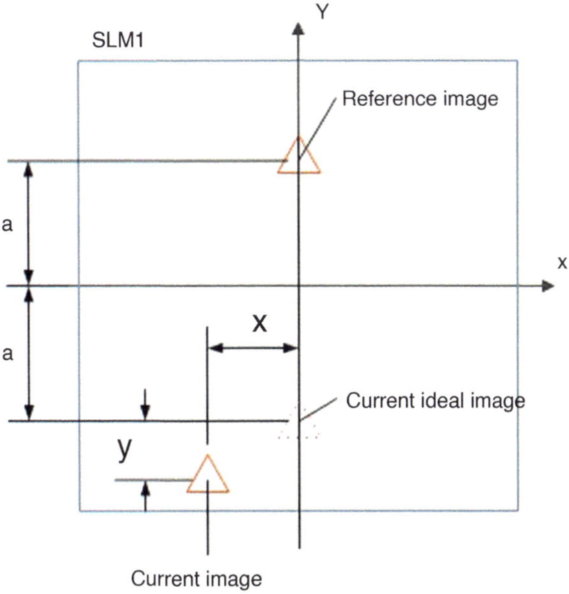

$$\left|T(u,v)\right|^2 = \left|F_{i-1}(u,v)\right|^2 + \left|F_i(u,v)\right|^2 + F_{i-1}F_i^* \exp\left\{-2i\pi\left[u\delta x + (2a+\delta y)v)\right]\right\} \\ + F_i F_{i-1}^* \exp\left\{2i\pi\left(u\delta x + (2a+\delta y)v\right]\right\}. \tag{3}$$

$\left|T(u,v)\right|^2$ is again subjected to a Fourier transform (Fourier lens implementation), and the correlation output is

$$c(x,y) = f_{i-1}(x,y) \otimes f_{i-1}(x,y) + f_i(x,y) \otimes f_i(x,y) + f_{i-1} \otimes f_i \\ * \delta(x - \delta x, y - 2a - \delta y) + f_i \otimes f_{i-1} * \delta(x + \delta x, y + 2a + \delta y), \tag{4}$$

where $*$ represents convolution and ⊗ represents correlation operation. The first two are autocorrelation phases, and the last two are cross-correlation terms, which are the points to be found and measured. The output correlation peak of the correlator is shown in Fig. 2, where the dotted circle in the figure represents the theoretical correlation peak position (the output image without image motion), and the solid circle represents the actual output correlation peak position corresponding to the actual scene position in Fig. 1. It can be seen that the image motion information is included in the related image, so this method can be used to measure the image motion of the space camera.

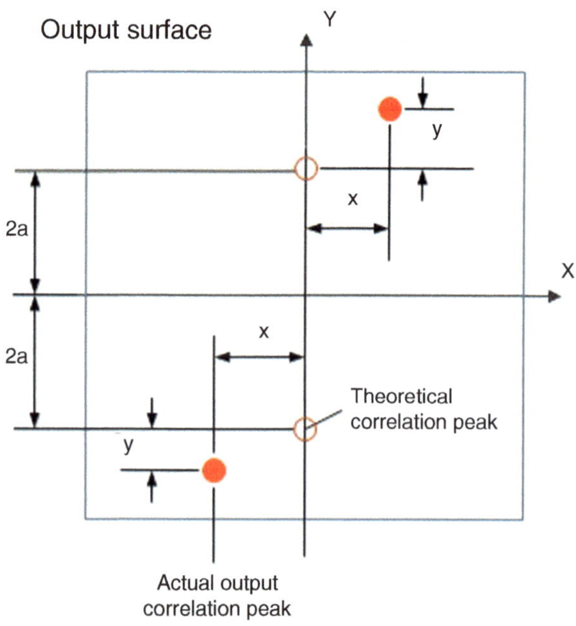


**Fig. 2** Output correlation peak position diagram

## 3 Correlated Peak Image Processing

According to the above principle analysis, the key to accurately calculate the image motion based on the obtained correlation peak is how to accurately calculate the position of the cross-correlation peak. The most commonly used method is the centroid method. However, the centroid method only achieves the desired effect on gray-scale symmetrical targets. In fact, due to the defocusing phenomenon in optical imaging, it is difficult to obtain high positioning accuracy[5, 6, 7].

In this paper, an adaptive global threshold segmentation method and a subpixel correlation method will be used to accurately identify the position of the correlation peak. As shown in Fig. 3, the peak of the useful signal of the correlation peak is much higher than the average value, so we use the global threshold method for threshold segmentation to separate the correlation peak from the background. Algorithms for determining thresholds in practice include a 2-mode method, a likelihood ratio detection method, and an iterative method. We use adaptive threshold method as the method to determine the actual image threshold. This method is very flexible, mainly based on real-time specific image data for threshold calculation, suitable for image processing with significant difference between useful signal and background signal. Its mathematical expression is

$$T = E + \lambda \times \delta, \tag{5}$$

where $E$ is the mean of the image, $\delta$ is the variance of the image, and $\lambda$ is the fixed weight. When $\lambda$ is too small, the detection probability of detecting the correlation peak signal will be large, but it also increases the false alarm probability; when $\lambda$ is

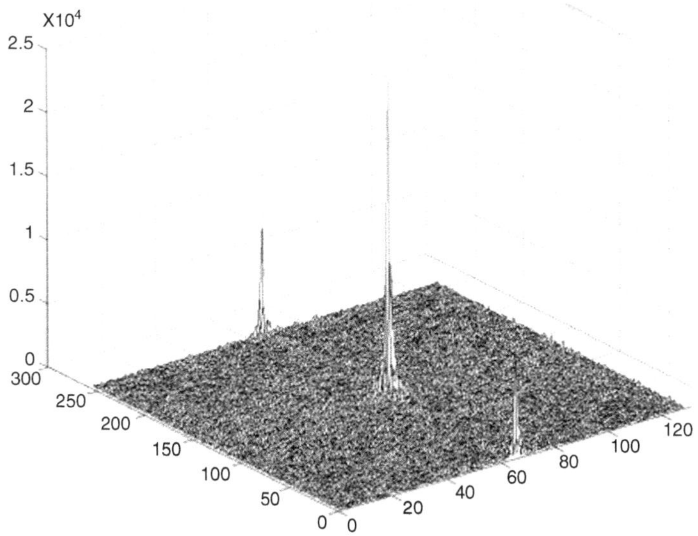


**Fig. 3** Power spectrum correlation peak map

too large, the threshold will increase, the false alarm probability will decrease, but the detection probability will also decrease. We have tested the statistics through a large number of tests, and the value of $\lambda$ is 6 under the condition of false alarm probability of the order of $10^{-3}$. The power spectrum correlation image that is thresholded is segmented using a subpixel correlation method to locate the position of the correlation peak. The subpixel correlation method is based on the correlation properties of the cross-correlation function:

$$C(x,y) = \sum_{(i,j)\in w} f(x+i, y+j)\, g(i,j), \tag{6}$$

where $C(x, y)$ is the correlation function, $f(x, y)$ is the source image of the target, $g(i, j)$ is the template, and $w$ is the template area. When $f(x, y)$ and $g(i, j)$ are determined, the greater the degree of overlap or similarity between the two in the gray level of the space, the larger $C(x, y)$. Therefore, the position of the target can be determined by determining the maximum position of the correlation function.

Since the correlation function matrix generally satisfies the Gaussian distribution on a single peak region centered on the maximum value, the analytical surface function of the region can be obtained by the fitting method, and the extreme point of the curved surface is taken as the subpixel position of the correlation peak. It is appropriate to fit the size of the fitting window by 3 × 3 or 5 × 5. Generally, the following parabolic equation is used to fit the relevant function surface:

$$f(i,j) = a_2 i^2 + a_1 i + b_2 j^2 + b_1 j + c. \tag{7}$$

Then, the best estimate of the target position deviation is the peak point of function $f(i, j)$ ($\partial f/\partial i = 0$, $\partial f/\partial j = 0$):

$$x_p = -\frac{a_1}{2a_2}, \qquad y_p = -\frac{b_1}{2b_2}. \tag{8}$$

Find the coefficients of the above formula by least squares method, set the correlation function to $R(i, j)$, and the coordinates of $Max(R(i, j))$ are (0, 0), and fit the data with 5 × 5 at the maximum value to minimize the sum of squares $E = \sum_i \sum_j [f(i, j) - R(i, j)]^2$ of the errors.

The following equations are available from $\partial E/\partial a_2 = 0$, $\partial E/\partial a_1 = 0$, $\partial E/\partial b_2 = 0$, and $\partial E/\partial c = 0$:

$$\sum_i \sum_j i^2 f(i,j) = \sum_i \sum_j i^2 R(i,j),$$

$$\sum_i \sum_j i f(i,j) = \sum_i \sum_j i R(i,j),$$

$$\sum_i \sum_j j^2 f(i,j) = \sum_i \sum_j j^2 R(i,j),$$

$$\sum_i \sum_j jf(i,j) = \sum_i \sum_j jR(i,j),$$

$$\sum_i \sum_j f(i,j) = \sum_i \sum_j R(i,j).$$

Solving the quaternion equations, we can find the coefficients $a_1$, $a_2$, $b_1$, and $b_2$.

## 4 Simulation Experiment

In the simulation, a 128 × 128 pixel dam map taken by the US Quickbird2 satellite was selected as the reference image. As shown in Fig. 4, the current image with a certain image motion is under the red line, so that a joint image of 128 × 256 pixels is used as a simulation input picture.

As shown in Fig. 5, in actual satellite motion, image motion is generally divided into forward image motion and lateral image motion, corresponding to the focal plane $x$ and $y$ directions of the space camera, the image motion on the imaging medium with components in both $x$ and $y$ directions. The image motion method is divided into two types: the first type of image motion is the same in the $x$ and $y$ directions from $T_0$ to $T_1$; the second type of image motion is different in the $x$ direction and the $y$ direction from $T_1$ to $T_2$.This is due to satellite attitude adjustment during satellite on-orbit imaging, and various movements of moving parts on the satellite cause camera lens shake, usually this gap is less than 3 pixels. As shown in

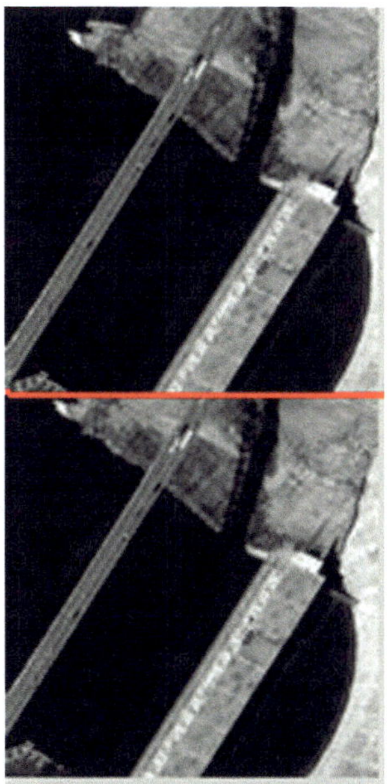

**Fig. 4** Simulates input

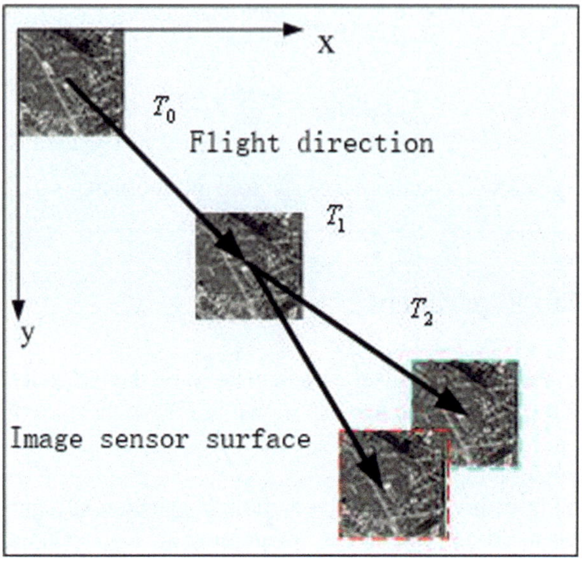


**Fig. 5** Image shifting diagram

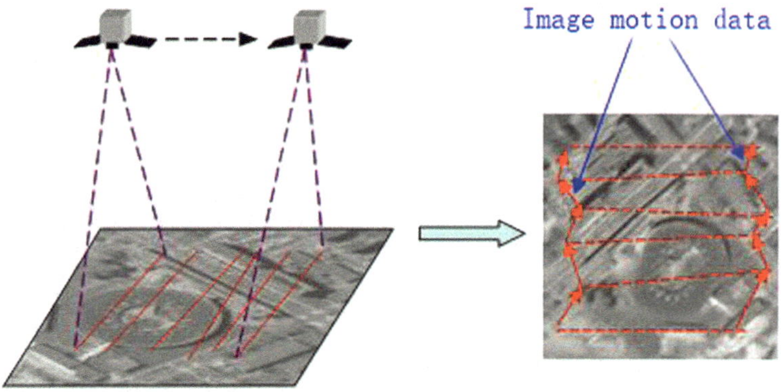


**Fig. 6** Space camera motion diagram of joint pictures

Fig. 6, the second image motion method is closer to the actual situation of the space camera. When the satellite moves in real time, the maximum image motion of adjacent two frames does not exceed 40 pixels.

Figure 7 shows the relationship between image motion and measurement error obtained by using the input image of Fig. 4 and the second image motion simulation. The abscissa indicates the amount of image motion in the *x* direction and the *y* direction, * and ◇ represent the measurement error values in the *x* and *y* directions, respectively. As shown in the Fig. 7, the measurement error increases with the increase of the image motion amount. When the image motion amount reaches 40 pixels, the measurement error reaches 0.25 pixels. As the image motion continues to increase, the correlation peak is overwhelmed because the correlation is too small.

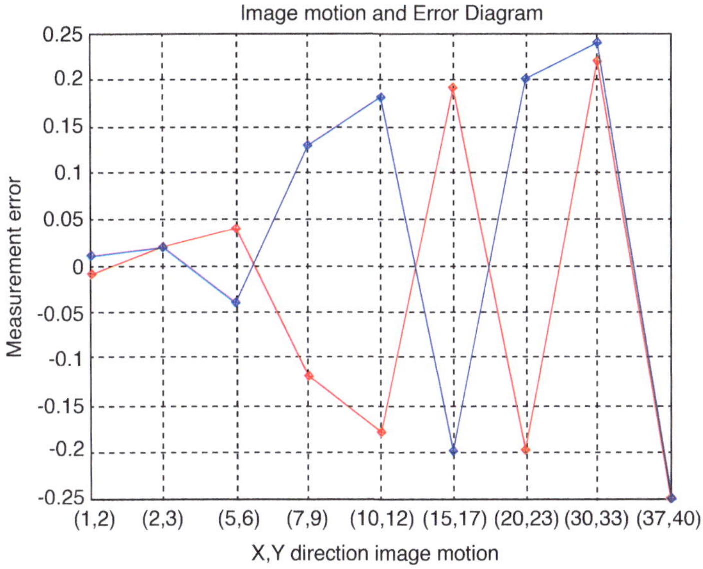


**Fig. 7** The relationship between image motion measurable image motion

In addition, when the camera is shooting on the ground, the image content collected by the image motion detector is constantly changing with the ground scene. In order to verify the accuracy of the algorithm, three different input images are used, and the image is from simple to complex. As shown in Fig. 9, the image contains a road with a simple structure and a complex urban community. Through simulation experiments, it is found that the algorithm has strong adaptability to the image, and the error value is almost unchanged from that of Fig. 7.

It is further considered that the device of the correlator itself is not an ideal device with no noise at all; for example, there will be scattering, spatial inhomogeneity, and surface error on the surface of the spatial light modulator, which modulates the input light intensity on the input surface. As a power spectrum and related output receiving device, CCD will inevitably have dark current noise, readout noise, etc., and these inherent defects will introduce background noise into the measurement. It is difficult to study the effect of these noises on the measurement accuracy when using the joint transform correlator to measure the image motion in actual situations, and the distribution form and size of these noises may vary depending on the device used. For the sake of simplicity, Gaussian white noise with a mean value of 0 different noise variance is added to the image 2 of Fig. 9 in the simulation. The influence of noise on measurement accuracy under different SNR is simulated. It is found that the measurement error is almost unchanged compared with Fig. 7.

At the same time, it is found that the noise is not the measurement accuracy but the maximum pixel value that can be measured. In the simulation, the image 3 in the most complicated image is taken as an example. Image 3 is shifted from the lower-left corner by 5 images to obtain image 4 with an SNR of 1. It can be seen from the

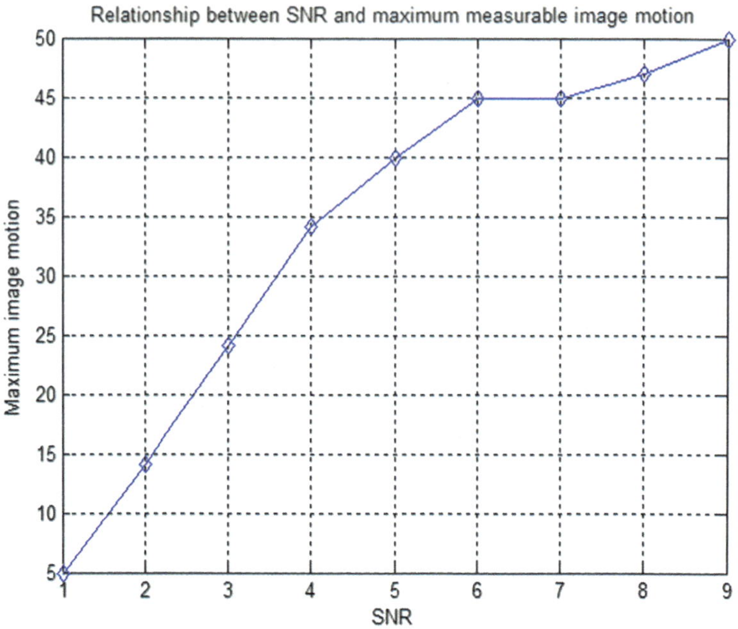


**Fig. 8** Relationship between SNR and maximum and measurement error

**Fig. 9** Simulation experiment picture (from left to right: image 1, image 2, image 3, image 4)

figure that the noise greatly reduces the correlation of the image, so the noise limits the maximum amount of image motion that can be detected. Figure 8 shows the maximum image motion relationship that image 3 can detect under different SNR conditions. When the SNR is 1, the maximum image motion is 5 pixels. When the SNR is 9, the maximum image motion is 50 pixels. The image motion is up to 60 pixels without adding noise. The same experiment was performed on image 1 and image 2 in Fig. 9, and the maximum image motion amount was 35 and 40 pixels, respectively, without adding noise. So in practice, when the input SNR is too low, it must be filtered to remove noise to ensure the effective range of image motion measurement.

From the above simulation experiments, we can draw the following conclusions:

1. The larger the image motion between the two related images, the lower the correlation between them and the smaller the correlation peak. When the image motion is large enough, the correlation peak is submerged. When the amount of image motion increases from small to large, the measurement error increases gradually when it is smaller than the maximum amount of image motion that can be detected, the maximum measurement error is 0.25 pixels. So, the measurement error depends largely on the quality of the power spectrum image peak position detection algorithm.
2. Image complexity also affects the measurement process of image motion, because the simpler the image, the lower the degree of correlation when there is image motion, and the more complex the image, the greater the degree of correlation. However, the complexity of the input image affects the intensity of the correlation peak and has little effect on the measurement accuracy of the image motion.
3. When the image SNR changes, the image motion error does not linearly increase or decrease. The SNR of the image can only affect the correlation of the image itself and does not affect the accuracy of the image motion measurement. When the image motion is the same, the SNR is large, the correlation peak is large, the SNR is small, and the correlation peak is small. The SNR only affects the maximum value of the image motion measurement. The more complex the image, the larger the SNR, and the larger the image motion maximum value. The simpler the image, the smaller the SNR, and the smaller the image motion maximum value.

## 5 Image Motion Measurement Test

As shown in Fig. 10, the spatial light modulator used in the experiment is an XGA3 type spatial light modulator produced by the optical center of the Central Research Laboratory of Scipher, UK, and the CCD is EL-400ME produced by DTA of Italy. The test device is mainly composed of a laser, a collimating optical system, a joint conversion correlator, a CCD camera, and a control computer [8, 9]. The joint transform correlator uses a single spatial light modulator, a single lens, and a computer-controlled opto-electric hybrid joint transform correlator. The computer simulates images of different image motion, injects spatial light modulators, and collects the relevant images output by the joint transform correlator with a CCD camera. The correlation peak image is processed by the automatic threshold and subpixel correlation power spectrum correlation peak method proposed in this paper, and the algorithm and performance of image motion measurement are evaluated.

In the experiment, the picture in Fig. 9 is used to verify the different input pictures, image signal-to-noise ratio, and image motion amount. The measured results

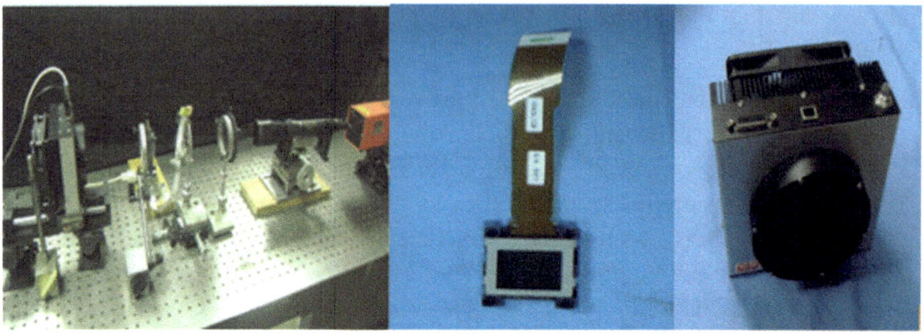

**Fig. 10** Image motion measurement test platform and main test device

are basically consistent with the simulation results, and the measurement error is less than 0.25 pixels. There are three issues to be aware of during the test:

1. Carefully adjust the position of the CCD, so that it is located on the back focal plane of the Fourier lens as much as possible, and rotate the position of the polarizer during the test, so that the light energy received by the CCD meets the test requirements.
2. The spatial light modulator itself may have defects such as scattering, spatial inhomogeneity, and surface error, which may affect the measurement accuracy.
3. When designing the optical path, pay attention to the uniformity of the laser beam intensity distribution. If the spatial light modulator is not uniformly illuminated, it will affect the measurement accuracy.

## 6 Conclusion

In summary, this paper uses optical correlation methods to analyze the image motion measurement of space cameras and several factors affecting measurement results. The simulation results show that the measurement accuracy depends largely on the accurate identification of the correlation peaks, which is linear with the image motion, and has no obvious relationship with the input image itself and the image signal-to-noise ratio. These factors only affect the maximum amount of image motion that the system itself can measure.

Finally, combined with the actual experiment, the measurement error is less than 0.25 pixels. The adaptive global threshold segmentation and subpixel correlation methods proposed in this paper have significant anti-interference to input and image noise, so the algorithm has strong robustness. In addition, because this method achieves correlation operation much faster than DSP and has high accuracy and flexible control, it has great prospects for real-time measurement on satellite in the future.

## References


1. Tchernykh, V., Dyblenko, S., Janschek, K.: Optical correlator based system for the real time analysis of image motion. Focal plane of an earth observation camera. Proc. SPIE. **4113**, 23–31 (2000)
2. Janschek, K., Tchernykh, V., Dybleenko, S.: Performance analysis of opto-mechatronic image stabilization for a compact space camera. Control. Eng. Pract. **15**, 333–347 (2007)
3. Janschek, K., Tchernykh, V., Dyblenko, S.: Integrated camera motion compensation by real-time image motion tracking and image deconvolution. Proc. IEEE/ASME. **2005**, 1437–1444 (2005)
4. Wen-Gang, H.U., Yong-Zhong, W.A.N.G., Wen-Shen, H.U.A.: The development and military application of optical correlator. Opt. Tech. **32**, 179–181 (2006)
5. Callagl, M.J.: The influence of SLM pixel size and shape on the performance of optical correlators and optical memories. SPIE. **4089**, 198–207 (2000)
6. Janschek, K., Tchernykh, V., Dybleenko, S.: Compensation of focal plane image motion perturbations with optical correlator in feedback loop. SPIE. **5570**, 280–288 (2004)
7. Kong-Bing, W.-Z., Tan, Y.-S.: Algorithm of laser spot detection based on circle fitting. Infrared Laser Eng. **31**, 275–279 (2002)
8. Tchernykh, V., Dyblenko, S., Janschek, K., Seifart, K., Harnisch, B.: Airborne test results for a smart pushbroom imaging system with optoelectronic image correction. SPIE. **5234**, 550–559 (2004)
9. Jutamulia, S., Guoguang, M., Zhai, H.: Use of laser diode in joint transform correlator. Opt. Eng. **43**(8), 1751–1758 (2004)

# Structure from Motion Using Homography Constraints for Sequential Aerial Imagery

**Banglei Guan, Zhang Li, Qifeng Yu, Yang Shang, and Xiaolin Liu**

**Abstract** With the omnipresence of unmanned aerial vehicle (UAV), sequential aerial imagery is becoming more widely available in both civilian and military programs. Considering that the monocular imagery is sequentially ordered according to its acquisition time and a dominant ground plane is often visible in the scene, we present a novel and complete automatic Structure from Motion system based on homography constraints. The homography induced by the ground plane is used to recover the camera pose. The windowed bundle adjustment and automatic loop closure are performed in sequential reconstruction. The global bundle adjustment is performed over the complete reconstruction. The proposed SfM system has successfully been used for sequential aerial imagery taken from a UAV.

**Keywords** Structure from motion · Homography · Ground plane · Sequential reconstruction

B. Guan · Z. Li (✉) · Q. Yu · Y. Shang
College of Aerospace Science and Engineering, National University of Defense Technology, Changsha, China

Hunan Provincial Key Laboratory of Image Measurement and Vision Navigation, Changsha, China

X. Liu
Hunan Provincial Key Laboratory of Image Measurement and Vision Navigation, Changsha, China

College of Intelligence Science and Technology, National University of Defense Technology, Changsha, China



H. P. Urbach, Q. Yu (eds.), *6th International Symposium of Space Optical Instruments and Applications*, Space Technology Proceedings 7,
https://doi.org/10.1007/978-3-030-56488-9_4

# 1 Introduction

Structure from Motion (SfM) has been an important field of research in computer vision and has been used successfully for a wide variety of applications, such as photogrammetry, visual navigation, and virtual reality. SfM involves recovering the 3D structure of scene and the camera motion. There are some open-source SfM systems that have been proposed recently. However, accuracy, computational complexity, and robustness of SfM algorithms are still a topic of research.

Basically, the existing SfM pipelines can be divided into three groups: sequential, incremental, and global SfM pipelines. Sequential SfM pipelines assume that the image sequence is ordered according to its acquisition time. By leveraging the temporal consistency of image sequence, the computational complexity of SfM is significantly reduced. Schönberger et al. [1] propose a SfM pipeline for close-distance and low-resolution image sequences, called MAVMAP. This SfM pipeline achieves high-resolution mapping with low-resolution cameras. Incremental SfM pipelines grow a 3D reconstruction by registering a new image at a time but do not require the image sequence's temporal consistency. Agarwal et al. [2] and Frahm et al. [3] have proposed incremental SfM pipelines for large-scale reconstruction of unordered photo collections. In addition, there are some open-source software for incremental SfM systems, for example, Bundler [4], VisualSFM [5], COLMAP [6], and so on. Global SfM pipelines take a fundamentally different approach to SfM by building and optimizing the viewing graph from unordered images [7]. Moulon et al. [8] propose a SfM pipeline based on the fusion of relative motions between image pairs, which has been implemented in OpenMVG.

At present, some SfM systems use the planar scene assumption as a prior to improve the accuracy and robustness of 3D reconstruction. Nicolas et al. [9] propose a linear method for computing a projective reconstruction based on planar homographies between views. Bartoli and Sturm [10] estimate the structure and motion parameters via user-provided geometry about a planer scene. Zhou et al. [11] detect and track the planes in the entire image sequence and then recover the camera motion and the scene structure using the homographies induced by the planes. Micusik and Wildenauer [12] use the planar structures of urban environments to improve overall SfM effectiveness. Raposo et al. [13] propose a SfM pipeline for jointly estimating the motion of stereo rig and the piecewise-planar structure. Collins and Bartoli [14] present a SfM system with planar scenes and affine cameras.

In this paper, we propose a novel and complete automatic SfM system for sequential aerial imagery based on homography constraints. The imagery is sequentially ordered according to its acquisition time. In addition, when an unmanned aerial vehicle (UAV) acquires the image sequence from a great height, a dominant ground plane is often visible in the scene. By taking the advantage of homography constraints induced by ground plane, our SfM system achieves more accurate and robust reconstruction.

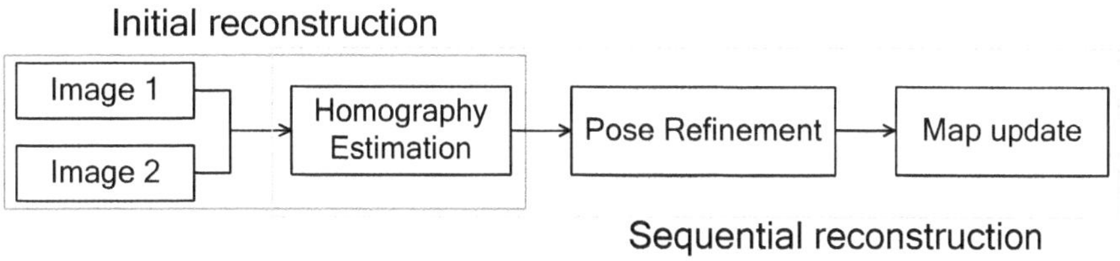


**Fig. 1** Overview of SfM pipeline

The remainder of the paper is structured as follows. Section 2 introduces our SfM pipeline. Section 3 describes the proposed algorithms. We evaluate our SfM pipeline using UAV dataset in Sect. 4. Finally, concluding remarks are given in Sect. 5.

## 2 SFM Pipeline

Our proposed SfM pipeline for sequential aerial imagery is shown in Fig. 1. We assume that a dominant ground plane is often visible in the scene. First, we choose a good initial image pair for initial reconstruction and estimate the homography between two images. The homography is decomposed into the relative rotation and translation of the initial image pair. The coordinates of 3D points are measured from the corresponding inliers of initial image pair by triangulation. Then, sequential reconstruction incrementally registers new image based on homography constraints. The translation can be only calculated up to one unknown scale factor. Considering the sequence's temporal consistency, the position of current image is directly set as the position of the nearest image and then a nonlinear pose refinement is employed by using all inliers from the RANSAC procedure.

## 3 Algorithms

### *3.1 Homography Estimation*

The general homographic relation for points belonging to a 3D plane and projected in two different views is defined as follows:

$$\lambda \mathbf{x}_j = \mathbf{H}\mathbf{x}_i, \tag{1}$$

where $\mathbf{x}_i = [x_i, y_i, 1]^T$ and $\mathbf{x}_j = [x_j, y_j, 1]^T$are the normalized homogeneous image coordinates of the points between the views $i$ and $j$, respectively, $\lambda$ is a scalar factor, and $\mathbf{H}$ is the homography between two views.

Let $\mathbf{X}$ be the coordinates of point in 3D plane relative to camera $i$ frame, $\mathbf{N} = [n_1, n_2, n_3]^T$ is the unit normal vector of the 3D plane with respect to the camera $i$ frame, and let $d > 0$ denote the distance from the 3D plane to the optical center of the camera $i$. Then, we have

$$\mathbf{N}^T\mathbf{X} = d. \tag{2}$$

The homography $\mathbf{H}$ can be given by:

$$\mathbf{H} = \mathbf{R} + \frac{1}{d}\mathbf{t}\mathbf{N}^T, \tag{3}$$

where $\mathbf{R}$ and $\mathbf{t}$ are, respectively, the rotation and the translation between views $i$ and $j$. In order to further eliminate the unknown scale in Eq. (1), multiplying both sides by the skew-symmetric matrix $[\mathbf{x}_j]_\times$, we obtain the equation:

$$\left[\mathbf{x}_j\right]_\times \mathbf{H}\mathbf{x}_i = \mathbf{0}, \tag{4}$$

where $[\mathbf{x}_j]_\times$ denotes the skew-symmetric matrix formed by $\mathbf{x}_j = [x_j, y_j, 1]^T$, which is expressed as follows:

$$\left[\mathbf{x}_j\right]_\times = \begin{bmatrix} 0 & -1 & y_j \\ 1 & 0 & -x_j \\ -y_j & x_j & 0 \end{bmatrix}.$$

Even though Eq. (4) has three rows, it only imposes two independent constraints on $\mathbf{H}$, because the skew-symmetric matrix $[\mathbf{x}_j]_\times$ is only of rank 2. Since each point correspondence gives rise to two independent equations, we require four point correspondences to solve for $\mathbf{H}$ up to one unknown scale factor.

## 3.2 *Homography Decomposition*

Based on the internal constraint of homography, the scale factor of the homography can be determined directly. We perform the SVD of homography matrix: $\mathbf{H} = \mathbf{U}\mathbf{\Sigma}\mathbf{V}^T$, where $\mathbf{U} \in SO(3)$, $\mathbf{V} \in SO(3)$, and $\mathbf{\Sigma} = \operatorname{diag}\{\sigma_1, \sigma_2, \sigma_3\}$. The scale factor of the homography is exactly the singular value $\sigma_2$. The homography $\mathbf{H}$ can be normalized as follows:

$$\mathbf{H} = \mathbf{H} / \sigma_2. \tag{5}$$

The rotation and translation between two different views are obtained by decomposing the homography $\mathbf{H}$. Note that the translation is only recovered up to scale, because the translation $\mathbf{t}$ and the distance between the camera $i$ frame and the 3D plane $d$ is coupled. Thus, when the homography $\mathbf{H}$ is decomposed to $\{\mathbf{R}, \mathbf{t}/d, \mathbf{N}\}$, there may be four sets of solutions [15]:

$$\begin{array}{ll} \{ \begin{array}{c} \mathbf{R}_1 = \mathbf{W}_1\mathbf{U}_1^T \\ \mathbf{N}_1 = [\mathbf{v}_2]_\times \mathbf{u}_1 \\ \mathbf{t}_1 / d_1 = (\mathbf{H} - \mathbf{R}_1)\mathbf{N}_1 \end{array} & \{ \begin{array}{c} \mathbf{R}_3 = \mathbf{R}_1 \\ \mathbf{N}_3 = -\mathbf{N}_1 \\ \mathbf{t}_3 / d_3 = -\mathbf{t}_1 / d_1 \end{array} \\ \{ \begin{array}{c} \mathbf{R}_2 = \mathbf{W}_2\mathbf{U}_2^T \\ \mathbf{N}_2 = [\mathbf{v}_2]_\times \mathbf{u}_2 \\ \mathbf{t}_2 / d_2 = (\mathbf{H} - \mathbf{R}_2)\mathbf{N}_2 \end{array} & \{ \begin{array}{c} \mathbf{R}_4 = \mathbf{R}_2 \\ \mathbf{N}_4 = -\mathbf{N}_2 \\ \mathbf{t}_4 / d_4 = -\mathbf{t}_2 / d_2 \end{array} \end{array} . \tag{6}$$

## 3.3 *Pose Refinement*

We can reconstruct the initial camera poses up to an unknown scale. We set Tx = 1 in the initial pair to unify scale in MAVMAP, in order to compare the reconstruction results of different methods. Based on the existing 3D point model, sequential reconstruction incrementally registers new cameras. But, the $\mathbf{t}$ can be only calculated up to one unknown scale factor. Considering the sequence's temporal consistency, the position of current image is directly set as the position of the nearest image and then a nonlinear pose refinement is employed by using all inliers from the RANSAC procedure.

All the parameters are optimized by a Levenberg–Marquardt algorithm that minimizes the reprojection errors. The minimized cost function $\varepsilon$ is described below. Under the assumption of Gaussian pixel noise, the proposed algorithm is a maximum-likelihood estimator.

$$\varepsilon = \min \sum_{m}^{k=1} \| \mathbf{x}_k - \tilde{\mathbf{x}} \left( \mathrm{K}, \mathbf{R}_W, \mathbf{T}_W, \mathbf{X}_W^k \right) \|, \tag{7}$$

where $k$ is the index of the 3D points in the existing 3D point model, $\mathbf{x}_k$ is the corresponding image coordinates, $\mathbf{x}\left(\mathrm{K}, \mathbf{R}_W, \mathbf{T}_W, \mathbf{X}_W^k\right)$ represents the reprojected image features of the 3D point $\mathbf{X}_W^k$, $\mathbf{K}$ denotes the intrinsic parameters of the camera, and $\mathbf{R}_W$ and $\mathbf{T}_W$ are the rotation and the translation from the 3D point model to the current image, respectively.

# 4 Experiment

In this experiment, we test our SfM pipeline on an image sequence acquired by UAV. The camera is typically looking toward the ground plane and 102 images are captured, as shown in Fig. 2. The image resolution is 4000 × 3000 pixels. We

**Fig. 2** UAV image

(a) (b)

**Fig. 3** SfM results. (**a**) Our SfM system (**b**) MAVMAP

compare our method against MAVMAP. The proposed algorithm is used to replace the 5 pt. essential matrix algorithm and the p3p algorithm in MAVMAP. Sequential reconstruction incrementally registers new image based on homography constraints. The other key steps of MAVMAP such as bundle adjustment are still retained in the experiment.

The camera poses and sparse 3D point cloud are shown in Fig. 3. The red mark indicates the recovered pose of camera, and the reconstructed point cloud is obtained from ours SfM pipeline at the same time. The poses of 102 images are recovered by both the proposed system and MAVMAP. Our SfM system using homography constraints achieves comparable results to MAVMAP.

In addition, we also make an analysis and statistics about the number of inliers obtained by our SfM pipeline and MAVMAP, which is shown in Table 1. It can be seen that our SfM pipeline obtains more inliers than the MAVMAP. Besides, the proposed SfM pipeline also obtains the higher ratio of inliers. The increase in the number of inliers can effectively improve the robustness of SfM system. This experiment demonstrates that our algorithm can successfully replace the 5 pt essential matrix algorithm and the p3p algorithm in the MAVMAP and obtain the robust reconstruction results.

**Table 1** Number of inliers obtained by our SfM pipeline and MAVMAP

| | MAVMAP | Our SfM pipeline |
|---|---|---|
| Average of the number of inliers | 110 | 132 |
| Ratio of inliers | 50.18% | 56.24% |

# 5 Conclusion

Considering that the monocular imagery is sequentially ordered according to its acquisition time and a dominant ground plane is often visible in the scene, we present a novel and complete automatic Structure from Motion (SfM) system based on homography constraints. The key technologies of our SfM system are as follows: For initial reconstruction, we estimate the homography induced by the ground plane and decompose the homography matrix to obtain the relative rotation and translation between initial image pair. For sequential reconstruction, the homography matrix between the current image and the previous image is estimated and the rotation matrix of the current image is recovered by homography decomposition. The position of current image can be directly set as the position of the nearest image by leveraging the temporal consistency. The rough pose of the current image is optimized using all 3D points. The proposed SfM system takes advantage of the presence of ground plane in the scene and has successfully been used for sequential aerial imagery taken from a UAV.

# References

1. Schönbergera, J.L., Fraundorfera, F., Frahmb, J.: Structure-from-motion for MAV image sequence analysis with photogrammetric applications. In: ISPRS-International Archives of the Photogrammetry, Remote Sensing and Spatial Information Sciences, vol. 1, pp. 305–312 (2014)
2. Agarwal, S., Furukawa, Y., Snavely, N., Simon, I., Curless, B., Seitz, S.M., Szeliski, R.: Building Rome in a day. Commun. ACM. **54**, 105–112 (2011)
3. Frahm, J.M., Fitegeorgel, P., Gallup, D., Johnson, T., Raguram, R., Wu, C., Jen, Y.H., Dunn, E., Clipp, B., Lazebnik, S.: Building Rome on a cloudless day. In: European Conference on Computer Vision, pp. 368–381 (2010)
4. Snavely, K.N.: Scene Reconstruction and Visualization from Internet Photo Collections. University of Washington, Washington (2008)
5. Wu, C.: Towards linear-time incremental structure from motion. In: International Conference on 3dtv-Conference, pp. 127–134 (2013)
6. Schönberger, J.L., Frahm, J.: Structure-from-motion revisited. In: IEEE Conference on Computer Vision and Pattern Recognition (2016)
7. Sweeney, C., Sattler, T., Hollerer, T., Turk, M., Pollefeys, M.: Optimizing the viewing graph for structure-from-motion. In: IEEE International Conference on Computer Vision, pp. 801–809 (2015)
8. Moulon, P., Monasse, P., Marlet, R.: Global fusion of relative motions for robust, accurate and scalable structure from motion. In: IEEE International Conference on Computer Vision, pp. 3248–3255 (2013)

9. Kaucic, R., Hartley, R., Dano, N.: Plane-based projective reconstruction. In: IEEE International Conference on Computer Vision, pp. 420–427 (2001)
10. Bartoli, A., Sturm, P.: Constrained structure and motion from multiple Uncalibrated views of a piecewise planar scene. Int. J. Comput. Vis. **52**, 45–64 (2001)
11. Zhou, Z., Jin, H., Ma, Y.: Robust plane-based structure from motion. In: IEEE Conference on Computer Vision and Pattern Recognition, pp. 1482–1489 (2012)
12. Micusik, B., Wildenauer, H.: Plane refined structure from motion. In: Scandinavian Conference on Image Analysis, pp. 29–40 (2017)
13. Raposo, C., Antunes, M., Barreto, J.P.: Piecewise-planar stereoscan: structure and motion from plane primitives. In: European Conference on Computer Vision, pp. 48–63 (2014)
14. Collins, T., Bartoli, A.: Stratified structure-from-motion for planar scenes and affine cameras. In: IEEE Transactions on Pattern Analysis and Machine Intelligence, p. 1 (2016)
15. Hartley, R., Zisserman, A.: Multiple View Geometry in Computer Visions. Cambridge University Press, Cambridge (2003)

# Simulation Samples Generation Method for Deep Learning in Object Detection of Remote Sensing Images

**Weichang Zhang, Wei Li, and Ningjuan Ruan**

**Abstract** As a typical data-driven method, deep learning relies on the availability of a huge amount of labeled data. Due to the serious lack of available samples, especially remote sensing images containing target information, the application of deep learning methods on object detection in remote sensing images is limited. Different from the common approach that tuning the deep learning model parameters to improve the detection performance, in this paper, we propose a simulation samples generating method for training the deep neural network, which can promote object detection ability for arbitrary deep learning framework. Based on the 3D model of target recovered by shape from laser scan technique and remote sensing background images, some image synthesis algorithms are employed for synthesizing multiple view 2D target images, which extends the number of images used in the training stage of a 2D object detection system. Recognition experimental results, using the detection framework of single shot multibox detector (SSD), demonstrate that Poisson image editing and Gaussian blur rendering can improve the quality of simulation remote sensing images, which significantly increases the detection performance of SSD on the various public remote sensing datasets.

**Keywords** Image synthesis · Deep learning · Remote sensing · Poisson image editing · Object detection

## 1 Introduction

As a powerful method for learning feature representations automatically from data, deep learning technique has attracted enormous attention in various fields, including data mining, natural language detection, and image recognition [1]. In particular, this technique has provided significant improvement for object detection, which is

W. Zhang · W. Li (✉) · N. Ruan
Beijing Key Laboratory of Advanced Optical Remote Sensing Technology, Beijing Institute of Space Mechanics and Electricity, Beijing, China



H. P. Urbach, Q. Yu (eds.), *6th International Symposium of Space Optical Instruments and Applications*, Space Technology Proceedings 7,
https://doi.org/10.1007/978-3-030-56488-9_5

expected to be combined with space remote sensing. With the rapid development of the deep learning detection framework, some previous works attempt to improve the accuracy of detection and recognition aiming at the specific targets such as ships and aircraft [2–6].

However, all these methods are necessary to face the same problem that the lack of available real remote sensing samples for training deep learning models. Although space remote sensors produce a large number of remote sensing images every day, researchers are still hard to acquire sufficient effective samples, especially remote sensing images containing target information. It limits the application of deep learning methods on object detection in remote sensing images.

To address it, this paper proposes the simulation samples generating method to solve the problem of small data challenges that have emerged in deep learning. Similar ideas have been studied for training deep neural networks to achieve Synthetic Aperture Radar (SAR) image recognition and camouflage target detection [7, 8]. The goal of this paper is to study feasible target images generating techniques to satisfy the requirements of training deep learning model used for space remote sensing object detection. Based on the 3D model of target recovered by shape from laser scan technique and remote sensing images, we first synthesize the foreground target images and background remoter sensing images through direct pixel value replacement. Then Poisson image editing and Gaussian blur rendering are employed to improve the quality of the synthetic images. Finally, we exemplify the use of simulation samples for training SSD model and demonstrate that the detection performance can be remarkably increased on publicly available datasets, including NWPU, DIRO, and RSOD.

## 2 Simulation Samples Generation

In this paper, the simulation samples generating method is based on the synthesis of front and back scenes, for which the 3D model of the aircraft at different angles and the background of sea surface are used to generate the simulation samples, as shown in Fig. 1.

### *2.1 Preliminary Synthesis Images*

In the first step, the foreground image in a completely black background is selected, when its RGB value is not equal to 0. Then these selected foreground images are embedded into the sea surface background. The foreground material images are directly inserted into the background to a random location, replacing the RGB value of the original background. A visualization of the synthesis images is shown in Fig. 2. In addition, the size of the target can be adjusted in the image by changing the downsampling parameters of the foreground material to restore remote sensing

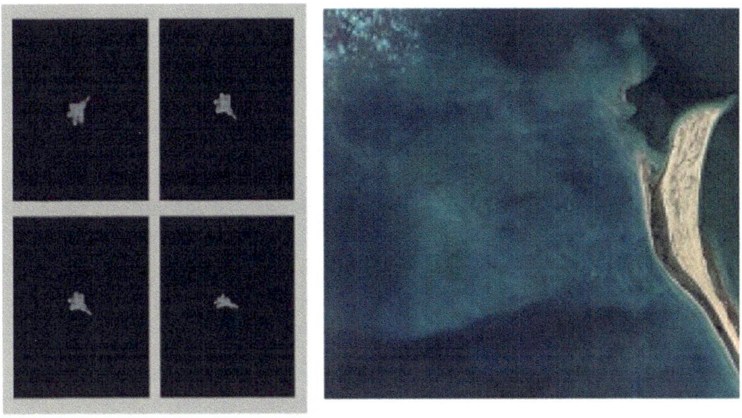

**Fig. 1** Material of foreground and background

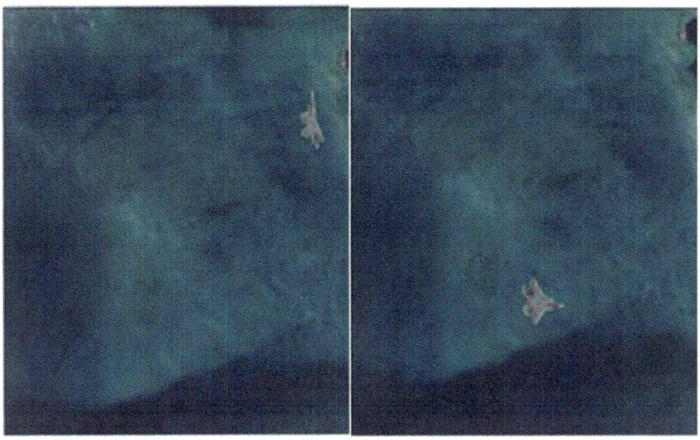

**Fig. 2** Samples of preliminary images

images in different situations. The background and foreground images are synthesized according to the resolution, which is scaled by the same criterion.

## 2.2 *Gaussian Blur Rendering*

Gaussian blur rendering is a type of noise whose probability density function obeys a Gaussian distribution (1). $\sigma^2$ and $\mu$ represent the variance and expectation of the Gaussian function. $\mu$ affects the center position of the function, and $\sigma^2$ affects the magnitude of the distribution.

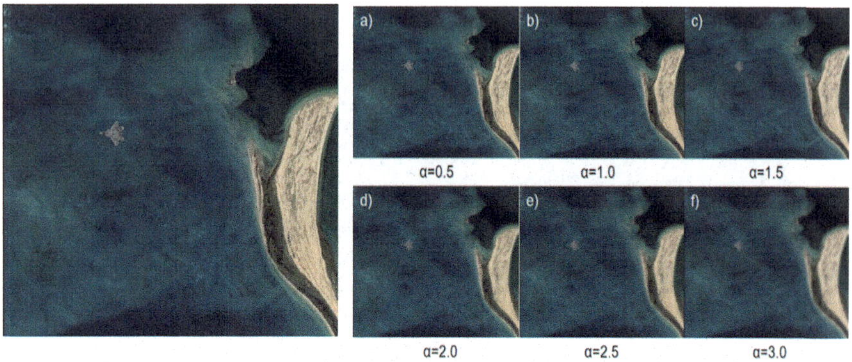


**Fig. 3** Original image (left), Gaussian blur in the order of 0.5–3.0 with 0.5 intervals (right)

$$p_{\mathrm{G}}(z)=\frac{1}{\sigma\sqrt{2\pi}}e^{-\frac{(z-\mu)^2}{2\sigma^2}}. \tag{1}$$

Introducing Gaussian blur on the generated image makes it be closer to the real situation. Its effect can be adjusted by changing the mean and variance of the Gaussian function to obtain better results. It can also be used to add noise to a solid-color background composite image to simulate a complex background. The following is a comparison between the image with different levels of Gaussian blur and the original image as shown in Fig. 3.

## 2.3 *Poisson Image Editing*

Poisson editing is a classic image synthesis method proposed by Perez et al. [9]. The mechanism of Poisson synthesis is to replace the traditional color intensity with color gradient. The color gradient in the synthetic region of the synthetic image edited by Poisson is the same as the background image, and the color intensity at the edge of the synthetic region is similar to the color intensity of the target region.

The core formula for Poisson editing is the Poisson Eq. (2) and Poisson partial differential equation with given Dirichlet boundary condition (3):

$$-\Delta\phi = f, \tag{2}$$

$$\begin{cases} -\Delta\phi(x,y) = f(x,y) & x,y \in \Omega \\ \phi(x,y) = g(x,y) & x,y \in \partial\Omega \end{cases} \tag{3}$$

where $\Delta=\frac{\partial^2}{\partial x^2}+\frac{\partial^2}{\partial y^2}$ is Laplacian operator, and $\Omega \in R^n$ is Bounded open set. InEqs. (2) and (3), $f$ and $\phi$ are real or complex-valued equations on the manifold, $f$ represents a known function, $\phi$ represents an unknown function, and $g$ is a given boundary condition.

Poisson image editing mainly involves three objects: source image, background image, and region of interest (ROI). It is divided into the following steps in applications:

(i) Solve the gradient of the source image.
(ii) Solve the gradient of the background image.
(iii) Solve the gradient field of the fused image and cover the background with the gradient field of the ROI region.
(iv) Take the partial derivative of gradient field to get the divergence of the region.
(v) Solve pixel values in ROI regions based on divergence and edge boundary conditions to reconstruct image.

Poisson image editing can realize different functions according to the operation process and adjustment of parameters, such as seamless clone, color change, illumination change, and texture flattening. In this paper, the seamless clone and its three modes are mainly used:

(i) Normal cloning: Preserve the gradient of the source image in the synthetic area.
(ii) Mixed cloning: The gradient of the synthetic area is determined by the source image and the background image together.
(iii) Monochrome transfer: Preserve the gradient of the source image and discard its color intensity, so that the color of the synthetic area is consistent with the background image.

Because normal cloning and monochrome transfer will produce obvious blur at the edges, the mixed cloning method is chosen for rendering the preliminary synthesis images. The results by three modes can be seen in the Fig. 4.

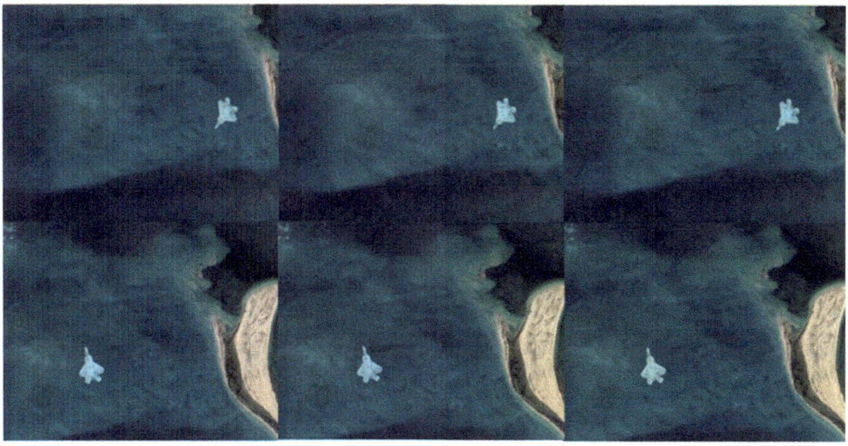

**Fig. 4** Two sets of pictures generated in three modes, normal cloning (left), mixed cloning (middle), monochrome transfer (right)

## 3 Training and Results

According to the simulation remote sensing target images, in this section the effect of simulation samples generating method for training the deep learning model is tested in object detection. Nowadays, there are such several derivative neural network models that have been proposed, such as R-CNN [10], SPP-Net [11], YOLO [12], SSD [13], etc. The network used in this training is a derivative network of Single Shot MultiBox Detector (SSD) model, whose basic structure is shown in Fig. 5. Based on VGG-16 model, SSD removes all dropout layers and combines selective search with multiscale features. The evident difference between SSD and other deep learning algorithms is that feature maps of different scales are applied to the final prediction stage, which improves its understanding of various levels of features to ensure accuracy and detection speed at the same time. It performs well on both complex targets and small targets because it learns features on different scales. SSD simultaneously acquires 6 different bounding boxes on each point of the different feature layers' feature map. Then Non-Maximum Suppression (NMS) is used to remove overlapping or incorrect bounding boxes to get the final bounding boxes.

Simulation image datasets based on preliminary synthesis images, Poisson image editing, and Gaussian blur rendering are used for training the SSD model. The detection performance of these models is evaluated by three public remote sensing datasets, including NWPU, DIRO, and RSOD, respectively.

Among them, the NWPU dataset is a public dataset containing aircraft targets on the ground, and we use more than 600 pictures of aircraft targets of NWPU as the detection set (Fig. 6). RSOD dataset is annotated by the research team of the Wuhan University, including 1000 images of four types of targets: such as airplane,

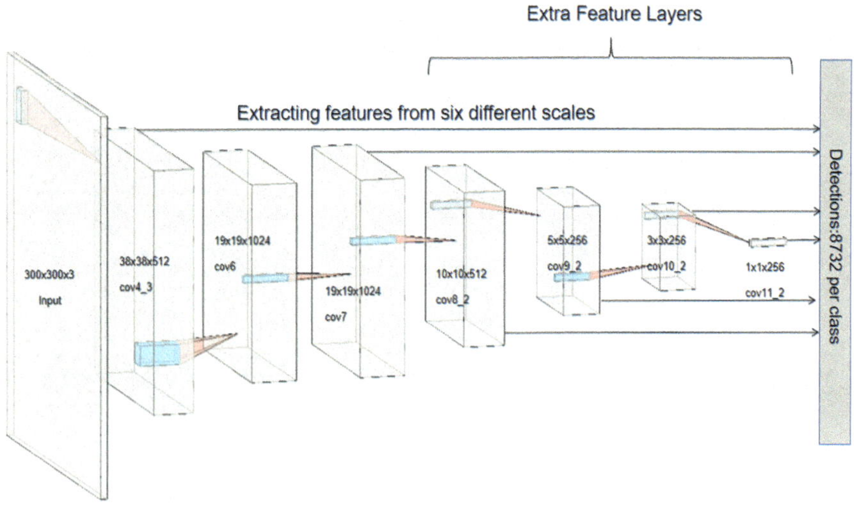


**Fig. 5** The basic structure of SSD [12]

**Fig. 6** NWPU dataset (left), DIRO dataset (middle), and RSOD dataset (right)

playground, overpass, and oil drum. In the dataset used in this study, each image contains more than ten airplane targets. DIRO dataset is a large-scale benchmark dataset for object detection in optical remote sensing images, including 23,463 images and 192,472 instances, covering 20 object classes. The part of this dataset that contains the images of airplanes is used in this study.

In this paper, three evaluation indicators are used to judge the target recognition ability of the model:

(i) Precision: The proportion of True positives in the identified pictures (4).
(ii) Recall: Proportion of all positive samples in the test set that are correctly identified as positive samples (5).
(iii) F1: Harmonic mean of precision and recall (6):

$$\text{Precision} = \frac{\text{TP}}{\text{TP}+\text{FP}}, \tag{4}$$

$$\text{Recall} = \frac{\text{TP}}{\text{TP}+\text{FN}}, \tag{5}$$

$$\text{F1} = 2\times\frac{\text{Precision}\times\text{Recall}}{\text{Precision}+\text{Recall}}, \tag{6}$$

where four possibilities True Positives (TP), True Negatives (TN), False Positives (FP), and False Negatives (FN) represent:

(i) TP: Positive samples are correctly identified as positive samples.
(ii) TN: Negative samples are correctly identified as negative samples.
(iii) FP: Negative samples are misidentified as positive samples.
(iv) FN: Positive samples are misidentified as negative samples.

Table 1 summarizes the results. It can be seen that the models trained on Poisson image editing and Gaussian blur rendering with alpha parameter range between 0 and 2 achieve the best performance on all public datasets, in precision, recall, and F1.

**Table 1** Effect of models trained on the generated dataset

| Model | Criterion | NWPU | DIRO | RSOD |
|---|---|---|---|---|
| Preliminary synthesis images | Precision | 0.2655 | 0.6373 | 0.1463 |
| | Recall | 0.1316 | 0.0858 | 0.0011 |
| | F1 | 0.1762 | 0.1512 | 0.0022 |
| Poisson image editing | Precision | 0.6733 | 0.7199 | 0.0647 |
| | Recall | 0.4529 | 0.2916 | 0.0071 |
| | F1 | 0.5416 | 0.4150 | 0.0127 |
| Gaussian blur rendering (alpha 0 ~ 1) | Precision | 0.3784 | 0.2882 | 0.0424 |
| | Recall | 0.2561 | 0.1095 | 0.0035 |
| | F1 | 0.3055 | 0.1587 | 0.0065 |
| Gaussian blur rendering (alpha 0 ~ 2) | Precision | 0.5069 | 0.4770 | 0.0594 |
| | Recall | 0.3519 | 0.1781 | 0.0067 |
| | F1 | 0.4154 | 0.2594 | 0.0120 |
| Gaussian blur rendering (alpha 0 ~ 3) | Precision | 0.4007 | 0.4225 | 0.0174 |
| | Recall | 0.3118 | 0.1583 | 0.0026 |
| | F1 | 0.3507 | 0.2303 | 0.0045 |

The Preliminary Synthesis Images present the worst training results. The reason may be the single material and background making the network overfit, or the complexity of the real image background is ignored. Although the model trained on Gaussian Blue Rendering is less effective than Poisson Image Editing, it still improves the detection performance of SSD model. It should be noticed that the detection ability relies on the variance of Gaussian blur. With the value of variance growing, the performance of SSD model first increases and reduces subsequently, because that Gaussian blur with the proper variance as the inherent feature of target images has been learned by the detection model.

Furthermore, the P–R curves are plotted in Fig. 7, with the corresponding cases of preliminary synthesis images, Poisson image editing, and Gaussian blur rendering ($\alpha \propto (0, 2)$), respectively.

As can be seen from the P–R curves, models trained on primary synthesis images perform worse than other methods. On all the simulation datasets, the curves have large fluctuation, leading to the low value of precision and F1. In contrast, the P–R curves of the models trained on the datasets generated by the other two methods are relatively flat. Besides, it can be noticed that the curves of models trained on the dataset which generated by Poisson image editing are closer to the parabola and the maximum recall value, as shown in the Fig. 7. This is the reason that datasets generated by Poisson image editing have the better performance.

Similar to what we observed on NWPU and DIRO datasets, RSOD dataset also shows the regularity of models trained on different simulation samples generation methods. But RSOD dataset contains a large number of targets on the same image, and all models have a poor recognition effect on it.

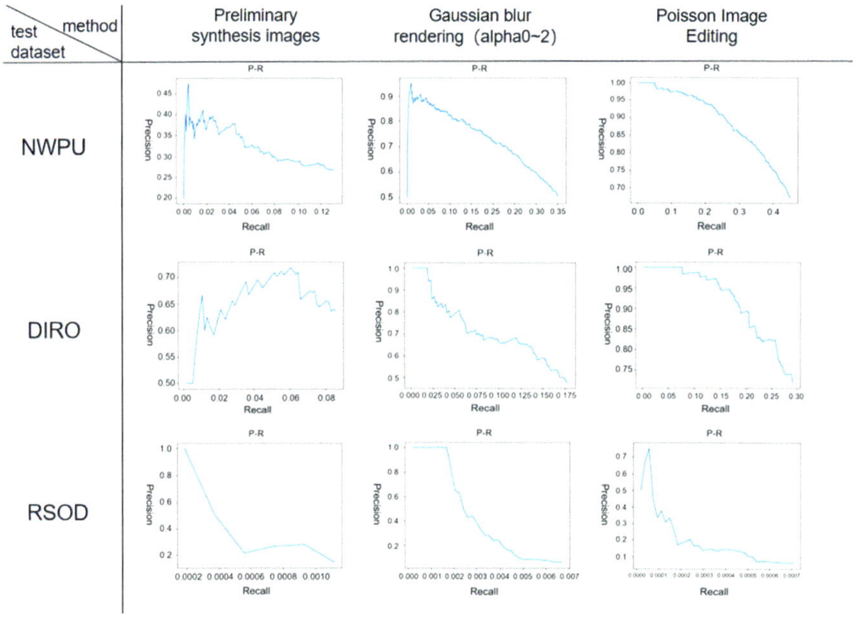


**Fig. 7** P–R curves of several groups of experiments

# 4 Conclusion

In this paper, several different methods are used to generate virtual image datasets and to train models through SSD deep learning algorithms for remote sensing detection. Based on the results of several models trained on the test dataset, it is found that Poisson image editing and Gaussian blur can improve the training effect. The original intention of using generated virtual images to train the model is to improve the performance for intelligent remote sensing detection. And image generation can also be used in many other aspects.

However, at this stage, the F1 evaluation index is far from satisfying the requirement of general purpose practical applications, so it is necessary to make further improvement and execution. More attempts, such as combining Poisson editing and Gaussian blur rendering, searching the perfect variance of Gaussian blur, optimizing the synthesis algorithm for Poisson image editing, and adding more realistic environmental effects and noise. In addition, increasing the amount and type of material to expand the target to images be identified is also a possible way.

## References

1. Li, S.X., Zhang, Z.L., Li, B.: A plane target detection algorithm in remote sensing images based on deep learning network technology. J. Phys. Conf. Ser. **960**, 012025 (2018)
2. Wang, H.Z., Gong, Y.C.: DeepPlane: a unified deep model for aircraft detection and recognition in remote sensing images. J. Appl. Remote. Sens. **11**(4), 1 (2017)
3. Jubelin, G., Khenchaf, A.: Multiscale algorithm for ship detection in mid, high and very high resolution optical imagery. In: 2014 IEEE Geoscience and Remote Sensing Symposium. Igarss IEEE, Quebec City, QC (2014)
4. Sun, Z.C., Tan, X.C., Hong, Z.H., Dong, H.P., Sha, Z.Y., Zhou, S.Z., et al.: Remote sensing image object detection based on deep convolution neural network. Aerosp. Shanghai. **35**(5), 18–24 (2018)
5. Huang, J., Jiang, Z., Zhang, H., Yao, Y.: Ship object detection in remote sensing images using convolutional neural networks. J Beijing Univ. Aeronaut. Astronaut. **43**(9), 1841–1848 (2017)
6. Wang, H.L., Zhu, M., Lin, C.B., Chen, D.B., Yang, H.: Ship detection of complex sea background in optical remote sensing images. Opt. Precis. Eng. **26**(3), 723–732 (2018)
7. Liu, C.Q., Chen, B., Pan, Z.H., Wang, W.H., Tang, X.B.: Research of target recognition technique via simulation SAR and SVM classifier. J. China Acad. Electron. Inform. Technol. **11**(3), 257–262 (2016)
8. Liu, Z., Chen, X.Q., Xie, Z.P., Jiang, X.J., Bi, D.K.: Simulation learning method for discovery of camouflage targets based on deep neural networks. Laser Optoelectron. Prog. **56**(7), 154–160 (2019)
9. Pérez, P., Gangnet, M., Blake, A.: Poisson image editing. ACM Trans. Graph. **22**(3), 313 (2003)
10. Girshick, R., Donahue, J., Darrell, T., Malik, J.: Region-based convolutional networks for accurate object detection and segmentation. IEEE Trans. Pattern Anal. Mach. Intell. **38**(1), 142–158 (2015)
11. He, K.M., Zhang, X.Y., Ren, S.Q., Sun, J.: Spatial pyramid pooling in deep convolutional networks for visual recognition. IEEE Trans. Pattern Anal. Mach. Intell. **37**(9), 1904–1916 (2014)
12. Redmon, J., Divvala, S., Girshick, R., Farhadi, A.: You only look once: unified, real-time object detection. In: IEEE Conference on Computer Vision and Pattern Recognition. IEEE, Las Vegas, NV (2016)
13. Liu, W., Anguelov, D., Erhan, D., Szegedy, C., Reed, S., Fu, C., et al.: SSD: single shot multibox detector. In: Leibe, B., Matas, J., Sebe, N., Welling, M. (eds.) Computer Vision – ECCV 2016. ECCV 2016, Lecture Notes in Computer Science, vol. 9905. Springer, Cham (2015). https://doi.org/10.1007/978-3-319-46448-0_2

# Aircraft Target Recognition in Optical Remote Sensing Image with Faster R-CNN

**Hu Tao, Li Runsheng, Hu Qing, and Ke Qingqing**

**Abstract** In order to detect and recognize aircraft targets in remote sensing images quickly and accurately, this paper proposes a method based on Faster R-CNN convolution neural network. Aircraft targets can be recognized and located rapidly through sample labeling, candidate region generation, aircraft model training, and target detection. The results show that it is feasible to use Faster R-CNN convolution neural network for target recognition in optical remote sensing images. It has significant advantages and potential in computational efficiency and accuracy.

**Keywords** Remote sensing image · Deep learning · Aircraft recognition · Faster R-CNN

## 1 Introduction

Optical remote sensing image contains abundant information and possesses fast update speed. It is obvious that the use of artificial methods to interpret optical remote sensing image and target recognition has long been unable to meet the needs of modern society in various fields. Therefore, real-time and efficient recognition of interested objects from optical remote sensing images is of great significance.

Deep learning was first proposed by Hinton et al. in 2006 [1]. Through the research, the deep learning model of convolutional neural network has been successfully applied to face recognition with its excellent performance. In 2012, Professor Geoffrey Hinton reduced the top-5 error rate of traditional detection algorithm to 15.3% [2] in ILSVRC (ImageNet Large Scale Visual Recognition Challenge) based on convolutional neural network in deep learning. In 2014, the algorithmic framework emerged, which made a significant breakthrough in the performance of deep learning in target detection. The average detection accuracy of region proposal + CNN algorithm on PASCAL VOC 2012 data set is 53.3% [3]. In 2015, the spatial pyramid pooling network [4] structure completed the input of any image size. In the

H. Tao · L. Runsheng (✉) · H. Qing · K. Qingqing
Information Engineering University, Henan, Zhengzhou, China



H. P. Urbach, Q. Yu (eds.), *6th International Symposium of Space Optical Instruments and Applications*, Space Technology Proceedings 7,
https://doi.org/10.1007/978-3-030-56488-9_6

same year, the accelerated version of R-CNN, Fast R-CNN, completed the end-to-end and loss function of the single-stage training process, which greatly accelerated the training speed. In 2016, Ren Shaoqing and others proposed a further improved network architecture Faster R-CNN in order to solve the problem of generating candidate box regions for long time in target detection [5]. In addition, some other algorithms based on convolutional neural network have also achieved good results [6–9].

Aiming at aircraft target recognition in remote sensing image, based on Faster R-CNN, an experimental environment for aircraft target detection and location in optical remote sensing image is built. The detection and location of aircraft target in optical remote sensing image are realized through sample labeling, candidate region generation, aircraft model training, and target detection and recognition, which is artificial intelligence. Energy technology provides theoretical and experimental support in the field of remote sensing target recognition.

## 2 Faster R-CNN Network Model

The process of target detection algorithm based on candidate region is candidate region generation, feature extraction, and detection and recognition of interested objects. Faster R-CNN has been developed by R-CNN and Fast R-CNN. The whole process has been unified into a deep learning network framework, which can be completed in GPU. The training speed and target detection speed of the neural network have been greatly improved. The structure of Faster R-CNN network model is shown in Fig. 1. The whole calculation process is as follows: (1) input any size

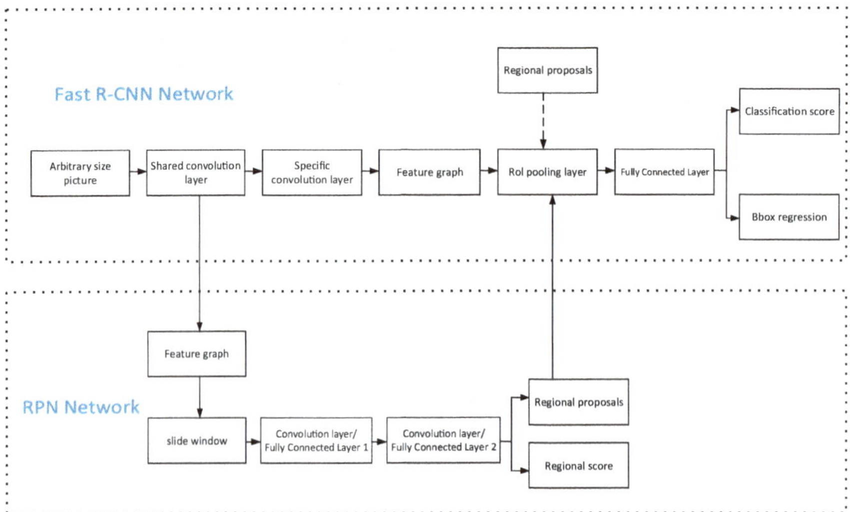


**Fig. 1** Faster R-CNN network model structure

picture to ZF network, and (2) propagate forward to the final shared convolution layer through ZF network and divide it into two paths. One way continues to propagate forward to the specific convolution layer to produce higher dimensional feature maps; the other way transmits to the input feature maps of RPN network; and (3) The feature maps transmitted to the RPN can get region proposals and region scores through the RPN, and non-maximum suppression is used for the region scores. Then, the region proposals with top-N scores are recommended to the ROI Pooling layer. (4) The high-dimensional feature maps obtained in Step 2 and the region proposals obtained in Step 3 are sent to the ROI Pooling layer simultaneously to extract the features of the corresponding region proposals. (5) The features of region proposals obtained in Step 4 pass through the full connection , then export the score of the region and the bounding boxes after regression [10].

## 2.1 Recommended Region Extraction

Faster-RCNN model uses RPN to generate target candidate regions. The core idea of RPN is to use convolution neural network to generate the recommendation area directly. This method only needs to slide on the final convolution layer once. By using anchor mechanism and border regression, the multiscale and multiaspect ratio recommendation area can be obtained ingeniously. The network structure is shown in Fig. 2 (with ZF network as the reference model). Above the dotted line is the structure in front of the last convolution layer of ZF network, and below the dotted line is the unique structure of RPN network. First, the convolution of $3 \times 3$ is performed, and then the output of $1 \times 1$ convolution is divided into two paths, one of which outputs the probability of target and nontarget, and the other outputs four parameters related to box, including the center coordinates $x$ and $y$ of box, the width $W$ of box, and the length $h$ of box.

Anchor is the core of RPN network. It takes the form of $(x1, y1, x2, y2)$ to represent the coordinates of the points in the upper-left and lower-right corner of the rectangle. In fact, RPN eventually sets up dense candidate anchors on the scale of the original map. Then CNN is used to determine which anchors are foreground anchors with targets and which are background anchors without targets.

One reshape before and after software max, combined with the form of Caffe basic data structure blob, is more conducive to the classification of software max. Box regression is used to map the original anchor to a more realistic box after a certain transformation. In this way, the modified anchor position is obtained. After getting the anchor correction parameters of each candidate area, the more accurate anchor is calculated. Then the anchor is ranked according to the regional score of the aircraft from large to small, and some obviously inconsistent anchors are eliminated, as well as other filtering conditions are added. The anchor is then output as a candidate box.

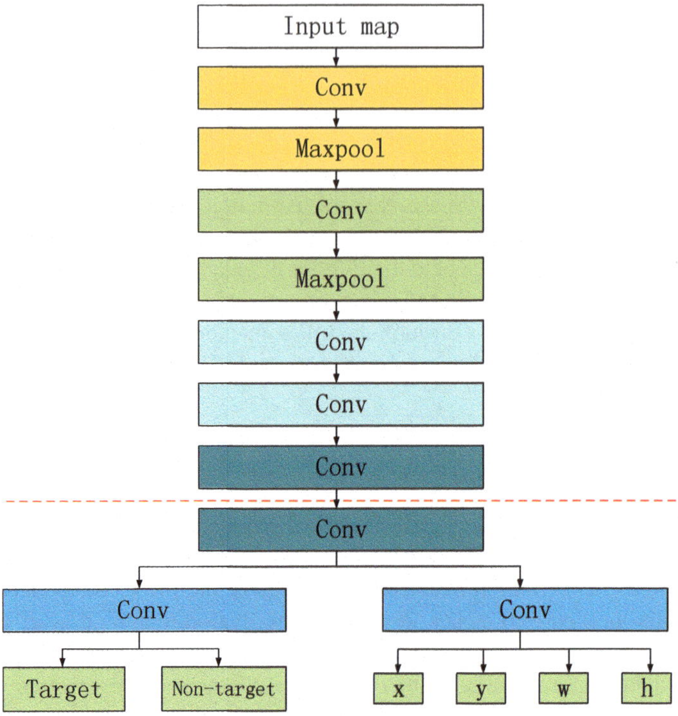


**Fig. 2** RPN network architecture (ZF network as reference model)

## *2.2 Training Network Model*

Faster R-CNN currently has three training network models: ZF, VGG_CNN_M_1024, and VGG16. This paper chooses ZF model. (1) Feature extraction process, input image→convolution→ReLU activation→maximum pooling→feature map. (2) The process of feature restoration, feature map anti-pooling anti-Relu activation deconvolution visual image. The size and step size of convolution core need to be set artificially, and many experiments are needed to find the best design. Because ZF network realizes feature visualization, we can analyze the visual image and find the optimal design parameters. At the same time, adjusting the parameters can reduce the scale of the data needed by the training network, so that the high-resolution remote sensing images with small data can also be classified as Faster R-CNN.

# 3 Experimental Design and Result Analysis

The experiment was completed under the framework of the deep learning platform Caffe. The high-speed operation is realized by GPU. The GPU used is NVIDIA GeForce GTX 1080ti, and the processor is Intel (R) Core (TM) i7-4790. The CUDA is used to drive the GPU for parallel computing to improve the processing speed [11].

IoU, accuracy, recall, AP, and P–R curves are used to evaluate the performance of the model. The IoU standard is used to measure the similarity between reality and prediction. In the training image of target detection, there are a large number of artificial annotation frames for certain types of targets. After processing by neural network, the prediction frame can be generated. The similarity or overlap rate of the two boxes is the IoU value.

P (Precision) is the accuracy:

$$\text{precision} = \frac{\text{TP}}{(\text{TP} + \text{FP})}. \tag{1}$$

R (Recall) is the recall rate:

$$\text{recall} = \frac{\text{TP}}{(\text{TP} + \text{FN})} = \frac{\text{TP}}{\text{P}}. \tag{2}$$

AP (Average Precision) is the average accuracy. Mean Average Precision (MAP) is the average AP value and the average accuracy of multitarget average.

P–R curve is a two-dimensional curve with accuracy as the vertical axis and recall rate as the horizontal axis.

## 3.1 *Experimental Data*

The experimental data set was taken from Google Earth. Six types of targets were selected: bomber (B52 and B1B), transporter (C17 and C130), tanker (KC135), fighter (multiple models), helicopter (various models), and passenger aircraft (various models). The data set contains 1409 remote sensing images in various scenarios. The resolution is 0.5 m.

According to VOC2007 format, batch renaming is carried out, and then the annotation program is designed to annotate the image one by one. After the annotation is completed, the XML file is generated. Then the samples are divided into training set, test set, and verification set. Among them, the test set accounts for about 50% of the total data set, and the test set and training set account for about 25% of the total data set, thus constituting the best network. Finally, the VOC2007 format data set is generated [8].

## 3.2 *Analysis of the Experimental Process and Results*

From a large number of images in the sample library, the aircraft detection model is trained and the P–R curves of various targets are obtained.

By collecting the images of all interested objects in a certain location, the recognition accuracy of the whole region can be analyzed and estimated through the IoU

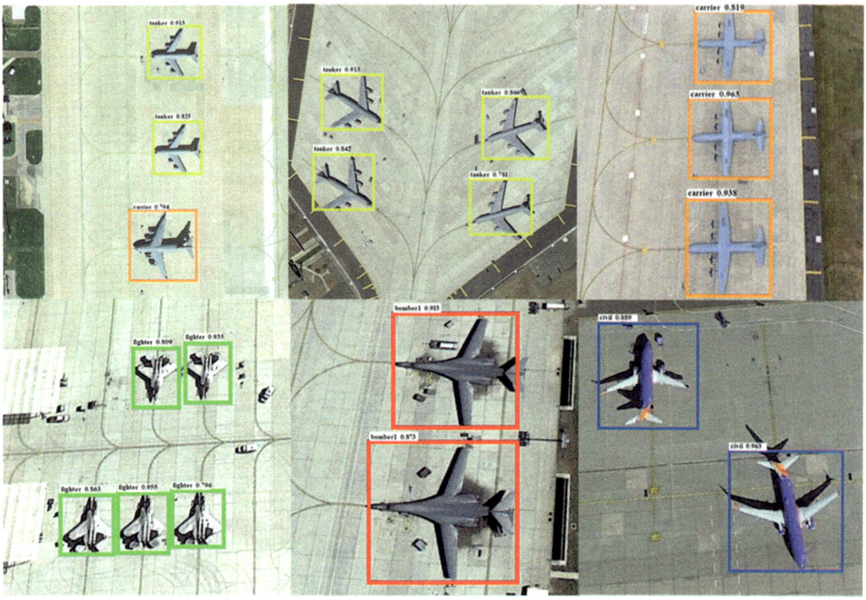


**Fig. 3** Target detection results of an airport

values of all targets in the whole region. Taking an airport as an example, the aircraft target is identified. The test results are shown in Fig. 3. There are six types of targets in the airport, and all of them are detected and recognized. Most of the IoU values are above 0.8. The results show that the model can estimate the detection accuracy of the target of interest.

A large number of B52 bomber images with different texture, shade, and material background are searched for experiment. The experimental results are shown in Fig. 4. Most of the IoU values in each scenario are above 0.9. The results show that the target detection and recognition model in this paper can achieve fast and accurate detection and recognition of targets in different scenarios.

The time for image detection and recognition is positively correlated with the size of the image. The larger the image, the more time will be required for detection and recognition, and the relationship curve will be more intuitive. Take the size of each image as the X-axis and the detection and recognition time as the Y-axis and draw a scatter diagram. Then, you fit it, and the curve you get is the relationship between them. The relationship between image size and required time is shown in Fig. 5.

For large-scale image, block, and scale processing should be performed, and avoid target areas when blocking [6]. The result is shown in Fig. 6. Every target is detected and located, and the IoU values are all above 0.7, most of which are above 0.9, and there is no error recognition. The results show that the model can detect and identify the objects of interest quickly and accurately in a large range of complex scenarios.

The remote sensing image of an airport is shown in Fig. 6. The airport has a single target, and the aircrafts are arranged orderly. From the perspective of algorithm recognition effect, the integrity and accuracy are high, and only one target is not recognized (Mark 1 in the figure). We can see the superiority of the algorithm.

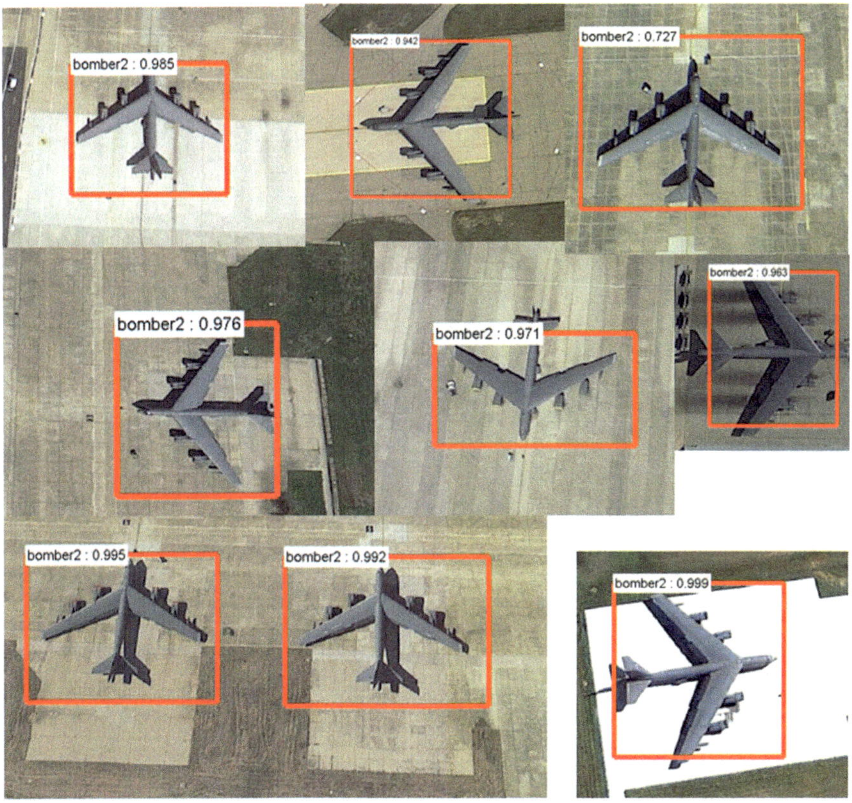


**Fig. 4** B52 bomber result in different scenes

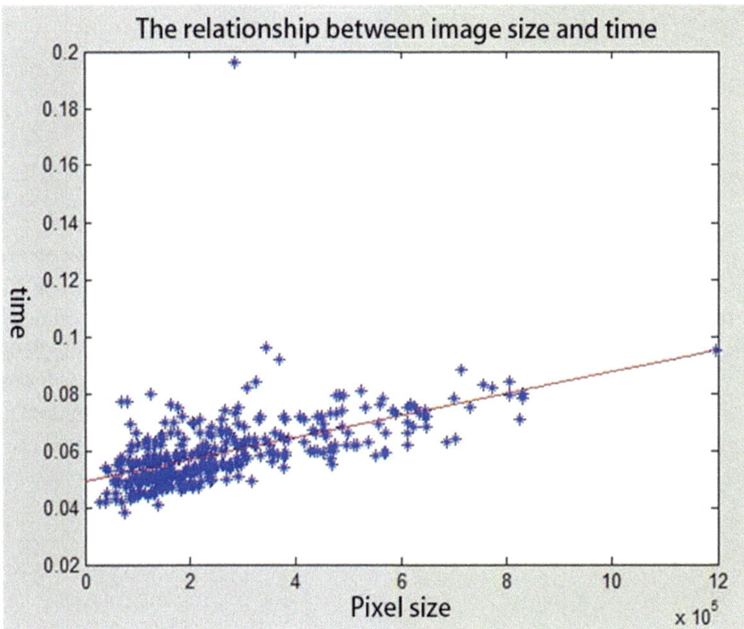


**Fig. 5** The relationship between image size and detection speed

**Fig. 6** Experimental results and local magnification of aircraft targets in large scale

# 4 Conclusion

In this paper, a Faster R-CNN network model based on ZF network is constructed on the Caffe deep learning framework. Through the experiment of aircraft target recognition in optical remote sensing images, the fast and accurate recognition of the model is analyzed. Experiments show that the model can estimate the detection accuracy of the target of interest in a certain location and can accomplish fast and accurate detection and recognition of the target in different scenarios, as well as fast and accurate detection and recognition of the target of interest in a large range of complex scenarios.

## References


1. Hinton, G., ESalakhutdinov, R.R.: Reducing the dimensionality of data with neural networks. Science. **313**(5786), 504–507 (2006)
2. Krizhevsky, A., Sutskever, I., Hinton, G.E.: Imagenet classification with deep convolution neural networks. In: Advances in Neural Information Processing Systems, pp. 1097–1105 (2012)
3. Girshick, R., Darrell, J., Darrell, T., Malik, J.: Rich feature hierarchies for accurate object detection and semantic segmentation. In: 2014 IEEE Conference on Computer Vision and Pattern Recognition, pp. 580–587. IEEE, Columbus, OH
4. Mingwei, H.: Convolutional neural network image classification algorithms based on spatial pyramid pooling. Master's degree thesis, Wuhan University, Wuhan (2018)
5. Shaoqing, R.: Efficient object detection based on feature sharing. Doctoral dissertation, China University of Science and Technology, Hefei (2016)
6. Zhiyuan, Z.: Optical remote sensing image aircraft detection based on deep learning. Master's degree thesis, Xiamen University, Xiamen (2016)
7. Xiaorui, L.: End-to-end target detection and attribute analysis algorithm based on deep learning and its application. Master's degree thesis, South China University of Technology, Guangzhou (2018)
8. Danxin, Z.: Research on deep learning method for aircraft target extraction in remote sensing images. Master's degree thesis, University of Chinese Academy of Sciences, Beijing (2018)
9. Wenbin, Y.: Research on application of convolutional neural network in remote sensing target recognition. Master's degree thesis, University of Chinese Academy of Sciences, Beijing (2018)
10. Feng-xiao, L.: Research on rapid target detection method based on deep learning. Master's degree thesis, Hangzhou University of Electronic Science and Technology, Hangzhou (2018)
11. Jia, Y., Shelhamer, E., Donahue, J., et al.: Caffe: convolutional architecture for fast feature embed ding. In: Proceedings of the 22nd ACM International Conference on Multimedia, pp. 675–678 (2014)

# On-Orbit SNR Measurement Using Staring Image Series

**Dianzhong Wang**

**Abstract** Signal to noise ratio is a critical parameter to assess the performance of a remote sensing satellite. On-orbit signal to noise ratio measurements have been dominated by homogeneous area method that requires homogeneous areas no less than 100 × 100 pixels within an image. This is not difficult to meet for high-resolution satellites in low Earth orbit. When the spatial resolution rises to 50 m × 50 m at nadir, just as the case of GF-4 satellite, it becomes hard to find out such a homogeneous area as large as 5 km × 5 km in most on-orbit images. Considering that the satellite has an observation mode called staring, that is acquiring image series in a short duration, a new method is proposed in this paper for measuring on-orbit signal to noise ratio through staring image series. The full measurement procedure includes on-orbit imaging, image sampling, calculating signal to noise ratio at pixel level and data fitting. A series of 22 images are acquired from which a homogeneous area with 728 pixels is determined. Each pixel series is calculated for a signal to noise ratio, and then their distribution is fitted by Gaussian model. The result shows that on-orbit signal to noise ratio could reach an expectation as high as 115 at a 95% confidence interval. This procedure could avoid impact of pixel's radiometric non-uniformity so as to give a rational assessment for satellite with staring mode. Therefore, it is recommended in future satellite project.

**Keywords** Signal to noise ratio · On-orbit measurement · Geo-stationary orbit Image series

## 1 Introduction

Signal to noise ratio (SNR or S/N) is defined as the ratio of signal and noise, which is a critical parameter to assess image quality of remote sensing. There are two major types of signal to noise ratio measurement: one is homogeneous area [1], and

D. Wang (✉)
Beijing Key Laboratory of Advanced Optical Remote Sensing Technology, Beijing Institute of Space Mechanics and Electricity, Beijing, China



H. P. Urbach, Q. Yu (eds.), *6th International Symposium of Space Optical Instruments and Applications*, Space Technology Proceedings 7,
https://doi.org/10.1007/978-3-030-56488-9_7

the other is geostatistics [2] or improved geostatistics [3], which was proposed to consider impacts of land cover types on system noise and measured signal to noise ratio with a function containing wavelength and land cover types. Other references proposed standard deviation of grey level between homogeneous area and nonhomogeneous area [4, 5]. All of the above researches were concerned with airplane platform or satellites in low Earth orbit that was no higher than several 100 km. While geostationary orbit is 35,786 km away from the Earth, about dozens times of Sun-synchronous orbit's height, extra distance leads to relatively lower spatial resolution for remote sensing [6]. GF-4 satellite was launched in December 2015 and known as the highest resolution in geostationary orbit at that time. This satellite owns an unprecedented observation mode called staring mode, which means successive monitoring in a short duration [7–10]. As a geostationary orbit satellite, GF-4 can avoid the deficiency of low Earth orbit satellite in successive monitoring and improve regional monitoring capability, which is badly needed in disaster and emergency monitoring [11–14]. However, spatial resolution of GF-4 satellite is only 50 m at nadir, means a 100 × 100 pixel homogeneous area equal to 5 km × 5 km large, which is practically hard to find out in practical image. As it can only use laboratory relatively calibration for on-orbit relatively radiometric correction, the spatial non-uniformity could not be eliminated ideally. In this case, a larger homogeneous area with heterogeneity obviously will lead to underestimation of signal to noise ratio. In order to solve this problem, a research was carried out on measurement of on-orbit signal to noise ratio through staring imaging series.

## 2 Data and Method

The full measurement procedure included on-orbit imaging, image sampling, calculating signal to noise ratio at pixel level and data fitting, as shown in Fig. 1.

Staring observation request was first submitted to China Centre for Resources Satellite Data and Application (CCRSDA), which was in charge of GF-4 satellite daily operation. The experiment site located in northeast China and their centre location was 126.0° E, 46.2° N, as shown in Fig. 2.

The experiment was conducted on July 26, 2016, and a series of 22 images were found and downloaded from CCRSDA's online data inquiry system as listed in Table 1.

The next step was to determine the homogeneous area from the series. In order to meet the requirement of 100 × 100 pixels, the least amount of pixels of the homogeneous area should not be smaller than the quotient of 10,000 and image numbers. In this case, a sample no less than 455 pixels was needed as the image number was 22. Larger homogeneous area and more images would be preferred. As these images were acquired within 10 min, the scenarios did not differ much between images. So we picked out an image from the series, searched throughout and found a lake to be homogeneous and big enough. Within the lake, a sample of 728 pixels was carefully chosen to avoid emergency of cloud, island and mixed pixels in any of the series, as shown in Fig. 3.

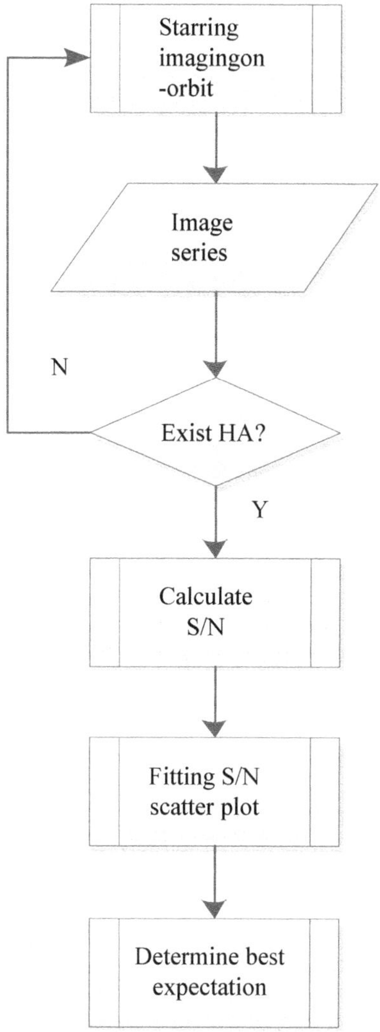


**Fig. 1** Flow of the proposed procedure

Obviously, sample from one image could not meet the requisite of the traditional homogeneous area method, which depended on the spatial mean $\overline{x_{i,j}}$ and standard deviation $\sigma_{i,j}$ of an $M \times N$ sample, as shown by the following equation:

$$S/N = \frac{\overline{x_{i,j}}}{\sigma_{i,j}}, i \leq M, j \leq N. \tag{1}$$

However each pixel in this sample stood for a position where there would be a series of pixel in other 21 images. Therefore, total sample series had 16,016 pixel

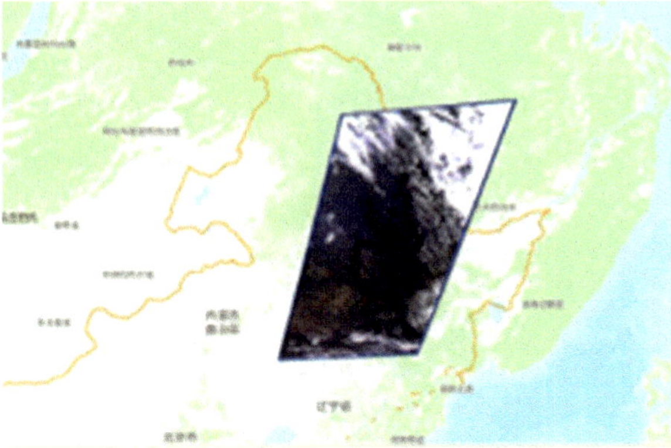

**Fig. 2** Location of the image series

**Table 1** Information of image series

| No. | Image name |
|---|---|
| 1 | GF4_B1_E126.0_N46.2_20160726_L1A0000124120 |
| 2 | GF4_B1_E126.0_N46.2_20160726_L1A0000124122 |
| 3 | GF4_B1_E126.0_N46.2_20160726_L1A0000124123 |
| 4 | GF4_B1_E126.0_N46.2_20160726_L1A0000124124 |
| 5 | GF4_B1_E126.0_N46.2_20160726_L1A0000124125 |
| 6 | GF4_B1_E126.0_N46.2_20160726_L1A0000124126 |
| 7 | GF4_B1_E126.0_N46.2_20160726_L1A0000124128 |
| 8 | GF4_B1_E126.0_N46.2_20160726_L1A0000124129 |
| 9 | GF4_B1_E126.0_N46.2_20160726_L1A0000124130 |
| 10 | GF4_B1_E126.0_N46.2_20160726_L1A0000124131 |
| 11 | GF4_B1_E126.0_N46.2_20160726_L1A0000124133 |
| 12 | GF4_B1_E126.0_N46.2_20160726_L1A0000124134 |
| 13 | GF4_B1_E126.0_N46.2_20160726_L1A0000124135 |
| 14 | GF4_B1_E126.0_N46.2_20160726_L1A0000124136 |
| 15 | GF4_B1_E126.0_N46.2_20160726_L1A0000124137 |
| 16 | GF4_B1_E126.0_N46.2_20160726_L1A0000124138 |
| 17 | GF4_B1_E126.0_N46.2_20160726_L1A0000124140 |
| 18 | GF4_B1_E126.0_N46.2_20160726_L1A0000124142 |
| 19 | GF4_B1_E126.0_N46.2_20160726_L1A0000124144 |
| 20 | GF4_B1_E126.0_N46.2_20160726_L1A0000124145 |
| 21 | GF4_B1_E126.0_N46.2_20160726_L1A0000124146 |
| 22 | GF4_B1_E126.0_N46.2_20160726_L1A0000124151 |

records in total and was big enough to be taken as valid sample for signal to noise ratio measurement.

For each pixel location in the sample series, there were 22 records in all 22 images. Therefore, the mean and the standard deviation of every pixel location could be calculated. Since the values of digital number were from the same pixel for the same land cover, the ratio above could reflect the temporal variation of each pixel.

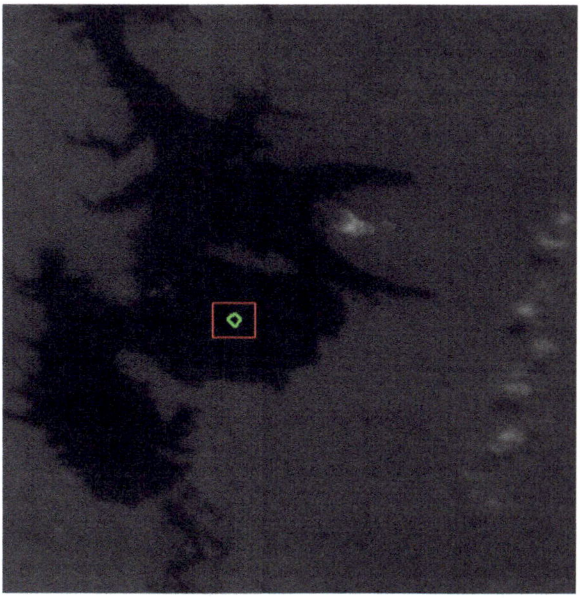

**Fig. 3** Sample location in the image (within red box)

The mean of a pixel could be taken as the signal and the standard deviation could be taken as noise, and the signal to noise ratio could be calculated with the following equation:

$$S/N = \frac{\overline{x_k}}{\sigma_k}, k \leq K, \quad (2)$$

where $K$ was the image number of the series, $\overline{x_k}$ was the mean and $\sigma_k$ was the standard deviation. In this way, all 728 pixels were calculated and then processed to find out the expectation in statistics.

## 3 Result and Analysis

Signal to noise ratio results were calculated for all 728 pixels. Obviously, the results were not of the same value but attributed to 119 values ranging from 72 to 275. Some value of signal to noise ratio was commonly shared by pixels while some was not. The situation commonness at pixel level was counted in 20 detailed levels for all these 119 values and is shown in Table 2.

From Table 2, we could see that counts of signal to noise ratio densely appeared between a much narrower range from 98 to 124. If the distribution was demonstrated in scatter plot as shown in Fig. 4, it was clear that they could be fitted by a certain trending line model. Several models were then tried to fit the scatter and finally Gaussian model was chosen as the best.

**Table 2** Counts of signal to noise ratio value

| Number of sharing pixels | Signal to noise ratio |
|---|---|
| 1 | 72,74,76,79,80,81,82,84,158,166,174,175,177,178,179,180,183,184,187,188, 191,193,196,206,218,220,223,232,265,275 |
| 2 | 75,83,153,163,164,172,176 |
| 3 | 145,151,162,167,171,173,182,186 |
| 4 | 85,86,87,91,125,138,143,146,148,149,154,159,168,170 |
| 5 | 134,152,157,169 |
| 6 | 104,136,137,139,140,144,150 |
| 7 | 93,94,100,112,128,129,147,155 |
| 8 | 117,119,120,127,156 |
| 9 | 88,89,108,130,142 |
| 10 | 97,116,131,135,141 |
| 11 | 92,95,103,110,113,114 |
| 12 | 106,109,118,123,126,133 |
| 13 | 96,105,115 |
| 14 | 90,101,102,132 |
| 15 | 107,111,122 |
| 16 | 99 |
| 18 | 121 |
| 19 | 98 |
| 20 | 124 |

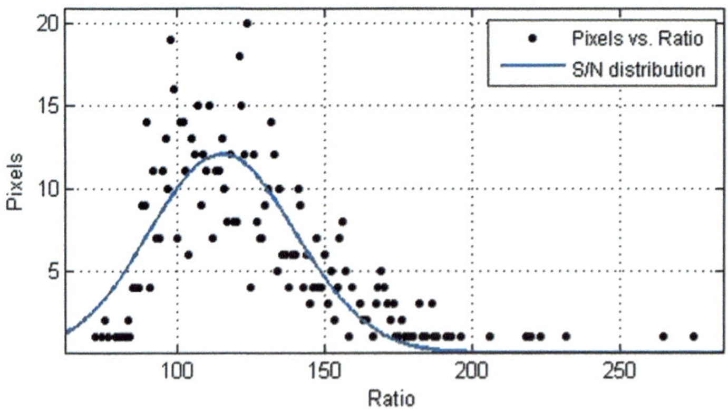


**Fig. 4** Scatter plot of signal to noise ratio

Gaussian model in this case had three coefficients as shown in the following equation:

$$f(x) = \frac{a}{e^{\left(\frac{x-b}{c}\right)^2}}. \tag{3}$$

At a 95% confidence interval, coefficient *a* equaled 12.01, *b* equaled 115.2 and *c* equaled 35.1. And *b* was the expectation in statistics of signal to noise ratio. It demonstrated that Gaussian model fitting could characterize the trending of scatter distribution. Every scatter had been integrated 22 statistics from the series, and all 728 scatters had been used for fitting, so that the information of the homogeneous area had been fully used. Moreover, as lab coefficients were still used in relative calibration due to the lack of on-orbit calibration, pixel's radiometric non-uniformity changes after launch could not be fully reflected. This procedure could avoid the problem of pixel's radiometric non-uniformity.

It could be transformed to 41.2 dB in decibel representation according to the following equation:

$$\mathrm{SNR} = 20 \times \lg\left(\mathrm{S/N}\right). \tag{4}$$

here S/N =115.2 as determined above.
Referring to Xu et al. who once reported a measurement of 44.0 dB in ocean and 42.7 dB in sand for on-orbit signal to noise ratio, the result in this paper was fairly close and proved rational [15].

## 4 Summary

In this paper, a method was proposed for measuring on-orbit signal to noise ratio. This procedure reflected pixel's temporal signal and noise character by calculated mean and standard deviation of each pixel series. The case in the paper used more than 20 pixels for each signal to noise ratio statistics and hundreds of pixels for fitting to get an expectation of 41.2 dB at a high confidence interval. Since it could avoid impact of pixel's radiometric non-uniformity and give a rational assessment for geostationary orbit satellite, it was recommended to use in relevant satellite projects.

This work was funded by NSFC: 41871278 and CHEOS: 50-Y20A07-0508-15/16.

## References

1. Duggin, M.J., Sakhavat, H., Lindsay, J.: Systematic and random variations in thematic mapper digital radiance data. Photogram. Eng. Remote Sens. **51**, 1427–1434 (1985)
2. Curran, P.J., Dungan, J.L.: Estimation of signal-to-noise: a new procedure applied to AVIRIS data. IEEE Trans. Geosci. Remote Sens. **27**, 620–628 (1989)
3. Atkinson, P.M., Sargent, I.M., Foody, G.M., Williams, J.: Exploring the geostatistical method for estimating the signal-to-noise ratio of images. Photogram. Eng. Remote Sens. **73**, 841–850 (2007)
4. Baltsavias, E.P., Pateraki, M., Zhang, L.: Radiometric and geometric evaluation of IKONOS geo images and their use for 3D building modeling. In: Proceedings of Joint ISPRS Workshop on High Resolution Mapping from Space, Hannover, Germany, September 19–21 (2001)

5. Crespi, M., De Vendictis, L.A.: Procedure for high resolution satellite imagery quality assessment. Sensors. **9**, 3289–3313 (2009)
6. Meng, L.J., Guo, D., Tang, M.H., Wang, Q.: Development status and prospect of high resolution imaging satellite in geostationary orbit. Spacecr. Recover. Remote Sens. **37**(4), 1–6 (2016) (in Chinese)
7. Wang, D.: Research on basic observation mode for high resolution satellite at geo-stationary orbit. Space Int. 52–54 (2015) (in Chinese)
8. Wang, D., He, H.: Observation capability and application prospect of GF-4 Satellite. In: Proceedings of 3rd International Symposium of Spaces Optical Instrument and Application, Beijing, China (2016)
9. Li, G., Kong, X.H., Liu, F.J., Lian, M.L.: GF-4 satellite remote sensing technology innovation. Spacecr. Recover. Remote Sens. **37**(4), 7–15 (2016) (in Chinese)
10. Ma, W.P., Lian, M.L.: Technical characteristics of the staring camera on board GF-4 satellite. Spacecr. Recover. Remote Sens. **37**(4), 26–31 (2016) (in Chinese)
11. Liu, M., Wu, W., Shu, Y., Fan, Y.D., Li, S.J., He, H.: Evaluation of water extraction based on GF-4 satellite data. Spacecr. Recover. Remote Sens. **37**(4), 96–101 (2016) (in Chinese)
12. Wu, W., Qin, Q.M., Fan, Y.D., Liu, M., Shu, Y.: Timeliness testing of GF-4 satellite data product and disaster reduction application service. Spacecr. Recover. Remote Sens. **37**(4), 101–109 (2016) (in Chinese)
13. Jing, F., Xu, Y.R., Zhang, X.Y., Shen, X.H., Chen, L.Z.: Potential application in earthquake research using data from GF-4 satellite. Spacecr. Recover. Remote Sens. **37**(4), 110–115 (2016) (in Chinese)
14. You, J., Zhang, H.Q., Chen, Y.F., Liu, H., Gao, Z.H.: Comparative study of dongting lake wetland information extraction based on GF-4 satellite image. Spacecr. Recover. Remote Sens. **37**(4), 116–122 (2016) (in Chinese)
15. Xu, W., Long, X.X., Li, Q.P., Cui, L., Zhong, H.M.: Image radiometric and geometric accuracy evaluation of GF-4 satellite. Spacecr. Recover. Remote Sens. **37**(4), 16–25 (2016)

# Estimation of Economic Parameters in Yangtze River Delta Using NPP-VIIRS and Landsat8 Data

**Bing Zhou, Xiaoli Cai, Yunfei Bao, and Jiaguo Li**

**Abstract** In view of traditional survey method cannot accurately and efficiently reveal the economic parameter spatial and temporal information, a new nighttime light index-based VIIRS surface reflectance data is proposed to simulate economic parameters. The built-up area was used for spatial scope of economic parameter spatial processing, acquired by the method of enhanced built-up and bareness index based on Landsat8 remote sensing data. Gross regional domestic product, secondary industry, tertiary industry, and gross industrial production at municipal scale were selected, and regression models were established between total nighttime light index and economic parameter. Then the nighttime light value using the VIIRS surface reflectance data and the economic parameter were compared and analyzed. The light-economic sensitive parameters at night were selected, and the grid data of sensitive economic parameters at 0.5 km × 0.5 km spatial scale were obtained by inversion. The research results show that the use of surface reflectance data to process the light values can improve the accuracy of the simulated economic parameter model. The average value of $R^2$ was increased from 0.798 to 0.888, and the average error has decreased from 0.516 to 0.354. GDP and tertiary industry are sensitive parameters, as the economic parameter rapidly increased and formed stretch area around the Shanghai, Nanjing, and other first-tier cities. This result is consistent with the actual situation. Meantime, the GDP density map and the tertiary industry density map can reflect the economic distribution details and macro distribution characteristics of the Yangtze River Delta, which can provide some basis for the

B. Zhou
Henan Key Laboratory of Big Data Analysis and Processing, Henan University, Kaifeng, China

X. Cai
School of Computer and Information Engineering, Henan University, Kaifeng, China

Y. Bao (✉)
Beijing Institute of Space Mechanics and Electricity, Beijing, People's Republic of China

J. Li
Aerospace Information Research Institute, Chinese Academy of Sciences, Beijing, China



H. P. Urbach, Q. Yu (eds.), *6th International Symposium of Space Optical Instruments and Applications*, Space Technology Proceedings 7,
https://doi.org/10.1007/978-3-030-56488-9_8

study of the economic strategy and sustainable development of the Yangtze River Delta.

**Keywords** Nighttime light data · NPP/VIIRS · Gross domestic product Economic parameter model

## 1 Introduction

The twenty-first century is a period of rapid development of human civilization, and economic parameters are one of the key indicators for measuring the development of human civilization. At present, most countries in the world independently formulate socioeconomic statistical calibers based on local needs. As a result, statistical data between countries cannot be directly compared, which affects the efficiency of use, and it is difficult to objectively and comprehensively reflect the laws of time and space of economic parameters. Luminous remote sensing provides a relatively reliable global indiscriminate data source, which can be regarded as a proxy variable for economic development. At present, night light data have become one of the important data sources for socioeconomic parameter space simulation, which provides us with timely and accurate spatial measurements of human activity.

Many scholars at home and abroad have studied the simulation or estimation methods of economic parameters. Nordhaus et al. [1] divided the grid of 1 ° × 1 ° around the world on the basis of previous studies and plotted the global GDP spatial distribution map. Zhou et al. [2] used a linear regression model to quantitatively study the relationship between nighttime light (NTL), GDP, and population and obtained a significant regional difference in the central and western regions of China through a series of inequality coefficients. Levin et al. [3] used night-time light data and population density, road length, road density, weekly traffic, and distance from intersections to build regression models and draw confidence ellipses and showed that night lighting in the Jewish quarters was positively correlated with population density and average weekly passenger flow. Nataliya et al. [4] used DMSP/OLS data research to confirm the feasibility of reconstructing the geographic pattern of economic activity using night light satellite measurement data, and the night light intensity often increased with population density and per capita GDP. Eleni [5] used NPP/VIIRS and DMSP/OLS data from 2012 and 2013 and found that night light radiation is closely related to tourism activities.

This paper selects 27 cities in the Yangtze River Delta region as the research object, introduces VIIRS reflectivity product data, divides the night light image and its reflectivity product pixel by pixel and rebuilds the model with economic parameters. It is found that the average error of the estimation can be reduced and the accuracy of the estimation can be improved. Then, using the simulation model, we obtained the GDP and tertiary industry grid data with a resolution of 0.5 km × 0.5 km in the Yangtze River Delta region in China in order to provide support for expanding the range of luminous data to simulate the socioeconomic parameters and provide a basis for drawing economic strategies and development routes in the Yangtze River Delta region.

# 2 Study Area and Data

## 2.1 *Study Area*

The Yangtze River Delta urban agglomeration is centered on Shanghai, located in the impact plain before the Yangtze River enters the sea, with a total population of about 243 million people. The economy is dominated by the secondary and tertiary industries. According to the outline of the Yangtze River Delta regional integrated development plan for 2019, the Yangtze River Delta includes Shanghai, Jiangsu, Zhejiang, and Anhui, covering an area of 225,000 square kilometers. This article uses 27 prefecture-level cities in the region. A study area is shown in Fig. 1.

## 2.2 *Data*

First, download the 2018 VIIRS monthly synthetic global image from the NOAA National Geographic Data Center website. Data of four socioeconomic parameters of 27 central cities in the Yangtze River Delta in 2018 are collected from the 2019 China urban statistical yearbook. The administrative boundary data of China's cities are from the National Basic Geographic Information Center. The data used are shown in Table 1.

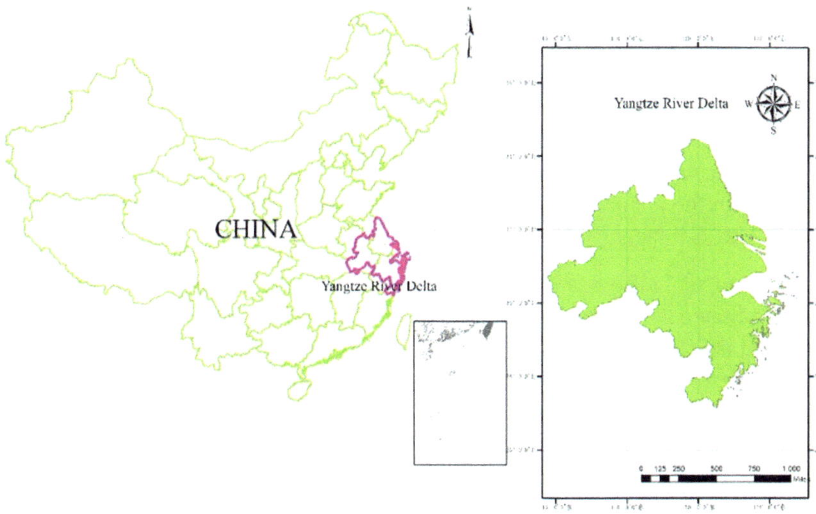


**Fig. 1** The location of study area

**Table 1** List of data sources

| Type of data | Date acquired | Source | Scale/spatial resolution |
|---|---|---|---|
| Administrative division map | 2017 | National Basic Geographic Information Center | 1:250,000 |
| Landsat 8 | 2018 | http://glovis.usgs.gov | 30 m × 30 m |
| Economic parameter | 2018 | Statistical yearbook and statistical bulletin by region | City-level division |
| NPP/VIIRS | 2018 | https://www.ngdc.noaa.gov/eog/viirs/download_dnb_composites.html | 0.5 km (after resampling) |
| VIIRS surface reflectance datasets | 2018 | https://search.earthdata.nasa.gov/ | 1 km |

## 2.3 *Data Preprocessing*

1. Preprocessing of night light data: The original data of VIIRS use the WGS1984 coordinate system. The luminous image acquisition time is about 1:30 in the morning and the spatial resolution of VIIRS is 15 arc seconds, which will cause the problem that the image decreases with the increase of latitude. In order to avoid the deformation of the image due to the coordinate system and to facilitate the calculation of the image area, we converted all the NPP/VIIRS image data into the Albers equal volume projection coordinate system. Finally, the image is resampled to a grid of 0.5 km × 0.5 km size. The first version of the monthly VIIRS synthetic data used in the study did not deal with the effects of aurora, fire, and other transient light sources, so there was background noise, and the light emissivity values were negative, minimum, and high. The presence of noise and outlier data can affect data simulation accuracy. In response to this situation, we did the following according to the method of Shi [6]: set the value less than zero to 0.001 (approximately 0, no effect on statistical analysis), set the background value to 0, set the value greater than 235, and set the value to 235 to remove some outliers. Since the original timescale of the NPP/VIIRS night light data products obtained is monthly products, it is necessary to use the synthesized annual data when spatializing. This article uses the average method to synthesize and obtain the annual data of NPP/VIIRS in the Yangtze River Delta region in 2018. VIIRS images of the study area are shown in Fig. 2.

2. Extraction of built-up areas: The extracted data of the built-up area use the United States Landsat8 remote sensing image. The transit time of the image is selected in 2018. The extraction method in this paper uses the enhanced built-up area and exposed index method (EBBI) proposed by As-syakur et al. [7]. This method uses a band operation to select a suitable threshold to extract the target. The equation for the EBBI index is

$$\mathrm{EBBI} = \frac{\left(\rho_{\mathrm{MIR}} - \rho_{\mathrm{NIR}}\right)}{10\sqrt{\rho_{\mathrm{MIR}} + \rho_{\mathrm{TIR}}}}, \tag{1}$$

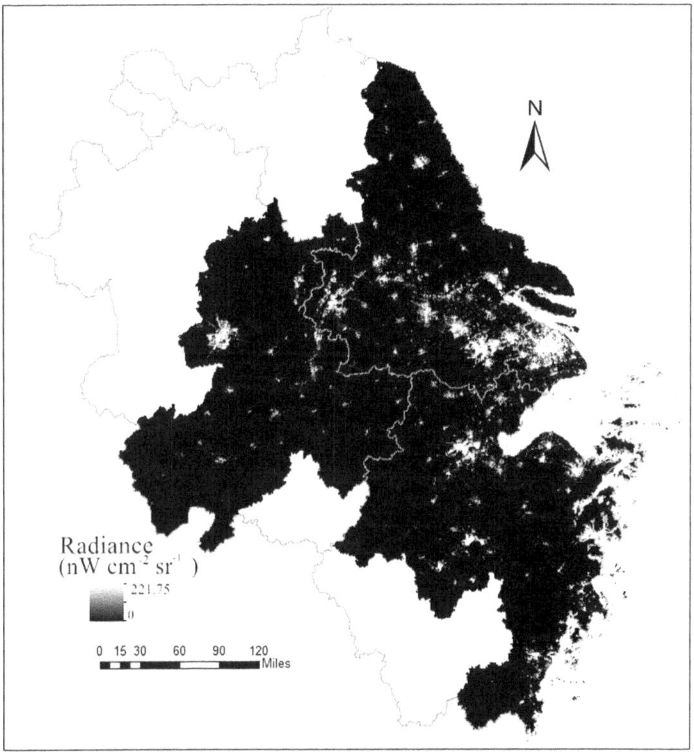


**Fig. 2** VIIRS Nighttime Light image for study areas

where $\rho_{NIR}$, $\rho_{MIR}$, and $\rho_{TIR}$ represent the near-infrared, mid-infrared, and the reflectivity of ground objects in the thermal infrared band, respectively, corresponding to the 5, 6, and 10 bands in the Landsat 8 OLI_TIRS image. It is worth noting that after extracting the built-up area through EBBI, it is necessary to change the misleading information according to the visual interpretation and overlay the corresponding remote sensing image. In order to analyze the extraction accuracy, the accuracy analysis was performed using the actual built-up area and extraction area of 7 cities in Zhejiang province in 2018. The results are shown in Table 2. According to Table 2, the overall accuracy is 86.85%.

3. Suomi-NPP VIIRS Surface Reflectance Processing: The VIIRS surface reflectance data are obtained after atmospheric correction, thin cloud processing, cloud detection, and aerosol correction. It is divided into 8-day composite data and daily composite data with a resolution of 1 km. In this study, the 8-day composite surface reflectance product is used. In the VIIRS surface reflectance data, each HDF5 file includes 9 bands of data, 1 solar azimuth data, and 1 observation angle data, and the data type is int16. In order to facilitate data processing, the first nine surface reflectance bands used in the HDF5 format data extraction research are stored in GeoTIFF format and given with corresponding coordi-

**Table 2** Accuracy table of built-up area

| Region | Extraction area (km²) | Actual area (km²) | Accuracy (%) |
|---|---|---|---|
| Ningbo | 600.74 | 512.26 | 85.27 |
| Wenzhou | 421.34 | 362.18 | 85.96 |
| Jiaxing | 423.76 | 359.83 | 84.91 |
| Shaoxing | 356.95 | 359.79 | 99.21 |
| Jinhua | 429.5 | 378 | 88.00 |
| Zhoushan | 61.44 | 76.3 | 80.52 |
| Taizhou | 253.34 | 301.47 | 84.03 |
| Total | 2349.83 | 2547.07 | 86.85 |

nates and projections. Due to the influence of thin clouds and haze, the reflectance received by satellite sensors will be different, which will affect the judgment of surface categories in the process of classification. In order to improve the quality of information extraction, the 8-day synthesized VIIRS data of 12 months and one period of each month is synthesized into annual data.

# 3 Modeling and Analysis

The main technical route of this article is shown in the Fig. 3.

## 3.1 Building a New NightTime Light Index

According to the statistical yearbook, the economic parameters of 27 central cities in the Yangtze River Delta were collected: GDP, total value of the secondary industry (SI), total value of the tertiary industry (TI), and total industrial product (IP). Use VIIRS images to count the total amount of nighttime light (NTL) in the built-up area of each prefecture-level city. The equation is

$$\mathrm{NTL}_n = \Sigma X_i, \tag{2}$$

where is the radiance value of the *i*th pixel and *n* is the *n*th prefecture-level city.

Establish a regression model of the four socioeconomic parameters of the 27 central cities in the Yangtze River Delta region and the total amount of lights in the built-up area, as shown in Fig. 4, which is the fitting effect of the four economic parameters and the total lights in the built-up area. The absolute coefficient ($R^2$) ranges from 0.6591 to 0.956, all of which passed the significance test. The higher the absolute coefficient, the better the fitting effect with NTL. According to Fig. 4, it is known that the regression model of GDP and the tertiary industry and total light has a better relationship. This is consistent with the findings of Zhao [8].

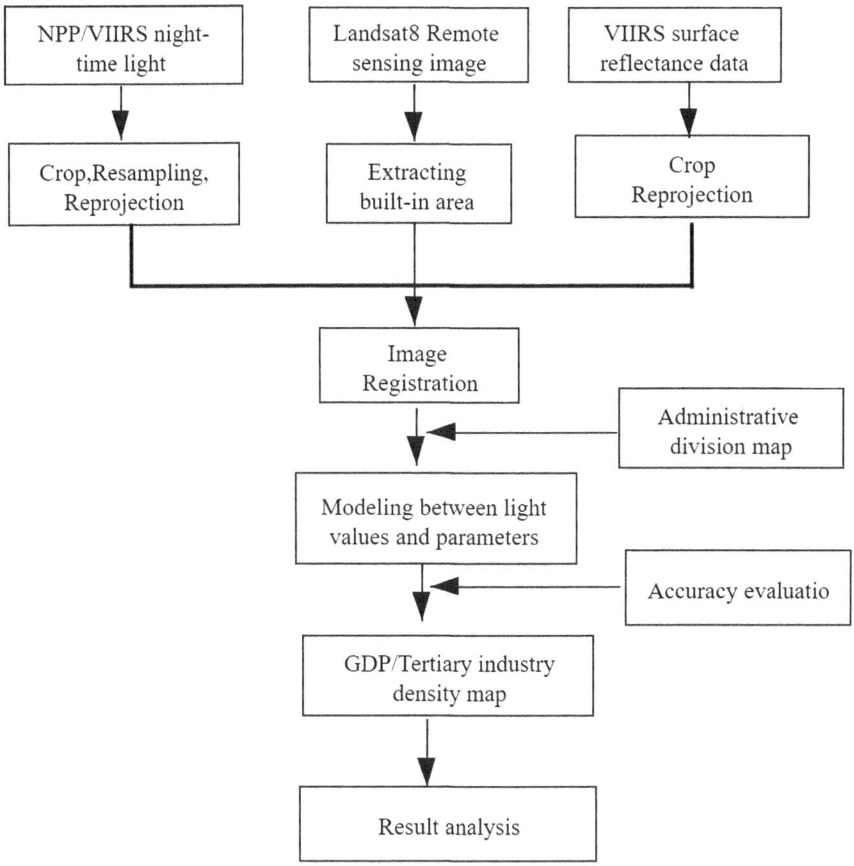


**Fig. 3** Main technical route

Use the VIIRS surface reflectance product to process the night light value and divide the night light image and its reflectance product pixel by pixel, and the calculation formula is

$$X_i^{'} = \frac{X_i}{\text{SR}}, \tag{3}$$

$$\text{NTL}_n^{'} = \sum X_i^{'}. \tag{4}$$

where $X_i$ is the radiance value of the $i$th pixel, SR is the surface reflectance value, $X_i^{'}$ is the radiance value of each pixel after the surface reflectance product is processed, $n$ is the $n$th ground level city, and $\text{NTL}_n^{'}$ is the total amount of light retrieved.

The total amount of light achieved by the surface reflectance product is used to establish a regression model with four economic parameters, as shown in Fig. 5.

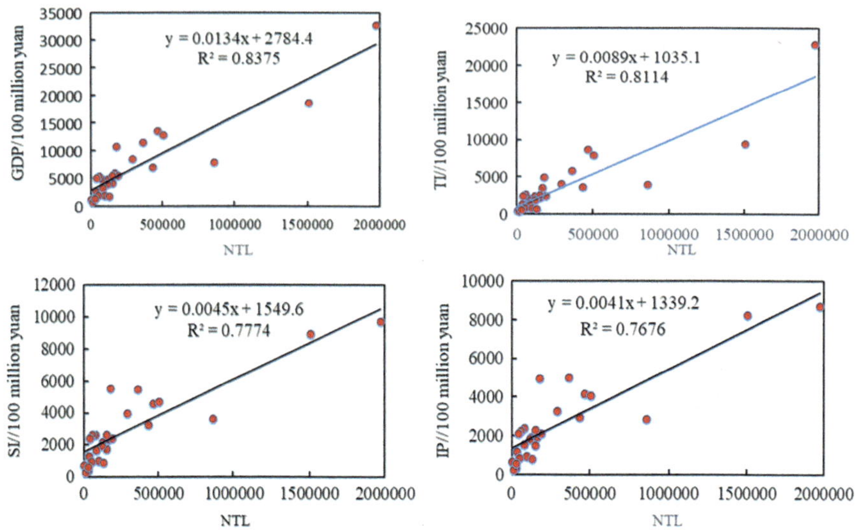


**Fig. 4** Comparison of 4 economic parameters and NTL regression model results

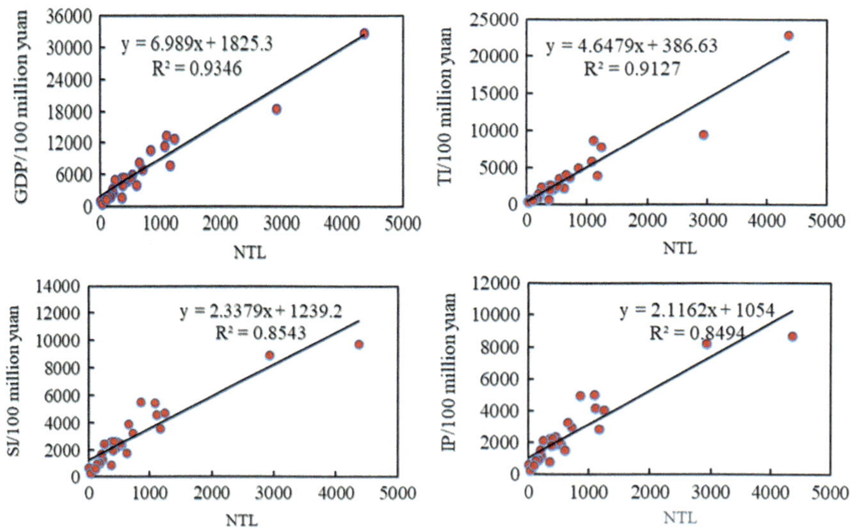


**Fig. 5** Comparison of 4 economic parameters and NTL regression model results

## 3.2 *Comparative Analysis*

Use the relative error (RE) and $R^2$ to compare the simulation results of the four economic parameters. The calculation formula is

$$\mathrm{RE} = \frac{\left|\mathrm{EP}_{\mathrm{est}} - \mathrm{EP}_{\mathrm{real}}\right|}{\mathrm{EP}_{\mathrm{real}}} \times 100\%, \tag{5}$$

$$\overline{\mathrm{RE}} = \frac{\sum \mathrm{RE}_i}{n}, \tag{6}$$

where RE is the relative error, $\overline{\mathrm{RE}}$ is the average relative error, $\mathrm{EP}_{\mathrm{real}}$ is the estimated economic parameter value, $\mathrm{EP}_{\mathrm{real}}$ is the actual economic parameter value, that is, the true value found in the statistical yearbook, and $n$ is the number of data.

Accuracy evaluation shows that the relative error of the VIIRS surface reflectance product has reduced from 0.516 to 0.454, and $R^2$ has increased from the original 0.798 to 0.888. The smaller the relative error, the closer the $R^2$ is to 1, indicating that the simulation effect is better. It can be seen that the accuracy of the model can be improved by introducing VIIRS surface reflectance data.

## 3.3 *Spatialization of GDP/Tertiary Industry Data*

According to Table 3, it is known that GDP and the tertiary industry output value have the greatest relationship with the night light value in the economic parameters. Therefore, according to the GDP and tertiary industry model in Fig. 5, the GDP and the tertiary industry are spatialized. Get the initial GDP and initial tertiary industry value on each grid cell.

For the regression results of GDP and the tertiary industry total value, the correction method for each city is adopted, and the correction coefficients $K_j$ of each city are constructed (refer to Eq. (7)). The regression results of each grid unit are adjusted, so that the simulation result data completely match the actual statistical data.

$$\mathrm{GDP}_{ji}^{'} = \mathrm{GDP}_{ji} \times K_j, \tag{7}$$

**Table 3** Comparison of average relative error and $R^2$ between the two methods

| | Model 1 | | Model 2 | |
|---|---|---|---|---|
| Variable | SE | $R^2$ | SE | $R^2$ |
| GDP | 0.484 | 0.838 | 0.300 | 0.935 |
| Total value of tertiary industry | 0.537 | 0.811 | 0.299 | 0.913 |
| Total value of secondary industry | 0.504 | 0.777 | 0.395 | 0.854 |
| Gross industrial production | 0.538 | 0.768 | 0.423 | 0.849 |
| Average | 0.516 | 0.798 | 0.354 | 0.888 |

Notes: Model 1: Total amount of nighttime light without using surface reflectance data; Model 2: Total value of nighttime light data with surface reflectance data

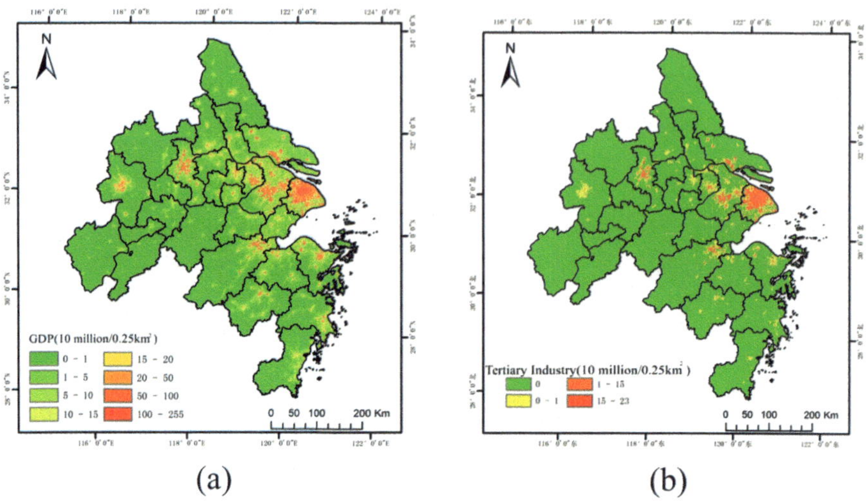


**Fig. 6** Simulation results of economic parameters. (**a**) GDP spatialization simulation results (**b**) Tertiary industry spatialization simulation results

$$K_j = \frac{\mathrm{GDP}_j}{\overline{\mathrm{GDP}_j}}, \tag{8}$$

where $\mathrm{GDP}'_{ji}$ represents the revised GDP data of the *i*th grid unit of the *j*th city, $\mathrm{GDP}_{ji}$ is the GDP data of the *i*th grid unit of the *j*th city obtained by regression, $K_j$ is the correction coefficient of the *j*th city, $\mathrm{GDP}_j$ is the actual statistical GDP data of the *j*th city, and $\overline{\mathrm{GDP}_j}$ is the GDP regression data of the *j*th city. The principle of the spatialization of the tertiary industry value is the same as that of GDP. From this, the simulated GDP numbers of the 0.5 km × 0.5 km grid unit in the Yangtze River Delta and the number of tertiary industries can be obtained (Fig. 6).

## 4 Conclusion

In this paper, we combine Landsat8 remote sensing image and NPP-VIIRS night light image and its reflectance data; by comparing the regression model of night light value and economic parameters before and after the use of reflectance data, it was found that VIIRS surface reflectance products can greatly improve the accuracy of the model and select economic parameters. The average value of $R^2$ was increased from 0.798 to 0.888. The average error has decreased from 0.516 to 0.354. It demonstrates the advantage of using VIIRS surface reflectance data to simulate economic parameters. Among the selected economic parameters, the simulation results of GDP, tertiary industry, and total night light are better. Through the spatialization of GDP and the tertiary industry, the results show that the high-value areas of the GDP and tertiary industry of the Yangtze River Delta region mainly form high-value rolling areas around Shanghai, Nanjing, Suzhou, Hangzhou, and other first-tier cities. The pattern of high east and low west is in line with the actual situation.

## References

1. Chen, X., Nordhaus, W.D.: Using luminosity data as a proxy for economic statistics. Proc. Natl. Acad. Sci. U.S.A. **108**, 8589–8594 (2011)
2. Zhou, Y.K., Ma, T., Zhou, C.H., Xu, T.: Night-time light derived assessment of regional inequality of socioeconomic development in China. Remote Sens. **7**, 1242–1262 (2015)
3. Levin, N., Duke, Y.S.: High spatial resolution night-time light images for demographic and socio-economic studies. Remote Sens. Environ. **119**, 1–10 (2012)
4. Nataliya, A.R., Boris, A.P.: Mapping geographical concentrations of economic activities in Europe using light at night (LAN) satellite data. Int. J. Remote Sens. **35**, 7706–7725 (2014)
5. Eleni, K., Chrysovalantis, T., Christos, C.: Estimating the relationship between touristic activities and night light emissions. Eur. J. Remote Sens. **52**, 233–246 (2019)
6. Shi, K.F., Yu, B.L., Huang, Y.X., et al.: Evaluating the ability of NPP-VIIRS night time light data to estimate the gross domestic product and the electric power consumption of China at multiple scales: a comparison with DMSP-OLS data. Remote Sens. **6**, 1705–1724 (2014)
7. Arthana, I.W., As-syakur, A.R., Adnyana, I.W., Arthana, I.W., Nuarsa, I.W.: Enhanced built-up and bareness index (EBBI) for mapping built-up and bare land in an urban area. Remote Sens. **4**, 2957–2970 (2012)
8. Zhao, N., Currit, N., Samson, E.: Net primary production and gross domestic product in China derived from satellite imagery. Ecol. Econ. **70**, 921–928 (2011)

# A Novel Stitching Product-Creating Algorithm Based on Sectional RPC Fitting

**Weican Meng, Fei Zhao, Feilin Peng, Wen Cao, and Haoran Dai**

**Abstract** TDI-CCD is widely used in the high-resolution optical remote sensing satellite. Because of the operational principle of the TDI-CCD, it leads to the line integral time variantion in the push-broom imaging system. There are two shortages in the traditional image-space-oriented stitching algorithms. One is the lack of strict geometric model is not available; the other is that the inner geometric accuracy is lacking consistency. Because the raw images are resampled, the RPC fitting accuracy and the algorithmic efficiency are reduced in the object-space-oriented stitching algorithms. Aiming at the shortage of the stitching algorithms, a stitching product-creating algorithm is put forward based on the sectional RPC fitting. In the novel algorithm, line integral time variation detection and sectional RPC calculation schema are proposed to improve RPC fitting accuracy. A series of tests are designed to verify the proposed method in this paper using images of HR camera of Mapping Satellite-1. The results of simulation and experiment show that the seamless image of the multi-TDI CCD can be obtained and the RPC fitting accuracy can be improved. Finally, the inner geometric accuracy of the stitching product image is obviously improved.

**Keywords** Linear push-broom camera · Image-space-oriented stitching algorithm · Object-space-oriented stitching algorithm · Line integration time hopping · Rational polynomial coefficients (RPC)

---

W. Meng · F. Zhao
Beijing Institute of Tracking and Telecommunication Technology,
Beijing, People's Republic of China

F. Peng · W. Cao (✉) · H. Dai
The School of Geo-Science and Technology, Zhengzhou University,
Zhengzhou, People's Republic of China



H. P. Urbach, Q. Yu (eds.), *6th International Symposium of Space Optical Instruments and Applications*, Space Technology Proceedings 7,
https://doi.org/10.1007/978-3-030-56488-9_9

# 1 Introduction

The traditional linear CCD camera can produce image data blurred or nonimaging under low illumination and high-speed motion. At the same time, it cannot meet the problem of enough ground cover width due to the limitation of manufacturing process. Therefore, the traditional linear CCD camera has been unable to adapt to the rapid development of optical remote sensing technology. The technology of TDI CCD mosaic camera is very innovative, which has solved the problem of traditional linear CCD camera, and has become the mainstream spaceborne optical sensor at present.

The original data acquired by a mosaic TDI CCD camera is a multislice image. In order to make full use of the camera's field of view, the original segmented image needs to be geometrically stitched. Document [1–4] summarizes the structure and image processing technology of TDI CCD camera abroad, and the technology only stays in the translation method or affine transform image model level in the segmentation algorithm. With the successful application of sky drawing No. 1, CBERS-02B/02C, and three remote sensing satellites in China, many scholars have done a lot of research on the TDI CCD image mosaic algorithm. At present, domestic and foreign product generation algorithms can be divided into two kinds of image stitching algorithms, such as image stitching and object splicing [5].

In terms of the image mosaic algorithm, Li Shiwei [6] extracted the connection point by using the image matching algorithm based on gray scale, according to the calculation of horizontal offset and vertical offset point weighted connection, and finally realized the piecewise image mosaic of CBERS-02B satellite HR camera through the whole translation. Long Xiaoxiang [7] firstly analyzed the influence of satellite yaw angle, integral time and stability of the platform, attitude and orbit error factors on the push and sweep images of splicing TDI CCD camera, and then studied the geometric splicing of TDI CCD image with CBERS-02B satellite HR camera as an example, and finally found that the splicing accuracy of connection points reached subpixel level. Men Weican [8] extracted uniformly distributed interslice along the flight direction and realized splicing of segmented TDI CCD images by constructing piecewise affine transformation model. Chen Qi [9] carried out geometric correction and image mosaic and emergence processing on the segmented TDI CCD images of CBERS-02C satellite high-resolution camera, and then integrated three processes to avoid the secondary resampling of images, and finally realized seamless splicing of segmented TDI CCD images.

In terms of the object-side stitching algorithm, Zhang Guo [10] spliced the inner field of the spliced sensor by using the method of virtual array and contrasted the forward intersection accuracy of the images before and after stitching as a quantitative evaluation of the accuracy of the stitching algorithm, which can test the impact of splicing algorithm on the production accuracy of subsequent photograph. The experimental results show that multi-piece spliced ALOS/PRISM sensors and the virtual linear array splicing method can achieve visually seamless and no loss of spatial forward intersection accuracy. Pan Hongbo [11] made an in-depth analysis on the camera structure of multislice TDI CCD splicing resource No. 3 of three-line array camera and adopted the process of "sensor correction" to solve the problem of splicing TDI CCD image, which is essentially a method of virtual linear array. Pan Jun [12] conducted an in-depth study of two object-side stitching algorithms, the virtual linear array method and the object-side projection surface method, and analyzed various sources of errors in the stitching process. The image data obtained by the CBERS-02C satellite HR camera was used to verify the above two stitching algorithms, were carried out, and qualitative and quantitative evaluation of the splicing results was carried out to verify the correctness and feasibility of the above two methods. Jin Shuying [13] suggested a coordinate inverse transform method for TDI-CCD pushbroom Images, that calculated the omage-space coordinates of a corresponding ground. The virtual scan line integral time was equal to construct the virtual image in this method. Hence, the efficiency and precision of algorithm was verified using ZY-1 02C satellite HR1 camera data. 02C satellite HR1 camera to verify the efficiency and precision of algorithm is high. Cao Bincai [14] corrected line jump caused by the error of the image point to balance resampling of the TDI images and verified that the algorithm can obtain better fitting precision of 0.06 plane as in the positioning characteristics under the condition of keeping the original image according to the traditional method of solution of precise RPCs with high-resolution Mapping Satellite-1 image.

Image stitching algorithm principle is simple and efficient, but its excessive dependence between corresponding points of information will lead to a decrease in the stitching accuracy or even the problem of unstitching. Moreover, there is no clear relationship between geometric objects, geometric quality is difficult to guarantee, and its related research is gradually decreasing. Object stitching algorithm has become the mainstream stitching algorithm because of its strict principle, clear geometry, describing ability, and other factors. The line integral time variation on fitting accuracy of RPC model, the object is both the basic mosaic algorithm sampling method to solve the whole scene image by integral time can be normalized; although RPC fitting accuracy increases to normal levels, the resampling process will affect the geometric accuracy of sound to the final splicing products in theory. At the same time, the operation efficiency of the object connection algorithm will also be reduced.

## 2 Virtual Scan Image Mosaic Algorithm Based on RPC Model Fitting of Segmented Image

The virtual push-broom image stitching product generation algorithm based on the slice image RPC model fitting is to set an ideal distortion-free virtual camera according to certain rules, whose line length is the same as the effective length of splicing type TDI CCD. Make virtual camera and real camera start up at the same time and have the same external position element; based on the continuity of the object space,

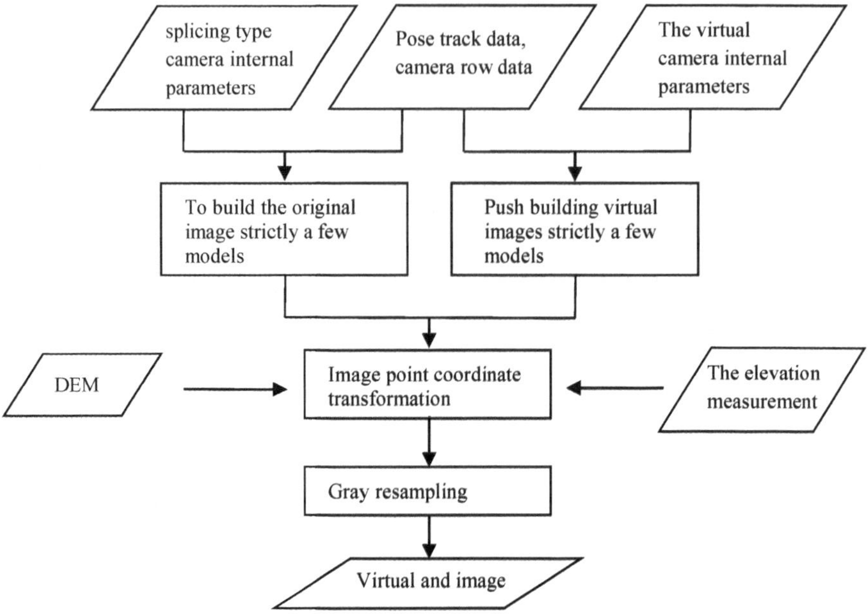


**Fig. 1** Diagram of virtual push-broom imagery production

the image point coordinate conversion relationship between the virtual push broom image and the original segmented image is constructed, and the splicing of TDI CCD segmented images can be realized by generating the virtual push broom image. The specific process is shown in Fig. 1:

1. On orbit geometric calibration of stitching TDI CCD camera in [13, 14], using the updated camera internal parameters to complete the internal orientation of each CCD agent in the camera coordinate system, and according to the image data, satellite data, camera placement data to complete the exterior orientation of the camera, and then build the rigorous geometric model of the original image;
2. The strict geometric imaging model in step (1) is a similar method of constructing virtual push-broom images;
3. For any image point $P(s, l)$ on the image of the virtual CCD camera, combining the strict geometric imaging model of the virtual push-scan image and the ground elevation information $H$, the image point $P$ is projected onto the ground to obtain its corresponding ground point coordinate $G(X, Y, Z)$, that is, the process of $P(s, l) + H \rightarrow G(X, Y, Z)$ is completed;
4. The ground points $G(X, Y, Z)$ can be mapped onto the original segmentation image by the back projection, and then the projection points $P'(x, y)$ can be obtained. In the strict geometric model, the efficiency is low due to iterative operation in the process of back projection of linear push-broom images. In

order to solve this problem, the RPC parameters of each split image are generated according to the single image geometric model before back projection. Thus, the back projection calculation is completed and the iterative process is avoided. At the same time, the problem that the accuracy of RPC model fitting is reduced due to line integral time hopping can be solved by using line integration time detection and RPC partition model;

5. After steps (3) and (4), the coordinate conversion relationship between the virtual push-broom image point $P(s, l)$ and the original fragmented image point $P'(x, y)$ is constructed, and the grayscale resampling algorithm can be used to obtain the grayscale of point $P'(x, y)$ on the original fragmented image and assign it to image point $P(s, l)$ on the virtual push image;
6. Go to the virtual push-broom images of all image points and repeat step (3) to (5), and you can get the whole push virtual imaging range of scan images, as the splicing TDI CCD image mosaic products.

The splicing product obtained based on the virtual push scan image generation algorithm has the geometric characteristics of the linear push scan image and clear geometric relationship of the object image, and the image is swept by the virtual camera in an ideal condition without optical distortion and CCD array distortion, and the images are considered as distortion-free images, when the influence of factors such as posture shake is not considered.

## 2.1 RPC Model

For linear array push-broom images, iterative calculation is needed based on the strict geometric model from the ground point back to the corresponding image point, and the efficiency is too low for the virtual projection image generation algorithm by point by point. The RPC model can achieve the same positioning accuracy as the strict geometric model [19], and more importantly, it can greatly improve the efficiency of the virtual push-broom image when the RPC model is used to calculate the image points from the ground surface without iterative calculation.

The RPC model links the image coordinates with the corresponding ground point coordinates in polynomial form. Generally, image coordinates and ground coordinates are regularized into $-1$ to 1, so as to enhance the stability of parameter solving.

The RPC model is defined as follows:

$$
\left\{
\begin{aligned}
X &= \frac{N_s\left(\tilde{B},\tilde{L},\tilde{H}\right)}{D_s\left(\tilde{B},\tilde{L},\tilde{H}\right)} \\
Y &= \frac{N_l\left(\tilde{B},\tilde{L},\tilde{H}\right)}{D_l\left(\tilde{B},\tilde{L},\tilde{H}\right)}
\end{aligned}
\right. \tag{1}
$$

In the last formula,

$$
\begin{aligned}
N_s\left(\tilde{B},\tilde{L},\tilde{H}\right) = {} & a_0 + a_1\tilde{L} + a_2\tilde{B} + a_3\tilde{H} + a_4\tilde{L}\tilde{B} \\
& + a_5\tilde{L}\tilde{H} + a_6\tilde{B}\tilde{H} + a_7\tilde{L}^2 + a_8\tilde{B}^2 \\
& + a_9\tilde{H}^2 + a_{10}\tilde{L}\tilde{B}\tilde{H} + a_{11}\tilde{L}^3 \\
& + a_{12}\tilde{L}\tilde{B}^2 + a_{13}\tilde{L}\tilde{H}^2 + a_{14}\tilde{L}^2\tilde{B} \\
& + a_{15}\tilde{B}^3 + a_{16}\tilde{B}\tilde{H}^2 + a_{17}\tilde{L}^2\tilde{H} \\
& + a_{18}\tilde{B}^2\tilde{H} + a_{19}\tilde{H}^3
\end{aligned},
$$

$$
\begin{aligned}
D_s\left(\tilde{B},\tilde{L},\tilde{H}\right) = {} & b_0 + b_1\tilde{L} + b_2\tilde{B} + b_3\tilde{H} + b_4\tilde{L}\tilde{B} \\
& + b_5\tilde{L}\tilde{H} + b_6\tilde{B}\tilde{H} + b_7\tilde{L}^2 + b_8\tilde{B}^2 \\
& + b_9\tilde{H}^2 + b_{10}\tilde{L}\tilde{B}\tilde{H} + b_{11}\tilde{L}^3 \\
& + b_{12}\tilde{L}\tilde{B}^2 + b_{13}\tilde{L}\tilde{H}^2 + b_{14}\tilde{L}^2\tilde{B} \\
& + b_{15}\tilde{B}^3 + b_{16}\tilde{B}\tilde{H}^2 + b_{17}\tilde{L}^2\tilde{H} \\
& + b_{18}\tilde{B}^2\tilde{H} + b_{19}\tilde{H}^3
\end{aligned},
$$

$$
\begin{aligned}
N_l\left(\tilde{B},\tilde{L},\tilde{H}\right) &= c_0 + c_1\tilde{L} + c_2\tilde{B} + c_3\tilde{H} + c_4\tilde{L}\tilde{B} \\
&+ c_5\tilde{L}\tilde{H} + c_6\tilde{B}\tilde{H} + c_7\tilde{L}^2 + c_8\tilde{B}^2 \\
&+ c_9\tilde{H}^2 + c_{10}\tilde{L}\tilde{B}\tilde{H} + c_{11}\tilde{L}^3 \\
&+ c_{12}\tilde{L}\tilde{B}^2 + c_{13}\tilde{L}\tilde{H}^2 + c_{14}\tilde{L}^2\tilde{B} \\
&+ c_{15}\tilde{B}^3 + c_{16}\tilde{B}\tilde{H}^2 + c_{17}\tilde{L}^2\tilde{H} \\
&+ c_{18}\tilde{B}^2\tilde{H} + c_{19}\tilde{H}^3
\end{aligned}
$$

$$
\begin{aligned}
D_l\left(\tilde{B},\tilde{L},\tilde{H}\right) &= d_0 + d_1\tilde{L} + d_2\tilde{B} + d_3\tilde{H} + d_4\tilde{L}\tilde{B} \\
&+ d_5\tilde{L}\tilde{H} + d_6\tilde{B}\tilde{H} + d_7\tilde{L}^2 + d_8\tilde{B}^2 \\
&+ d_9\tilde{H}^2 + d_{10}\tilde{L}\tilde{B}\tilde{H} + d_{11}\tilde{L}^3 \\
&+ d_{12}\tilde{L}\tilde{B}^2 + d_{13}\tilde{L}\tilde{H}^2 + d_{14}\tilde{L}^2\tilde{B} \\
&+ d_{15}\tilde{B}^3 + d_{16}\tilde{B}\tilde{H}^2 + d_{17}\tilde{L}^2\tilde{H} \\
&+ d_{18}\tilde{B}^2\tilde{H} + d_{19}\tilde{H}^3
\end{aligned}
.
$$

In the upper, middle, and regular geodetic coordinates, which are usually constant 1, are regularized image coordinates, respectively, by type calculations:

$$\tilde{B} = \frac{B - B_{\text{off}}}{B_{\text{scale}}}, \tag{2}$$

$$\tilde{L} = \frac{L - L_{\text{off}}}{L_{\text{scale}}}, \tag{3}$$

$$\tilde{H} = \frac{H - H_{\text{off}}}{H_{\text{scale}}}, \tag{4}$$

$$X = \frac{s - s_{\text{off}}}{s_{\text{scale}}}, \tag{5}$$

$$Y = \frac{l - l_{\text{off}}}{l_{\text{scale}}}. \tag{6}$$

In the last formula, $B_{\text{off}}$, $B_{\text{scale}}$, $L_{\text{off}}$, $L_{\text{scale}}$, $H_{\text{off}}$, and $H_{\text{scale}}$ are regularization parameters of the ground coordinates. $s_{\text{off}}$ and $s_{\text{scale}}$ are regularization parameters of the image coordinates.

## 2.2 PRC Model Parameter Solving

Under the condition of rigorous geometric model, the RPC parameter is generally used to solve the terrain-independent scheme. The basic principle is that, first, the virtual object grid is generated based on the rigorous geometric model, and the unknown parameters of the RPC model are computed based on the discrete virtual object points.

The main steps of virtual property grid construction are as follows: first, the image grid points ($s_i$, $l_i$) are uniformly distributed according to the image size; second, the maximum and minimum elevations of the image coverage area are obtained using SRTM DEM and other public DEM data; third, the elevation range is uniformly divided into several elevation plane $H_j$; fourth, the coordinates of corresponding ground points can be calculated according to the intersection of the elevation and sight vector determined by the strict geometric model; finally, the virtual object square grid is constructed, as shown in Fig. 2.

After obtaining the virtual square grid and its corresponding image points, 10 ground coordinates and image coordinate regularization parameters are calculated first, and the formulas are as follows:

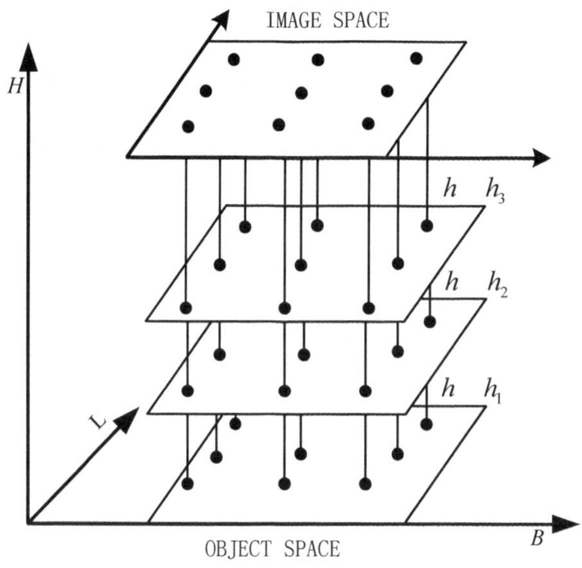


**Fig. 2** Diagram of virtual space grids

$$\left\{\begin{array}{l} B_{\text{off}} = \dfrac{\sum B}{n} \\ L_{\text{off}} = \dfrac{\sum L}{n} \\ H_{\text{off}} = \dfrac{\sum H}{n} \end{array}\right., \tag{7}$$

$$\left\{\begin{array}{l} B_{\text{scale}} = \max\left(\left|B_{\max} - B_{\text{off}}\right|, \left|B_{\min} - B_{\text{off}}\right|\right) \\ L_{\text{scale}} = \max\left(\left|L_{\max} - L_{\text{off}}\right|, \left|L_{\min} - L_{\text{off}}\right|\right) \\ H_{\text{scale}} = \max\left(\left|H_{\max} - H_{\text{off}}\right|, \left|H_{\min} - H_{\text{off}}\right|\right) \end{array}\right., \tag{8}$$

$$\left\{\begin{array}{l} s_{\text{off}} = \dfrac{\sum s}{n} \\ l_{\text{off}} = \dfrac{\sum l}{n} \end{array}\right., \tag{9}$$

$$\left\{\begin{array}{l} s_{\text{scale}} = \max\left(\left|s_{\max} - s_{\text{off}}\right|, \left|s_{\min} - s_{\text{off}}\right|\right) \\ l_{\text{scale}} = \max\left(\left|l_{\max} - l_{\text{off}}\right|, \left|l_{\min} - l_{\text{off}}\right|\right) \end{array}\right.. \tag{10}$$

According to the calculated ground coordinate and image coordinate regularization parameters, the coordinates are normalized and then according to Eqs. (2) to (10) the RPC parameters can be solved. The Eq. (1) is expressed as

$$\begin{aligned} F_X &= N_s\left(\tilde{B}, \tilde{L}, \tilde{H}\right) - X \times D_s\left(\tilde{B}, \tilde{L}, \tilde{H}\right) = 0, \\ F_Y &= N_l\left(\tilde{B}, \tilde{L}, \tilde{H}\right) - Y \times D_l\left(\tilde{B}, \tilde{L}, \tilde{H}\right) = 0. \end{aligned} \tag{11}$$

Then, the error equation can be expressed as

$$\mathbf{v} = \mathbf{Bx} - \mathbf{l},\ \mathbf{P}, \tag{12}$$

where

$$\mathbf{B} = \begin{bmatrix} \dfrac{\partial F_X}{\partial a_i} & \dfrac{\partial F_X}{\partial b_j} & \dfrac{\partial F_X}{\partial c_i} & \dfrac{\partial F_X}{\partial d_j} \\ \dfrac{\partial F_Y}{\partial a_i} & \dfrac{\partial F_Y}{\partial b_j} & \dfrac{\partial F_Y}{\partial c_i} & \dfrac{\partial F_Y}{\partial d_j} \end{bmatrix}, i \in [0,19], j \in [1,19] \tag{13}$$

$$\mathbf{x} = \begin{bmatrix} a_i & b_j & c_i & d_j \end{bmatrix}^T, i \in [0,19], j \in [1,19] \quad (14)$$

$$\mathbf{l} = \begin{bmatrix} -F_X^0 \\ -F_Y^0 \end{bmatrix} \quad (15)$$

According to the least square principle, the normal equation of the error equation can be defined as follows:

$$\left(\mathbf{B}^T\mathbf{P}\mathbf{B}\right)\mathbf{x} = \mathbf{B}^T\mathbf{P}\mathbf{l}. \quad (16)$$

In some cases, if least square method is used to calculate the RPC parameters directly, the normal matrix BTPB will have many conditions, which can lead to ill-conditioned normal equations and makes the inversion of normal matrix unstable; finally, it results in poor accuracy of RCP parameter fitting and even with no calculation result [17].

Therefore, the unbiased spectral correction iteration method is adopted to improve the normal equation state [18], and **x** is added simultaneously to the two sides of Eq. (16):

$$\left(\mathbf{B}^T\mathbf{P}\mathbf{B} + \mathbf{E}\right)\mathbf{x} = \mathbf{B}^T\mathbf{P}\mathbf{l} + \mathbf{x}. \quad (17)$$

In the equation, **E** is the unit matrix of order n, and because the function on both sides of the above formula has an unknown parameter **x**, it needs to be solved with a solution. The iteration formula is as follows:

$$\hat{\mathbf{x}}^{(n)} = \left(\mathbf{B}^T\mathbf{P}\mathbf{B} + \mathbf{E}\right)^{-1}\left(\mathbf{B}^T\mathbf{P}\mathbf{l} + \hat{\mathbf{x}}^{(n-1)}\right). \quad (18)$$

In the above formula, $x^{(n)}$ and $x^{(n-1)}$ are respectively the $n^{th}$ and $n\text{-}1^{th}$ spectral correction iterative estimated values of the position parameter.

## 2.3 *Line Time Hopping Detection and Partitioning Solution*

The study of [20] shows that the integral time jump can be divided into the category of high-frequency attitude errors, while the RPC model is difficult to fit the high-frequency errors. In this paper, it is found that the linear integral time hopping leads to the substitution error of RPC model over 1 pixel.

In view of this situation, this paper detects the line where the line integration time jump is based on file integration time jump and then performs the RPC parameter fitting calculation on the original TDI CCD segmented image in the flight direction. The specific process is shown in Fig. 3. Since the line integral time jump exists only

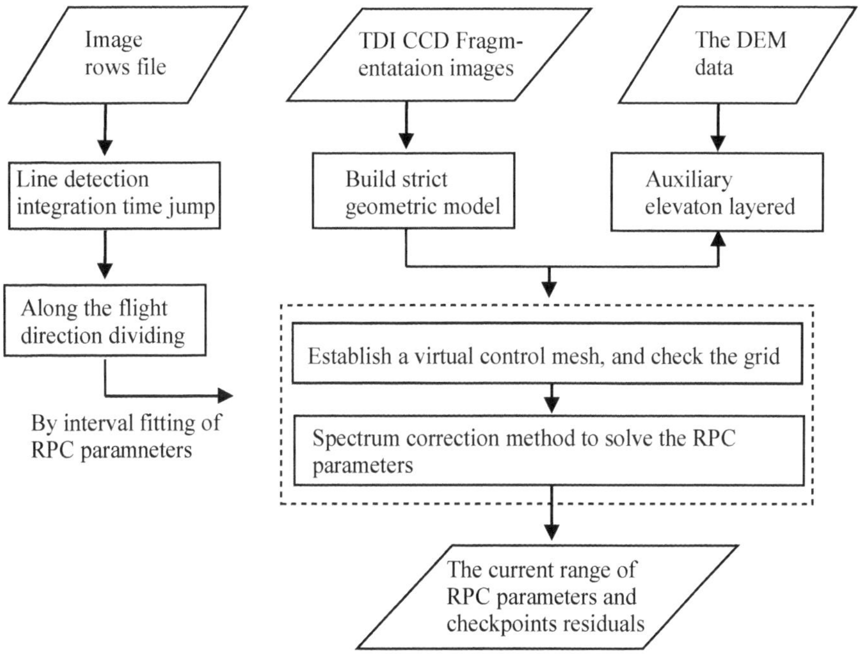


**Fig. 3** Diagram of sectional RPC calculation schema

1~2 times in a scene, the number of partitions will be less, and the algorithm process will not be too complicated while improving the accuracy of RPC fitting.

## 3 Experiment and Analysis

The original images of TH-1 01 and 02 satellite high resolution cameras are used as experimental data to verify the algorithm in this paper from three aspects: RPC fitting of sliced images, virtual push-broom image generation, and virtual push-broom image geometric quality evaluation.

### *3.1 Slice Image RPC Fitting Experiment*

In order to verify the influence of the integration time jump on the accuracy of the traditional RPC model, the 4-scene high-resolution camera original images of Mapping Satellite 01 were selected to generate the RPC parameters and verify the fitting accuracy. The number of images and the number of segments of 4-scene original images are 35,000 lines and eight pieces, respectively, and are named

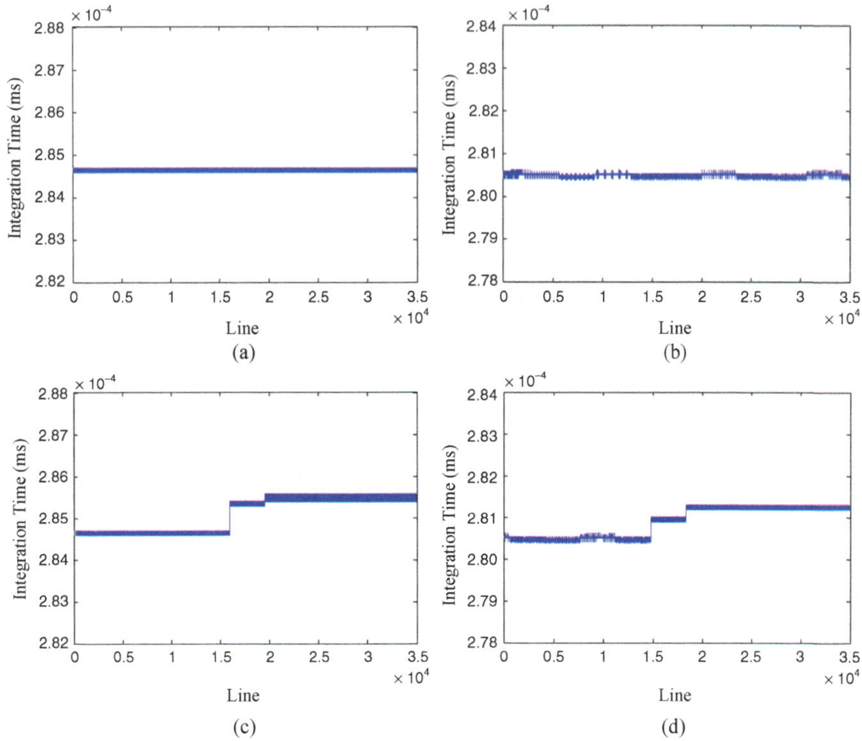


**Fig. 4** (**a**–**d**) Gives the trend graph of the line integral time jumps of 4 scene images with line number. (**a**) 01-005-134-A scene of constant integration time. (**b**) 01-005-134-B scene of constant integration time. (**c**) 01-005-135-A scene of non-constant integration time. (**d**) 01-005-135-B scene of non-constant integration time

01-005-134-A, 01-005-134-B, 01-005-134-A, and 01-005-134-B. In the first two images, there was no integral time hopping in November 28, 2010. The posterior two-scene imaging was performed in December 20, 2010, and there was an integral time jump. Among them, the 01-005-135-A scene image has the line integral time jump in the 16034th, 19539th row; the 01-005-135-B scene image has the line integral time jump in the 14824th, 18384th row (Fig. 4).

The experiment used third-order RPC model with different denominators in 32 pieces of 4 scenes of original image, which generates virtual control grid with intervals of 512 × 512 pixels and elevation layer of 5. Based on the virtual control grid, the inspection grid with 216 × 216 pixel interval and elevation layer of 10 is generated by encryption and translation. The virtual control grid is used to solve the RPC parameters, and the grid is checked for the RPC model instead of the accuracy verification.

First, the whole RPC is fitted to all the rows (1~35,000 rows) of each slice image, and its substitution accuracy is shown in Table 1.

From the results in Table 1, we can see that:

**Table 1** Statistics table of RPC fitting accuracy caused by line integration time hopping

| image | Patch number | Line direction | | Column direction | |
|---|---|---|---|---|---|
| | | Max/pixel | RMS/pixel | Max/pixel | RMS/pixel |
| 01-005-134-A The whole scene (1~35,000 line) | CCD1 | −0.009698 | 0.003871 | −0.000257 | 0.000091 |
| | CCD2 | −0.009709 | 0.003871 | −0.000276 | 0.000098 |
| | CCD3 | −0.009717 | 0.003871 | −0.000293 | 0.000105 |
| | CCD4 | −0.009726 | 0.003872 | −0.000312 | 0.000112 |
| | CCD5 | −0.009732 | 0.003872 | −0.000330 | 0.000119 |
| | CCD6 | −0.009749 | 0.003873 | −0.000355 | 0.000129 |
| | CCD7 | −0.009760 | 0.003874 | −0.000380 | 0.000138 |
| | CCD8 | −0.009773 | 0.003875 | −0.000407 | 0.000149 |
| 01-005-134-B The whole scene (1~35,000 line) | CCD1 | 0.022080 | 0.007954 | −0.000191 | 0.000068 |
| | CCD2 | 0.022106 | 0.007955 | −0.000209 | 0.000074 |
| | CCD3 | 0.022128 | 0.007955 | −0.000225 | 0.000081 |
| | CCD4 | 0.022153 | 0.007955 | −0.000243 | 0.000088 |
| | CCD5 | 0.022175 | 0.007955 | −0.000260 | 0.000094 |
| | CCD6 | 0.022229 | 0.007956 | −0.000283 | 0.000104 |
| | CCD7 | 0.022282 | 0.007956 | −0.000306 | 0.000112 |
| | CCD8 | 0.022350 | 0.007957 | −0.000331 | 0.000123 |
| 01-005-135-A The whole scene (1~35,000 line) | CCD1 | −2.306142 | 0.751749 | −0.000174 | 0.000063 |
| | CCD2 | −2.306142 | 0.751749 | −0.000192 | 0.000070 |
| | CCD3 | −2.306145 | 0.751749 | −0.000210 | 0.000077 |
| | CCD4 | −2.306146 | 0.751748 | −0.000229 | 0.000084 |
| | CCD5 | −2.306149 | 0.751748 | −0.000247 | 0.000092 |
| | CCD6 | −2.306153 | 0.751747 | −0.000269 | 0.000100 |
| | CCD7 | −2.306160 | 0.751747 | −0.000291 | 0.000109 |
| | CCD8 | −2.306167 | 0.751747 | −0.000315 | 0.000119 |
| 01-005-135-B The whole scene (1~35,000 line) | CCD1 | −1.247867 | 0.561882 | −0.000155 | 0.000056 |
| | CCD2 | −1.247868 | 0.561881 | −0.000173 | 0.000063 |
| | CCD3 | −1.247871 | 0.561881 | −0.000190 | 0.000069 |
| | CCD4 | −1.247873 | 0.561880 | −0.000209 | 0.000077 |
| | CCD5 | −1.247877 | 0.561880 | −0.000226 | 0.000084 |
| | CCD6 | −1.247881 | 0.561879 | −0.000248 | 0.000092 |
| | CCD7 | −1.247887 | 0.561879 | −0.000269 | 0.000101 |
| | CCD8 | −1.247893 | 0.561878 | −0.000293 | 0.000110 |

1. For the nonline jump in two images 01-005-134-A and 01-005-134-B, the model substitution accuracy of RPC parameters in row and column direction is better than that of $10^{-2}$ pixels;
2. For the existing line jump in two images 01-005-135-A and 01-005-135-B, the model substitution accuracy of the column direction model is better than $10^{-3}$ pixels; but, the RMS of line direction substitution is greater than 0.5 pixels, which is far greater than the normal substitution error of RPC model, indicating that line integral time hopping will greatly reduce the line direction fitting accuracy of RPC model.

According to the method mentioned above, the line integral time jump detection and interpretation of RPC fitting are carried out for 2-scene image with line integral time jump, and the image is divided into 3 regions in the flight direction according to the line where the line integral time jump. Table 2 gives the RPC parameters of each partition between alternative precision.

From the results in Table 2, we can see that:

1. The partition solution RPC parameters can effectively overcome the effect of integral time hopping on RPC model substitution precision;
2. The precise substitution of entire image RPC from 0.75 pixel level to $10^{-3}$ pixel level reaches the normal level of RPC model substitution precision.

## 3.2 Virtual Push-Broom Image Generation Experiment

For each original image (size 35,000 rows × 32,768 columns), according to the method described in this paper, the virtual push-broom image generation experiment was conducted by taking 01-005-134-A and 01-005-135-A as an example, Fig. 5 gives the experimental results.

Figure 5 shows that the virtual push image generation algorithm based on the slice image RPC model fitting can successfully complete the geometric splicing of the original splicing image of the TDI CCD (this paper only focuses on the image geometric processing technology research and the image in Fig. 5 is not dodging processed).

On the basis of correctly generating the pseudo push-broom image, the stitching area is partially enlarged (the actual pixel magnification of two times), and the splicing image before and after the in-orbit geometric calibration is qualitatively evaluated by visual observation. Figure 6 shows the partial enlargement of the CCD1~CCD8 section of the 01-005-135-A image before and after the geometric calibration in orbit.

As can be seen from Fig. 6, when virtual image is generated by sweeping with internal parameters before geometric calibration, splicing products can be successfully obtained, but there is a relatively obvious dislocation in each splicing area. The reason is that the intrinsic parameters before geometric calibration of geometric calibration cannot accurately describe the relative placement relationship of each CCD. After on-orbit geometric calibration, the geometric distortion of each CCD is accurately described, and the relative geometric accuracy is high among each CCD image, so it can realize the seamless stitching directly adjacent CCD slice images.

## 3.3 Virtual Push-Broom Image, RPC, Fitting Precision

On the basis of the rigorous geometric imaging model of the virtual push-broom image, the RPC model of the virtual push-broom image is constructed by using the method of solving the RPC parameter independent of the terrain, and the

**Table 2** Statistics table of sectional RPC fitting accuracy

| Image and Fitting interval | Patch number | Line direction | | Column direction | |
|---|---|---|---|---|---|
| | | Max/pixel | RMS/pixel | Max/pixel | RMS/pixel |
| 01-005-135-A (1~16,033 line) | CCD1 | −0.006447 | 0.002919 | 0.000002 | 0.000001 |
| | CCD2 | −0.006447 | 0.002919 | 0.000002 | 0.000001 |
| | CCD3 | −0.006447 | 0.002919 | 0.000002 | 0.000001 |
| | CCD4 | −0.006447 | 0.002919 | 0.000002 | 0.000001 |
| | CCD5 | −0.006447 | 0.002919 | 0.000002 | 0.000001 |
| | CCD6 | −0.006447 | 0.002919 | 0.000002 | 0.000001 |
| | CCD7 | −0.006447 | 0.002919 | 0.000002 | 0.000001 |
| | CCD8 | −0.006447 | 0.002919 | 0.000002 | 0.000001 |
| 01-005-135-A (16,034~19,538 line) | CCD1 | 0.000393 | 0.000192 | −0.000002 | 0.000001 |
| | CCD2 | 0.000393 | 0.000192 | −0.000002 | 0.000001 |
| | CCD3 | 0.000393 | 0.000192 | −0.000002 | 0.000001 |
| | CCD4 | 0.000393 | 0.000192 | −0.000002 | 0.000001 |
| | CCD5 | 0.000393 | 0.000192 | −0.000002 | 0.000001 |
| | CCD6 | 0.000393 | 0.000192 | −0.000002 | 0.000001 |
| | CCD7 | 0.000393 | 0.000192 | −0.000002 | 0.000001 |
| | CCD8 | 0.000393 | 0.000192 | −0.000002 | 0.000001 |
| 01-005-135-A (19,539~35,000 line) | CCD1 | 0.009161 | 0.004359 | 0.000002 | 0.000001 |
| | CCD2 | 0.009161 | 0.004359 | 0.000002 | 0.000001 |
| | CCD3 | 0.009161 | 0.004359 | 0.000002 | 0.000001 |
| | CCD4 | 0.009161 | 0.004359 | 0.000002 | 0.000001 |
| | CCD5 | 0.009161 | 0.004359 | 0.000002 | 0.000001 |
| | CCD6 | 0.009161 | 0.004359 | 0.000002 | 0.000001 |
| | CCD7 | 0.009161 | 0.004359 | 0.000002 | 0.000001 |
| | CCD8 | 0.009161 | 0.004359 | 0.000002 | 0.000001 |
| 01-005-135-B (1~14,823 line) | CCD1 | −0.007260 | 0.003412 | 0.000002 | 0.000001 |
| | CCD2 | −0.007260 | 0.003412 | 0.000002 | 0.000001 |
| | CCD3 | −0.007260 | 0.003412 | 0.000002 | 0.000001 |
| | CCD4 | −0.007260 | 0.003412 | 0.000002 | 0.000001 |
| | CCD5 | −0.007260 | 0.003412 | 0.000002 | 0.000001 |
| | CCD6 | −0.007260 | 0.003412 | 0.000002 | 0.000001 |
| | CCD7 | −0.007260 | 0.003412 | 0.000002 | 0.000001 |
| | CCD8 | −0.007260 | 0.003412 | 0.000002 | 0.000001 |
| 01-005-135-B (14,824~18,383 line) | CCD1 | 0.000334 | 0.000195 | 0.000002 | 0.000001 |
| | CCD2 | 0.000334 | 0.000195 | 0.000002 | 0.000001 |
| | CCD3 | 0.000334 | 0.000195 | 0.000002 | 0.000001 |
| | CCD4 | 0.000334 | 0.000195 | 0.000002 | 0.000001 |
| | CCD5 | 0.000334 | 0.000195 | 0.000002 | 0.000001 |
| | CCD6 | 0.000334 | 0.000195 | 0.000002 | 0.000001 |
| | CCD7 | 0.000334 | 0.000195 | 0.000002 | 0.000001 |
| | CCD8 | 0.000334 | 0.000195 | 0.000002 | 0.000001 |

(continued)

**Table 2** (continued)

| Image and Fitting interval | Patch number | Line direction | | Column direction | |
|---|---|---|---|---|---|
| | | Max/pixel | RMS/pixel | Max/pixel | RMS/pixel |
| 01-005-135-B (18,384~35,000 line) | CCD1 | 0.013960 | 0.006541 | 0.000002 | 0.000001 |
| | CCD2 | 0.013960 | 0.006541 | 0.000002 | 0.000001 |
| | CCD3 | 0.013960 | 0.006541 | 0.000002 | 0.000001 |
| | CCD4 | 0.013960 | 0.006541 | 0.000002 | 0.000001 |
| | CCD5 | 0.013960 | 0.006541 | 0.000002 | 0.000001 |
| | CCD6 | 0.013960 | 0.006541 | 0.000002 | 0.000001 |
| | CCD7 | 0.013960 | 0.006541 | 0.000002 | 0.000001 |
| | CCD8 | 0.013960 | 0.006541 | 0.000002 | 0.000001 |

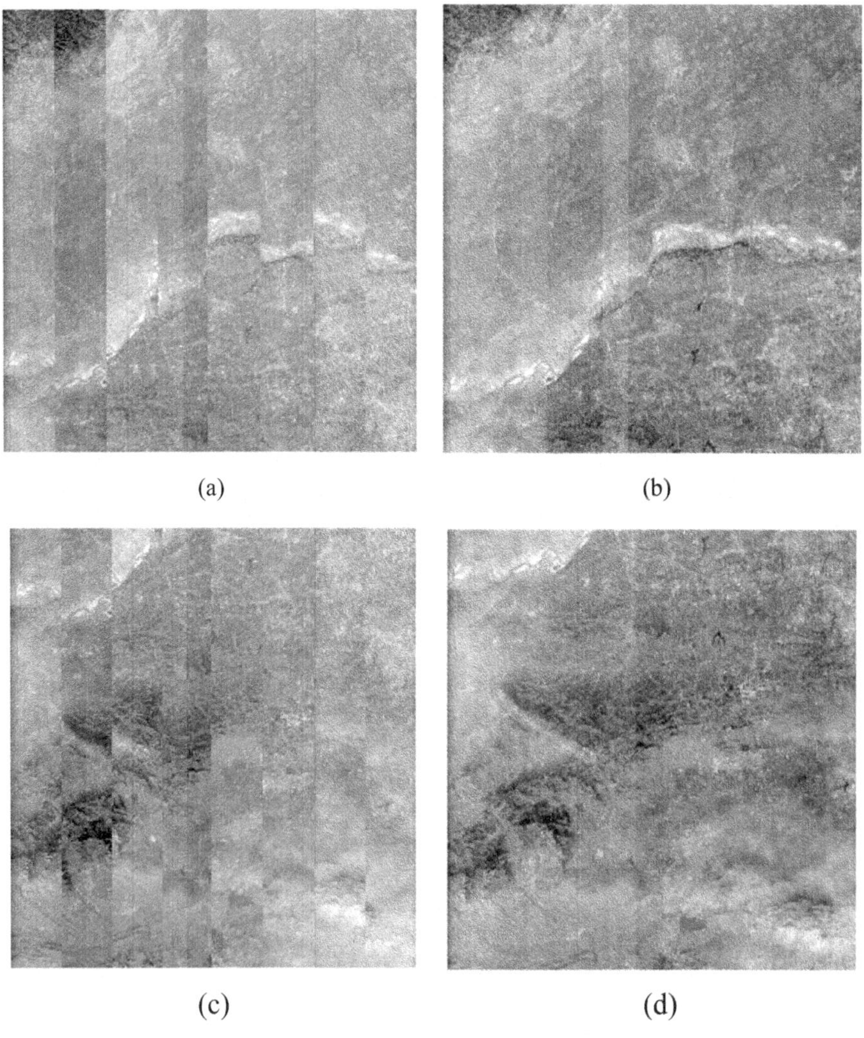


**Fig. 5** Comparison diagram of virtual push-broom images generated. (**a**) 01-005-134-Ascene original image. (**b**) 01-005-134-A scene virtual push-broom image. (**c**) 01-005-135-A scene original image. (**d**) 01-005-135-A scene virtual push-broom image

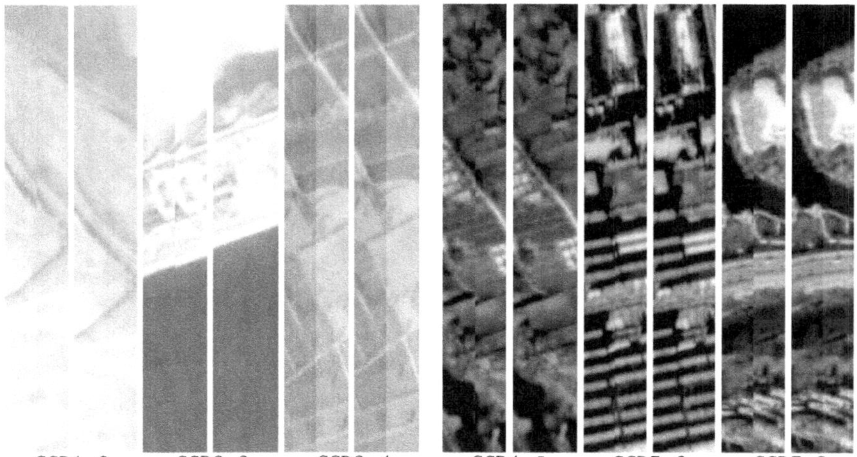


**Fig. 6** Comparison diagram of partial enlargement of virtual push-broom image generated

**Table 3** Statistics table of virtual push-broom image RPC fitting accuracy

| Image number | Line direction | | Column direction | |
|---|---|---|---|---|
| | Max/pixel | RMS/pixel | Max/pixel | RMS/pixel |
| 01-005-134-A | 0.000126 | 0.000048 | −0.000200 | 0.000027 |
| 01-005-134-B | −0.000219 | 0.000056 | −0.000330 | 0.000035 |
| 01-005-135-A | 0.000336 | 0.000072 | −0.000200 | 0.000025 |
| 01-005-135-B | −0.000626 | 0.000038 | −0.000300 | 0.000033 |
| 02-005-135-A | 0.000169 | 0.000023 | −0.000300 | 0.000031 |
| 02-005-135-B | 0.000183 | 0.000015 | −0.000310 | 0.000035 |

substitution accuracy of the model is evaluated. During the experiment, virtual control grid with intervals of 512 × 512 pixels and elevation layer of 5 was generated; based on the virtual control grid, the encryption and translation generated 216 × 216 pixel intervals and elevation layer of 10 inspection grid.

The virtual control grid is used to solve the RPC parameters, and the virtual check grid is used for the RPC model substitution precision verification. The model adopted is the RPC model with unequal 3 orders and unequal denominator. The RPC fitting accuracy of 6-scene virtual sweeping images was statistically analyzed, and the results were shown in Table 3. On the basis of RPC precision verification of virtual push-broom images, the corresponding coordinates of the image points were reversely calculated based on the ground coordinates of control points and the RPC model of virtual push image. The calculated values of the coordinates of the image points were subtracted from the measured values, and the errors were calculated to describe the internal geometric precision of the virtual pushover image (i.e., the internal geometric positioning accuracy consistency). Compared with the traditional image mosaic algorithm (global embedding + piecewise affine transformation), the results are shown in Table 4.

Table 4 Statistics table of stitched production image inner geometric accuracy

| Image number | Control points | Algorithm product in this paper | | | | Traditional image stitching algorithm product | | | |
|---|---|---|---|---|---|---|---|---|---|
| | | No control point | | Control point Affine transformation of image | | No control point | | Control point Affine transformation of image | |
| | | Line direction | Column direction | Line direction | Column direction | Line direction | Column direction | Line direction | Column direction |
| 01-005-134-A | 12 | 1.186 | 1.039 | 1.194 | 1.016 | 4.786 | 10.337 | 2.344 | 3.057 |
| 01-005-134-B | 12 | 1.296 | 1.339 | 1.045 | 1.173 | 5.213 | 9.291 | 2.469 | 3.540 |
| 01-005-135-A | 51 | 1.766 | 1.908 | 1.550 | 1.862 | 6.272 | 11.664 | 3.182 | 5.033 |
| 01-005-135-B | 51 | 1.953 | 1.877 | 1.904 | 1.931 | 5.964 | 10.873 | 2.990 | 5.165 |
| 02-005-135-A | 45 | 1.811 | 2.03 | 1.956 | 1.867 | 6.048 | 12.039 | 3.065 | 5.271 |
| 02-005-135-B | 35 | 1.711 | 1.639 | 1.622 | 1.752 | 6.319 | 11.988 | 3.248 | 5.372 |

From the available statistic data (Table 3), the virtual push-broom image RPC model fitting accuracy is better than $1 \times 10^{-4}$ pixels. The reason for its high accuracy is that the virtual camera has no geometric distortion and does not exist for the integral time jump, which makes the virtual push-broom imaging models smooth, and it lays the foundation for high-precision RPC model fitting. As seen from Table 4, under no control condition, the error in image positioning of traditional image mosaic products can reach a maximum of 12 pixels, while the maximum error in image positioning of the virtual push-broom image generated by the algorithm in this paper is only 2 pixels, indicating that the internal geometric accuracy of traditional image mosaic product is low, but the internal geometric accuracy of new algorithm is much higher than that of traditional image splicing products. Although a small number of control points are used to compensate the RPC model image-side affine transformation error, the internal geometric accuracy of traditional image mosaic products has improved, but the maximum error in image positioning remains in about 5 pixels, and the internal geometric accuracy of the virtual push-broom image generated by the algorithm in this paper is still significantly higher than that of the traditional image mosaic product. The new algorithm uses control points to compensate for the image-side affine transformation errors of the RPC models in each scene, and the internal geometric accuracy of the image has not been significantly improved. The reason is the internal parameters of on-orbit geometric calibration results better to complete the internal geometric distortion elimination, without control points to ensure that the virtual image and push internal geometric accuracy.

## 4 Conclusion

The problem of reducing the fitting accuracy of RPC model is studied by means of the core design idea of object stitching and the time jump of line integral. The following conclusions have been obtained:

1. The new algorithm can not only achieve seamless slice CCD images but also generate RPC parameter files based on its strict geometric model while generating the image product.
2. The integral time for detecting and solving the RPC model partition design can effectively solve the line integral time jump, thus reducing the fitting accuracy problems of RPC model.
3. Because of the attitude and orbit data of virtual push-broom images, the virtual camera is smooth and there is no internal distortion, so the RPC model substitution precision can reach a higher level and it fully meets the needs of practical application.
4. We compared with traditional stitching products, the internal geometric accuracy of splicing products obtained by the new algorithm is significantly increased.

5. When we compared with [13] algorithm, RPC model is roughly the same, but because of the lack of resampling steps, the computational efficiency is improved. However, the new algorithm generates multiple RPC parameter files when the row integral jump occurs, which is not convenient for users to generate image products using RPC parameter files.

The virtual push-broom image stitching algorithm based on the fitting of the RPC model of the segmented image can provide a certain technical foundation for the high-precision geometric processing of the stitched TDI CCD satellite image, but there are still some problems that need further research, including:

1. The high frequency error will have a greater impact on image quality, the positioning accuracy of submeter satellite images, the satellite submeter high-frequency error detection, and modeling, solving, and generating images without distortion are worthy of further study.
2. The internal geometric accuracy of qualitative and quantitative evaluation of the splicing products system still needs further research and exploration.

## References


1. Jacobsen, K.: Calibration of optical satellite sensors. http://www.isprs.org/proceedings/papers/Ca;SatJac_Jacobsen.pdf (2014)
2. Zhang, L., Gruen, A.: Multi-image matching for DSM generation from IKONOS imagery. ISPRS J. Photogramm. Remote Sens. **60**(1), 195–211 (2006)
3. Zhang, L., Gruen, A.: Automatic DSM generation from linear array imagery data. In: Proceedings of IAPRS Congress, pp. 128–133. IAPRS, Turkey (2004)
4. Jacobsen, K.: Calibration of IRS-1C PAN-camera. In: Joint Workshop "Sensors and Mapping from Space", Hannover, pp. 1–8 (1997)
5. Fen, H.: Research on Inner FOV Stitching Theories and Algorithms for Sub-images of Three Non-collinear TDI CCD Chips. Wuhan University, Wuhan (2010)
6. Shiwei, L., Tuanjie, L., Hongqi, W.: Image mosaic for TDICCD push-broom camera image based on image matching. Remote Sens. Techbol. Appl. **24**(3), 374–378 (2009)
7. Xiaoxiang, L., Xiaoyan, W., Huimin, Z.: Analysis of image quality and processing method of a space-borne focal plane view splicing TDI CCD camera. SCIENCE CHINA Inf. Sci. **41**, 19–31 (2011)
8. Weican, M., Shulong, Z., Baoshan, Z., Bincai, C.: TDI CCDs imagery stitching using piecewise affine transformation model. J. Geomat. Sci. Technol. **30**(5), 505–509 (2013)
9. Qi, C., MingWei, S.: Automated seamless mosaicking of multi-strip data from CBERS-02C imagery. In: International Conference on Remote Sensing, Environment and Transportation Engineering (RSETE), pp. 513–517 (2013)
10. Guo, Z., Bin, L., Wanshou, J.: Inner FOV stitching algorithm of spaceborne optiacal sensor based on the virtual CCD line. J. Image Graph. **15**(5), 1046–1052 (2011)
11. Hongbo, P., Guo, Z., Xinming, T., et al.: The geometrical model of sensor corrected products for ZY-3 satellite. Acta Geodaetica et Cartographica Sinica. **15**(5), 1046–1052 (2011)
12. Jun, P., Fen, H., Mi, W., Shuying, J.: Inner FOV stitching of ZY-1 02C HR camera based on virtual CCD line. Geo. Inf. Sci. Wuhan Univ. **40**(4), 71–77 (2015)

13. Shuying, J., Fen, H., Mi, W., Jun, P.: A novel coordinate inverse transform method for TDI CCD push-broom images. Geo. Inf. Sci. Wuhan Univ. **41**(5), 590–597 (2016)
14. Bincai, C., Zhenge, Q., Shulong, Z., Weican, M., Delin, M., Fang, C.: A solution to RPCs of satellite imagery with variant integration time. Surv. Rev. **46**, 392–399 (2016)
15. Weican, M., Shulong, Z., Wen, C., et al.: Establishment and optimization of rigorous geometric model of push-broom camera using TDI CCD arranged in an alternating pattern. Acta Geodaetica et Cartographica Sinica. **44**(12), 1340–1350 (2015)
16. Weican, M., Shulong, Z., Wen, C., et al.: High accuracy on-orbit geometric calibration of linear push-broom cameras. Geo. Inf. Sci. Wuhan Univ. **40**(10), 1392–1399 (2015)
17. Bin, L., Jianya, G., Wanshou, J., Xiaoyong, Z.: Improvement of the iteration by correcting characteristic value based on ridge estimation and its application in RPC calculating. Geo. Inf. Sci. Wuhan Univ. **37**(4), 399–403 (2012)
18. Xiaoyong, Z., Guo, Z., Xuwen, Q.: The formulation of RPC for domestic optical satellite imager. Remote Sens. Land Resour. **2**, 32–35 (2009)
19. Yongsheng, Z., Danchao, G., Jun, L., et al.: Application of High Resolution Remote Sensing Satellite. Science Press, Beijing (2014)
20. Xiuxiao, Y., Xiang, Y.: Calibration of angular systematic errors for high resolution satellite imager. Acta Geodaetica et Cartographica Sinica. **41**(3), 385–392 (2012)

# Research on Hexagonal Remote Sensing Image Sampling

**Mingyang Zheng, Jin Ben, Wen Cao, Rui Wang, and Jianbin Zhou**

**Abstract** Hexagonal discrete global grids can provide an excellent solution for the massive multi-source, multi-temporal, and multi-resolution raster data integration and management. Traditional images are rectangular pixels, and cannot be expressed on a hexagonal grid. Therefore, how to obtain images based on hexagonal pixels has attracted widespread academic attention. Combining current hexagonal sampling methods, this paper studies the evaluation criteria of hexagonal sampling accuracy, summarizes the previous research, and proposes a more general hexagonal sampling method for remote sensing images. This method mainly involves signal preprocessing, spectrum analysis, calculation of sampling interval, and establishment of accuracy evaluation standards. Finally, we verify the feasibility of the proposed hexagon algorithm to provide a reference for hexagon sampling.

**Keywords** Multi-source data · Hexagonal sampling · Nyquist sampling
Accuracy evaluation criteria

## 1 Introduction

With the successful implementation of the National Project "High-Resolution Earth Observation", China's ability to acquire geospatial data has been unprecedentedly increasing [1]. Faced with unprecedented massive, multi-source, multi-temporal, and multi-resolution spatial data, traditional spatial data models lack a global unified spatial reference [2], and exist problems such as difficulty in integrating multi-scale data [2], inaccurate large-scale analysis [3], and so on. It is not possible to

M. Zheng (✉) · J. Ben · R. Wang · J. Zhou
Information Engineering University,
Zhengzhou, Henan Province, People's Republic of China
e-mail: benj@lreis.ac.cn

W. Cao
School of Geo-Science and Technology, Zhengzhou University, Zhengzhou, Henan Province, People's Republic of China



H. P. Urbach, Q. Yu (eds.), *6th International Symposium of Space Optical Instruments and Applications*, Space Technology Proceedings 7,
https://doi.org/10.1007/978-3-030-56488-9_10

achieve unified integration and management of massive remote sensing data. The Discrete Global Grid System (DGGS) can provide a new data model for the integration and management of remote sensing data. DGGS takes the globe as the research object, establishes a globally unified spatial reference datum, and a multi-source data processing model that takes position as the object. In constructing common graphics for DGGS, hexagonal graphics have obvious advantages, compared with triangles and quadrilaterals, due to the consistent grid topology of hexagons, which is helpful for the implementation of spatial analysis algorithms [4]. The sampling efficiency for hexagons is 13% higher than that of the quadrilateral, which is helpful for the data visualization [5]. Therefore, the hexagonal DGGS is more conducive to the organization, processing, analysis, and visualization of remote sensing image data, and is expected to model the remote sensing data in a way that is more suitable for computer process.

Traditional remote sensing images are represented by quadrilateral pixels. Although it is convenient for computer storage, as the sampling dimension increases, the spatial sampling efficiency of the quadrilateral is far lower than the hexagonal sampling efficiency, which will greatly increase the computer storage and calculation costs. From the perspective of aliasing, the adjacent relationship of square pixels is inconsistent, and the pre-filtering effect in the axial direction is significantly different from that in the diagonal direction. For hexagonal pixels, their neighbor relationships are consistent, with consistent connectivity [5] and circular symmetry [6], which is more conducive to the design of symmetrical filters. In addition, hexagonal pixels have many other advantages, such as higher angular resolution [7] and more harmony with human vision [8]. Based on these advantages, it is of great research significance to realize hexagonal sampling of remote sensing images.

Hexagonal sampling has aroused widespread concern in academia, but the research on remote sensing image sampling lacks systematic review and summary. This paper analyzes the hexagonal samping technology and summarizes measurement of hexagonal sampling precision in China and overseas, and proposes a solution of obtaining hexagonal images. First, signal spectrum is analyzed, signal preprocessed, Second, sampling intervals is calculated. Third, measurement of hexagonal sampling precision is selected to evaluate the quality of the data. Finally, the results show that the hexagonal sampling algorithm based on equal density is achievable, and the Fourier transform of the hexagonal image and the visualization of the hexagonal image are possible.

## 2 Classification of Hexagonal Sampling Methods

Hexagonal sampling research started from the 1960s. Daniel P. Petersen [9, 10] extended Shannon's sampling theorem to N dimensions and mentioned that rectangular sampling is not the optimal sampling scheme. Middleton and Sivaswamy [11] showed that the hexagonal sampling rate is 13.4% higher than the quadrangular sampling. Hexagonal image research can be mainly divided into two categories:

One category is to obtain hexagonal images based on rectangular images, which is relatively simple but is very difficult to evaluate the accuracy of the data. The other is to obtain the hexagonal image using the hexagonal sensors directly, but relevant research based on this method are few.

## *2.1 Obtaining Hexagonal Images Based on Rectangular Pixels*

The methods for obtaining hexagonal data based on rectangular images include pseudo hexagonal pixels, oversampling, interpolation sampling, and resampling with equal pixel density.

### 2.1.1 Pseudo Hexagonal Pixel

Wüthrich and Stucki [12], Wu et al. [13] proposed a method to create pseudo hexagonal pixels from a cluster of square pixels. The resulting hexagonal pixels have no overlap or gap and the six sides of pixel cluster have equal lengths, such that it is similar to hexagons to a great extent.

The gray value of the pseudo hexagonal pixel is the average gray value of a square pixel cluster. This methods, as shown in Fig. 1a can obtain images with less distortion, but the image resolution is often greatly lost.

### 2.1.2 Interpolated Hexagonal Pixels

Xiaochong Tong [14] used interpolation sampling algorithms to create a hexagonal pixel. The six pixels obtained by interpolation form a hexagonal pixel, and the gray value of the hexagonal pixel is determined by the gray values of the center pixel and

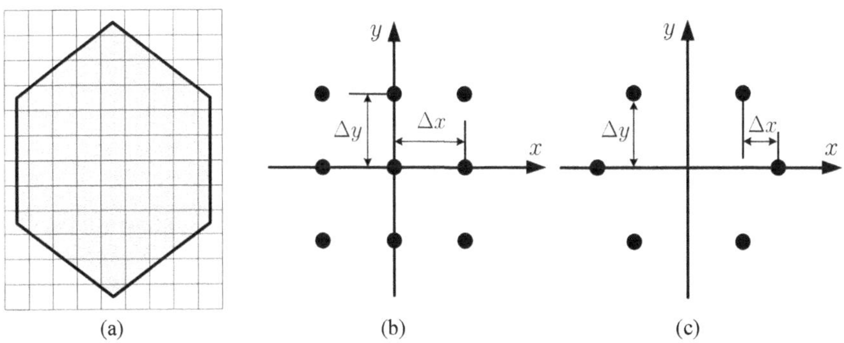


**Fig. 1** Get Hexagonal Pixel Schematic. (**a**) Pseudo Hexagonal Pixel, (**b**) Rectangular sampling, (**c**) Hexagonal sampling

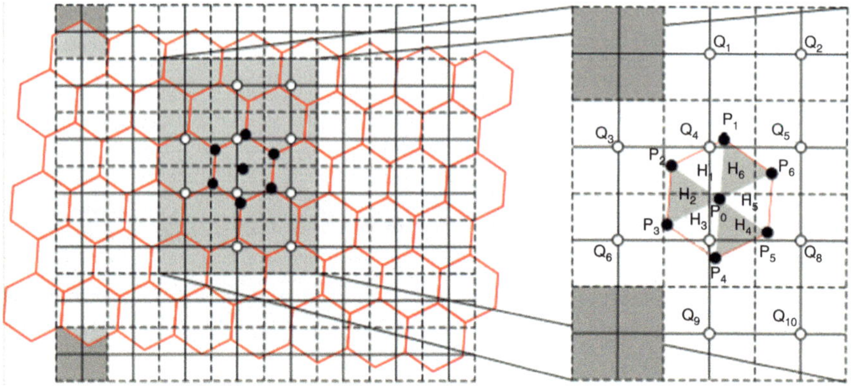


**Fig. 2** Multi-point interpolation hexagon sampling [14]

the six vertex pixels. This multi-point interpolation method not only guarantees the sampling efficiency, but also makes the image grayscale display uniform. However, the data accuracy evaluation for this method is more complicated (Fig. 2).

### 2.1.3 Resampling Based on Equal Density

Burton et al. [15, 16] calculated the hexagonal sampling interval based on the method of equal pixel density, as shown in Fig. 1b, c. Xiqun Lu and Chun Chen [8], Yalu Li [17] obtained hexagonal image data based on equal resampling of pixel area.

Assume that the frequency band of a two-dimensional continuous signal lies in a circular region with a radius $w$ in the frequency domain. Let $T_1$ and $T_2$ are the horizontal and vertical sampling interval of the quadrilateral. According to the Nyquist sampling theorem, it can be deduced that the sampling interval of the quadrilateral image is $T_1 < \frac{\pi}{w}, T_2 < \frac{\pi}{w}$, so the quadrilateral sampling matrix is

$$\begin{bmatrix} \frac{\pi}{w} & 0 \\ 0 & \frac{\pi}{w} \end{bmatrix} \tag{1}$$

Hexagonal sampling interval is represented by rectangular sampling interval $T_1$, $T_2$. Assuming $T_1 = \frac{\sqrt{3}T_2}{2}$, the image sampling density can be expressed by the inverse of the matrix determinant, the pixel densities result the same for the two lattices at the Nyquist limit [18], the hexagonal sampling matrix can be calculated as

$$\begin{bmatrix} \frac{\sqrt{3}\pi}{3w} & \frac{\sqrt{3}\pi}{3w} \\ \frac{\pi}{w} & -\frac{\pi}{w} \end{bmatrix} \tag{2}$$

Through Eqs. (1) and (2), we can prove that the hexagonal sampling efficiency is higher than that of the quadrilateral. This method mainly calculates the original pixel retention ratio to evaluate the accuracy of the hexagonal data.

## 2.2 *Direct Hexagonal Sampling*

Mersereau [19] proposed a hexagonal sampling method based on the Nyquist sampling theorem. The signal function of the two-dimensional continuous signal is $f(t_1,t_2)$ after preprocessing, and the signal function is $F(\xi,\eta)$ after Fourier transform. $\xi_{max}$, $\eta_{max}$ denote the horizontal cut-off frequency and vertical cut-off frequency of the hexagonal sampling. The value of the signal spectrum function $F(\xi,\eta)$ is limited to the following range:

$$\frac{\xi^2}{\xi_{max}^2}+\frac{\eta^2}{\eta_{max}^2}\le 1 \tag{3}$$

Because the circular structure has circular symmetry, a circular band filter is used to pre-process the signal, i.e. $\xi_{max} = \eta_{max} = w$. Hexagonal sampling is similar to rectangular sampling. Hexagonal pre-filtering is the same as quadrilateral pre-filtering. The value of the sampling sequence is the sampling value corresponding to the analog waveform, except that the sampling point positions are different. $n_1$ and $n_2$ are the number of horizontal and vertical sampling points of the quadrilateral, $T_1$ and $T_2$ are the horizontal and vertical sampling interval of the quadrilateral, then the preprocessing function $f(t_1,t_2)=f(n_1\times T_1,n_2\times T_2)$. According to the Nyquist sampling principle, $T_1<\frac{\pi}{w},T_2<\frac{\pi}{w}$ is calculated. Since two adjacent rows of the hexagonal sample differ by half an interval, the preprocessing function is $\tilde{f}(t_1,t_2)=f\left(\frac{2n_1-n_2}{2}\times V_1,n_2\times V_2\right)$, where $V_1$, $V_2$ are the horizontal sampling interval and vertical sampling interval of the hexagon, $n_1$ and $n_2$ are the number of horizontal sampling points and the number of vertical sampling points. According to the pre-filter function of the quadrangle and the hexagon, we can calculate $V_1<\frac{2\sqrt{3}\pi}{3w},V_2<\frac{\pi}{w}$.

## 3 Measurement of Hexagonal Sampling Precision

The amount of information loss of the hexagonal image data is used to evaluate the quality of the hexagonal data compared to the quadrilateral image. The existing accuracy evaluation methods can be roughly divided into four types: interpolation sampling accuracy evaluation method, mean square error evaluation accuracy method, Parseval's theorem evaluation accuracy method, and MTF value accuracy evaluation.

### *3.1 Evaluation of Interpolation Sampling Accuracy*

The multi-point interpolation sampling method is used to reconstruct the signal, and then a digital image is obtained based on the reconstructed signal [14]. However, it is necessary to consider linear interpolation, sampling, quantization, and the mean square error produced by multiple pixels. The gray levels of the reconstructed digital image and the original remote sensing image are compared to determine the sampling accuracy. This evaluation algorithm has a rigorous mathematical theoretical foundation, but the calculation is more complicated.

### *3.2 Mean Square Error Evaluation Accuracy*

Jeevan and Krishnakumar [20] introduced a Mean-Squared-Error (MSE) method to calculate the average difference of pixels in the three steps of pre-filtering, sampling, and reconstruction, respectively, and to calculate the MSE value. The higher the MSE value, the greater the difference between the original and processed images. The calculation formula is as follows:

$$\mathrm{MSE} = \frac{1}{N}\sum_{i}\sum_{j}\left(X_{ij} - V_{ij}\right)^{2} \tag{4}$$

where $X_{ij}$ and $V_{ij}$ are the processed image pixels and the original image pixels, and $N$ represents the number of pixels.

### *3.3 Parseval's Relation*

Brown [21] mentioned a method using Parseval's theorem to evaluate image accuracy, i.e. the sum of squares of a signal function is equal to the sum of squares of its Fourier transform. The energy of a digital image is equal to the sum of the squares of the amplitude values at each sample point. Therefore, the data accuracy was evaluated by calculating the amplitude of the hexagonal sampling points.

### *3.4 Modulation Transfer Function*

The value of modulation transfer function is an important evaluation index for optical image. The larger the Modulation Transfer Function (MTF) value, the better the optical performance. Bruno Aiazzia [18] evaluated the accuracy of hexagonal, octagonal, and quadrilateral data by calculating MTF values.

## 4 Hexagonal Sampling Method

The aperiodic signal is a two-dimensional signal, and its frequency domain is also continuous time aperiodic. In order to obtain a digital image, it is necessary to prefilter the continuous filtering.

### *4.1 Time-Frequency Analysis*

Analyzing the signal in the frequency domain has a very important effect on data processing. The frequency domain can clearly show features which are difficult to show in the space domain. According to the Fourier transform theorem, the aperiodic signal $f(t_1, t_2)$ that changes with time or space can be regarded as a superposition of a plurality of fundamental harmonic signals having different frequencies. $\delta(U)$ is the two-dimensional sampling pulse function, then the amplitude of the sampling point is $C = f(t_1, t_2) * \delta(U)$. Define the aperiodic signal function $f(t_1, t_2)$ whose Fourier transform is $F(\xi_1, \xi_2)$. If the spatial-domain coordinate axis is scaled and transformed, its frequency domain will also change. It is assumed that the two-dimensional scaling matrix of the function is $A$, then $f(A(t_1, t_2))$ Fourier transform is $\frac{1}{|A|}F\left(A^{-T}\left(\xi_1,\xi_2\right)\right)$, assuming $Z = (t_1, t_2)$, $\xi = (\xi_1, \xi_2)$, proved as follows:

$$F\left(f\left(AZ\right)\right) = \int e^{-2\pi i Z\xi)} f\left(AZ\right) d_Z$$

$$X = AZ$$

$$d_X = |A| d_Z$$

$$Z\xi = A^{-1} X\xi$$

$$F(f(X)) = \frac{1}{|A|} \int e^{-2\pi i X A^{-T} \xi} f(X) d_x$$

$$F(f(AZ)) = \frac{1}{|A|} F(A^{-T} \xi) \tag{5}$$

## 4.2 Processing and Sampling Signals

According to the second part of the sampling method, the signal is pre-processed using a circular filter, where $f(t_1, t_2)$ is the pre-processed signal function. According to the sampling theorem, the highest sampling frequency is not less than twice the cut-off frequency. Let the cut-off frequency be $w$, the sampling interval of the quadrilateral image is $T_1 < \frac{\pi}{w}, T_2 < \frac{\pi}{w}$, $\tilde{f}(t_1, t_2) = f_s(n_1 \times T_1, n_2 \times T_2)$, where $n_1$ and $n_2$ are the number of horizontal and vertical sampling points of the quadrilateral. The hexagonal pre-filtering is the same as the quadrilateral pre-filtering, and both the hexagonal and quadrangular sampling values correspond to the aperiodic waveform values, as shown in Fig. 3.

There is a half sampling interval between the two adjacent rows of the hexagonal sampling lattice, and the pre-processed signal is $\tilde{f}(t_1, t_2) = f\left(\frac{2n_1 - n_2}{2} \times V_1, n_2 \times V_2\right)$, where $V_1$ and $T_2$ are the horizontal sampling interval and vertical sampling interval of the hexagon. Spatial sampling interval is related to the number of sampling points and the cut-off frequency, according to the quadrangle pre-filtering function and the number of sampling points; the hexagonal sampling interval is calculated as $V_1 < \frac{2\sqrt{3}\pi}{3w}, V_2 < \frac{\pi}{w}$ (Fig. 4).

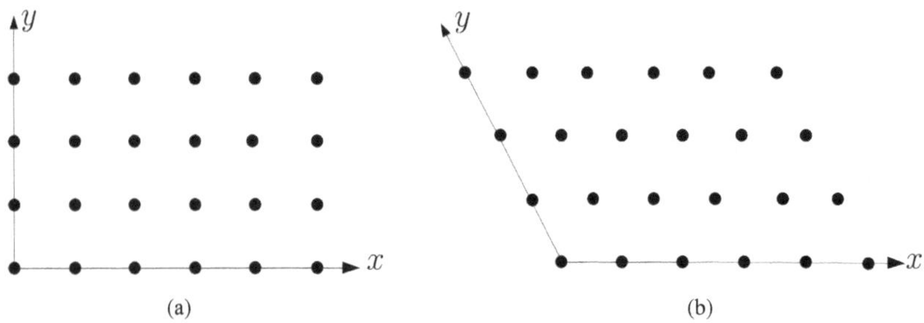


**Fig. 3** Two-dimensional sampling lattice. (**a**) Rectangular sampling, (**b**) Hexagonal sampling

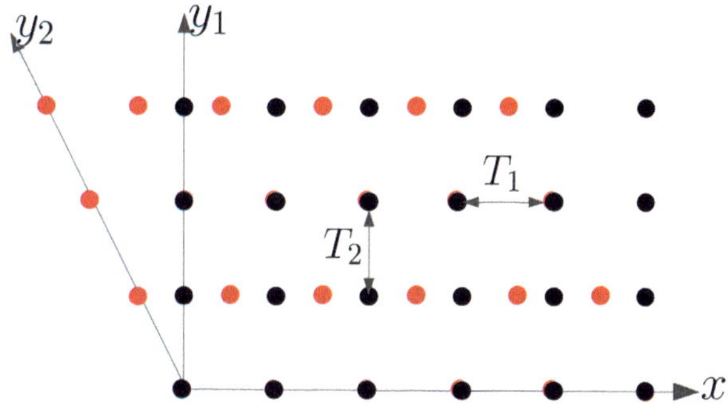


**Fig. 4** Samples on continuous signals

All sampling points in the figure are samples on continuous signals, where the red sampling points are hexagonal sampling points, and the black sampling points are quadrangular sampling points. Although the positions of the sampling points on the even rows are the same, the horizontal sampling interval of the hexagon is $\frac{2\sqrt{3}}{3}$ times the horizontal sampling direction of the quadrangle, which means that the sampling pulse function undergoes a stretching transformation in the horizontal direction. According to Eq. (5), the function undergoes a scaling transformation and the frequency spectrum also changes to $\frac{1}{|A|}F\left(A^{-T}\left(\xi_1,\xi_2\right)\right)$. The gray value of the sampling point also changes, so it is necessary to analyze the hexagonal pixel accuracy after sampling.

## *4.3 Accuracy Evaluation*

Kamgar-Parsi [22] used the average quantization error value to evaluate the relative noise sensitivity of the hexagonal and quadrilateral sampling grids. Experimental results show that there is almost no difference between the effects of quantization errors of the two systems, so the quantization errors can be ignored. Hexagonal sampling is obtained by horizontally shifting two adjacent rows of quadrilateral sampling samples by half of the sampling interval and stretching the sampling interval in the vertical sampling direction. Therefore, we consider using Parseval's theorem to evaluate the accuracy of hexagonal pixels. The gray value of the pixel obtained by calculating the hexagonal sampling is compared with the corresponding gray value of the initial quadrangular pixel to further evaluate the accuracy of the hexagonal data.

# 5 Obtaining and Application of Hexagon Remote Sensing Image

## 5.1 Obtaining Hexagonal Images

Traditional images are rectangular pixels. Each pixel not only has geographic attributes, and represents the spatial resolution of the image. In order to keep the geographical location and spatial resolution of hexagonal pixels the same as that of quadrilaterals, we can use the resampling method with equal cell area to obtain the hexagonal image. Assuming that the side length of the rectangular pixel is 1, the side length of the hexagonal pixel with equal area is calculated to be 0.6204, then, the coordinates of the sampling point are the same as the quadrilateral, and the gray value of pixel is unchanged. Figure 5a, b are the 10 × 10 rectangular remote pixel

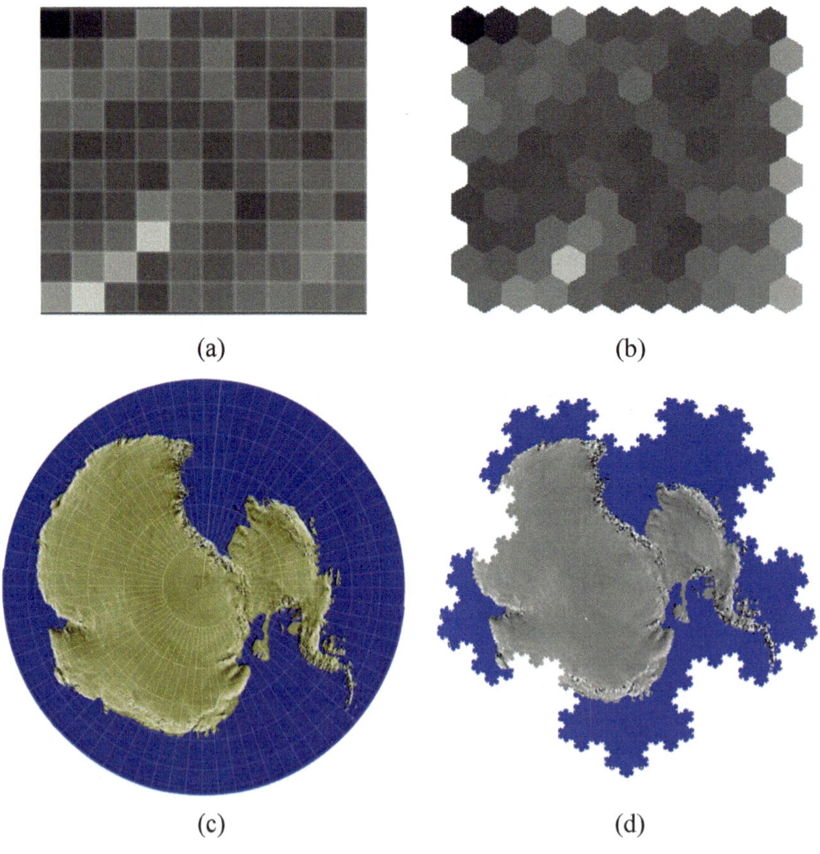

**Fig. 5** Hexagonal remote sensing image and quadrilateral remote sensing image. (**a**) 10 × 10 original image, (**b**) 10 × 10 hexagon image, (**c**) Remote sensing image of high-latitude longitude and latitude grid, (**d**) High-latitude hexagon remote sensing image

image and the $10 \times 10$ hexagonal pixel image. In addition, in order to show the real sampling result, we also transform the rectangular pixel data image of the gtopo30 data set at 60–90 south latitude (Fig. 5c) to the hexagon pixel image, as shown in Fig. 5d. The experimental results show that the rectangular pixels in the high-latitude area are squeezed and deformed. Since hexagonal pixels are equal area in the global scope, they will not be squeezed and deformed. Therefore, obtaining hexagonal pixel image solves the problem of large amounts of data redundancy caused by equally spaced latitude and longitude grids.

## 5.2 *Fourier Transform of Hexagon Image*

The frequency domain can clearly show features which are difficult to show in the space domain. The difficulty of the discrete Fourier transform based on hexagonal pixels is that the transform kernel cannot be separated, i.e. the two-dimensional discrete Fourier transform cannot be separated into two one-dimensional discrete Fourier transforms, so the butterfly algorithm cannot be used. Therefore, different Fourier transform algorithms are generated on different hexagonal grids. In this paper, the hexagonal two-dimensional spatial function is converted to a one-dimensional function, and then the one-dimensional function is Fourier transformed to obtain the hexagonal image spectrum. The hexagonal spectrum obtained by Fourier transform of a one-dimensional function is shown in Fig. 6b, the result graph for the frequency domain is rotated counterclockwise by $0.0330467^{o}$.

In order to verify the accuracy of the hexagonal Fourier transform, we set different cut-off frequencies, and use a high-pass filter to filter the hexagonal spectrum image. The result of the inverse Fourier transform of the spectrum image is shown in Fig. 7.

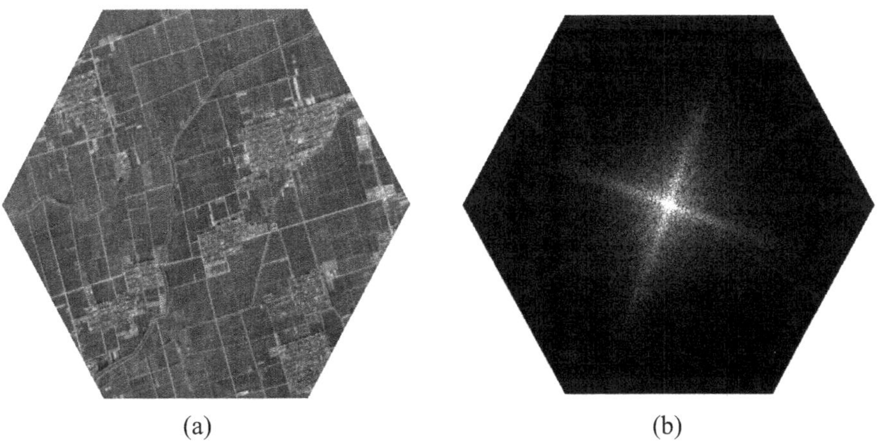

**Fig. 6** Samples on continuous signals. (**a**) Hexagonal pixel image, (**b**) Hexagonal pixel image spectrum

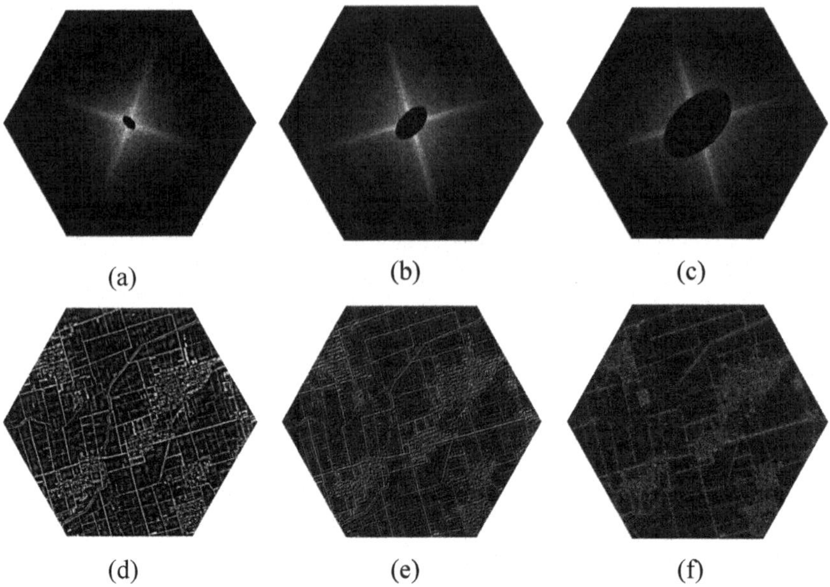

**Fig. 7** Comparison of filter effects with different cut-off frequencies. (**a**) Cut-off frequency 0.03, (**b**) Cut-off frequency 0.07, (**c**) Cut-off frequency 0.15, (**d**) Cut-off frequency 0.03, (**e**) Cut-off frequency 0.07, (**f**) Cut-off frequency 0.15

In fact, after performing high-pass filtering for the quadrilateral image, the filtered low-frequency shape is a circle, but after performing high-pass filtering for the hexagonal image, the filtered low-frequency shape is an ellipse, which is resulted from the difference between the base vectors of hexagonal pixels and that of rectangular pixels. Comparison results show that the filtering effect with a cut-off frequency of 0.03 is the best.

## 5.3 *Visualization of Hexagonal Image*

In order to visualize the hexagonal image on the discrete global grid system, we used vegetation coverage data of multiple provinces, which came from the Resource and Environmental Data Cloud Platform, as shown in Fig. 8.

Hexagon images obtained through resampling rectangular image can display various types of vegetation. Compared with remote sensing images at low and medium latitudes, the sampling advantages of hexagons are not obvious, but compared with rectangular images at high latitudes, the advantages of hexagonal images are obvious. Using sampling points with the identical number as rectangular images, hexagonal images can show more information about features. Figure 9 is the diagram of hexagonal image visualization.

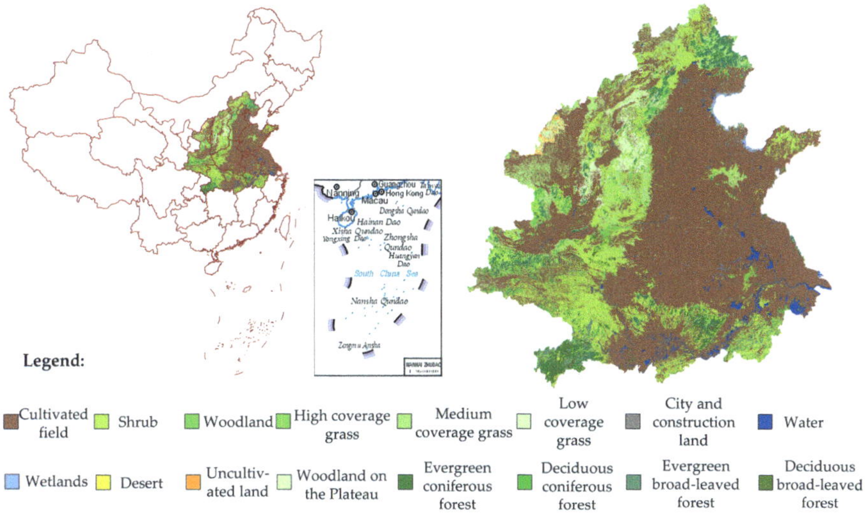


**Fig. 8** Rectangular image vegetation coverage diagram

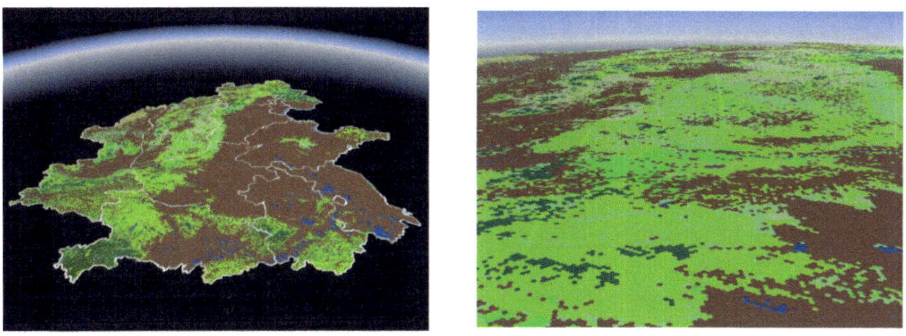


**Fig. 9** Schematic diagram of vegetation cover. (**a**) Global viewpoint, (**b**) Local viewpoint

# 6 Conclusion

The experimental results show that the hexagonal remote sensing image can be obtained by using the equal area sampling method. Hexagonal image not only solves the problem of high-latitude pixel distortion, but also improves the storage efficiency of the computer compared to remote sensing images of equal latitude and longitude. The extraction results of hexagonal linear features show that it is feasible to convert the two-dimensional function into one-dimensional to obtain the hexagonal spectrum, which overcomes the problem that the two-dimensional transformation kernel of the hexagon cannot be separated. At the same time, the visualization of the hexagonal image can verify that the hexagonal image can obtain more sampling information at the same sampling point. The next task is to quantify the accuracy of the hexagonal data.

## References


1. Deren, L., Peilaing, Z., Guisong, X.: Automatic analysis and data mining of remote sensing big data. Acta. Geod. Cartoraphica Sinica. **43**(12), 1211–1216 (2014)
2. Min, S., Xuesheng, Z., Renliang, Z.: Global GIS and its key technologies. Geomat. Inf. Sci. Wuhan Univ. **33**(1), 41–45 (2008)
3. Xuesheng, Z., Miaole, H., Jianju, B.: Spatial Digital Modeling of Global Discrete Grids. Surveying and Mapping Press, Beijing (2007)
4. Chenghu, Z., Ouyang, M.T.: Research progress of geographic grid model. Prog. Geogr. **28**(5), 657–662 (2009)
5. Jin, B., Xiaochong, T., Chenghu, Z., Kaixin, Z.: Hexagonal discrete grid system generation algorithm for regular octahedron. J. Geo-Inf. Sci. **17**(7), 789–797 (2015a)
6. Gardiner, B., Coleman, S., Scotney, B.: Comparing hexagonal image resampling techniques with respect to feature extraction. In: 14th International Machine Vision and Image Processing Conference. Cambridge Scholars Publishing, Newcastle upon Tyne (2011)
7. Singh, I., Oberoi, A.: Comparison between square pixel structure and hexagonal pixel structure in digital image processing. Int. J. Comput. Sci. Trends Technol. **3**, 176–181 (2015)
8. Lu, X., Chun, C.: Research on digital images of regular hexagon lattice structure. J. Image Gr. **06**, 82–88 (2004)
9. Petersen, D.P., Middleton, D.: Sampling and reconstruction of wave-number-limited functions in N-dimensional Euclidean spaces. Inf. Control. **5**(4), 279–323 (1962)
10. Lucas, D.: A multiplication in N-space. IEEE Trans. Image Process. Am. Math. Soc. **74**, 1–8 (1979)
11. Middleton, L., Sivaswamy, J.: Hexagonal Image Processing. Springer, London (2005)
12. Wüthrich, C.A., Stucki, P.: An algorithmic comparison between square- and hexagonal-based grids. CVGIP Gr. Model Image Process. **53**(4), 324–339 (1991)
13. Wu, Q., He, S., Hintz, T.B.: Bi-lateral filtering based edge detection on hexagonal architecture. In: Proc IEEE Int Conf Acoustics, Speech, and Signal Processing, pp. 713–716. IEEE (2005)
14. Xiaochong, T., Jin, B., Yongsheng, Z.: Construction and fast display of global multi-resolution data model. Sci. Surv. Mapp. **31**(1), 72–74, 79 (2006)
15. Burton II, J.C.: End-to-end analysis of hexagonal vs. rectangular sampling in digital imaging systems. Dissertation, College of William & Mary (1993)
16. Burton, J.C., Miller, K.W., Park, S.K.: Rectangularly and hexagonally sampled imaging-system-fidelity analysis. In: Visual Information Processing II, vol. 1961, pp. 81–92. International Society for Optics and Photonics, Bellingham, WA (1993)
17. Yalu, L.: Geometric Structure and Fourier Transform of Hexagon Global Discrete Grid System. Information Engineering University, Zhengzhou (2017)
18. Aiazzi, B., Baronti, S., Capanni, A., Santurri, L., Vitulli, R.: Advantages of hexagonal sampling grids and hexagonal shape detector elements in remote sensing imagers. In: 2002 11th European Signal Processing Conference, Toulouse, pp. 1–4. IEEE (2002)
19. Mersereau, R.: The processing of hexagonally sampled two-dimensional signals. Proc. IEEE. **67**(6), 930–949 (1979)
20. Jeevan, K.M., Krishnakumar, S.: Compression of images represented in hexagonal lattice using wavelet and Gabor filter. In: International Conference on Contemporary Computing & Informatics. IEEE (2015)
21. Brown, J.L.: On quadrature sampling of bandpass signals. IEEE Trans. Aerosp. Electron. Syst. **AES-15**(3), 366–371 (1979)
22. Kamgar-Parsi, B.: Dynamical stability and parameter selection in neural optimization. In: Neural Networks, 1992. IJCNN. International Joint Conference on. IEEE (1992)

# A Method to Determine the Rate Random Walk Coefficient of Fiber Optic Gyroscope

**Lijun Ye, Fucheng Liu, Yinhe Chen, Liliang Li, and Hexi Baoyin**

**Abstract** The mechanism of FOG's (fiber optic gyroscope) RRW (rate random walk) is not clear; there are few researches on the estimation of FOG's RRW coefficient based on orbit data. A data processing method based on the combination of on-board and on-ground processing is proposed; the integral angle of long period can be obtained by on-board processing; the RRW coefficient is estimated by Allan variance method by on-ground processing; the problem of large estimation error caused by long RRW correlation time is solved. The identification of RRW correlation time is given based on Allan variance curve; the calculation of the standard deviation of integral angle error is given; correctness of the coefficient estimation is proved by comparative analysis.

**Keywords** Rate random walk · Long correlation time · Fiber optic gyroscope Allan variance method

## 1 Introduction

As a key measurement component, gyroscope is widely used in the control system of satellite, launch vehicle, and UAV (unmanned aerial vehicle) [1]. The working principle of FOG (fiber optic gyroscope) is to sense the rotation angular rate by Sagnac Effect produced by the fiber coil. With the development of material science

---

L. Ye (✉)
School of Aerospace Engineering, Tsinghua University, Beijing, China

Shanghai Aerospace Control Technology Institute, Shanghai, China

Shanghai Key Laboratory of Aerospace Intelligent Control Technology, Shanghai, China

F. Liu · Y. Chen · L. Li
Shanghai Aerospace Control Technology Institute, Shanghai, China

Shanghai Key Laboratory of Aerospace Intelligent Control Technology, Shanghai, China

H. Baoyin
School of Aerospace Engineering, Tsinghua University, Beijing, China



H. P. Urbach, Q. Yu (eds.), *6th International Symposium of Space Optical Instruments and Applications*, Space Technology Proceedings 7,
https://doi.org/10.1007/978-3-030-56488-9_11

[2] and other aspects of technical improvements, the measurement accuracy of FOG has been greatly improved, and it is urgent to research all of the FOG error sources, mainly include five parts [3–8]: QN (quantization noise), ARW (angle random walk), the BI (bias instability), RRW (rate random walk), and RR (rate ramp).

Allan variance method [3] is widely used in the estimation of coefficients; the estimation of coefficients with the shorter correlation time (QN/ARW/BI) is easier, while the estimation of RRW coefficient is much more difficult because of its long correlation time. The mechanism of RRW is not clear [4–8], and also there are few researches on the accurate estimation of high-precision FOG's RRW coefficient.

# 2 FOG Noises Introduction and Analysis

## 2.1 FOG Noises Introduction

QN: It is directly caused by the output digital characteristics of fog, which is essentially the truncation error caused by the minimum resolution of digital quantity.

ARW: It is the result of the integration of the angular rate white noise, which causes the angular rate error of the FOG to be white noise, while the angular shows random walk, and it is the key technical parameter of the FOG.

BI: It shows that the zero bias fluctuates slowly with time.

RRW: It is the result of integration of the angular acceleration white noise, which causes the angular acceleration error of the FOG to be white noise, while the angular rate shows random walk, its mechanism is still uncertain.

RR: In essence, it is a kind of definite error; it shows the monotonous change of the light source very slowly and lasts for the whole lifetime of FOG.

## 2.2 Analysis

In practical application, Allan variance estimation is based on a set of limited data, increasing the correlation time, the number of separable independent subsets will be reduced, and then the accuracy of Allan variance coefficient estimation will be reduced; the unreliability is calculated as follows:

$$\sigma\left(\delta_{\mathrm{AV}}\right)=\frac{1}{\sqrt{2R/(r-1)}} \tag{1}$$

Among that, $R$ is the total number of data, $r$ is the number of data contained in each subset. The smaller $\sigma(\delta_{\mathrm{AV}})$ is, the higher confidence of the estimation.

For example, the duration of data is $10^4$ s, and one sample for 1 s, $R = 10000$, the correlation time is 5000 s, $r = 5000$, then $\sigma(\delta_{AV}) = 50\%$, and the unreliability is obviously high.

In order to make the $\sigma(\delta_{AV})$ lower without increasing the $R$ ($R = 10000$), $r$ should be reduced ($r = 2$), that means one sample for 5000 s, and the duration of data is $5 \times 10^7$ s, then we have $\sigma(\delta_{AV}) = 0.7\%$, and the unreliability is much lower.

# 3 Method

## 3.1 General Introduction

There are advantages to analyze the FOG RRW based on the on-orbit data, and the on-board computer is used to collect and process the data. In fact, the on-board computer can be seen as the data preprocessor, and the data with long correlation time are calculated and telemetered down by the on-board computer. With the help of on-board computer, we can increase the correlation time of data, and reduce the total amount of data.

However, the on-orbit FOG data analysis may be affected by the working mode of spacecraft. In order to ensure the purity of the data, we need to eliminate the data during the unsteady working mode such as spacecraft maneuver. Also, it is necessary to eliminate the influence of the orbit period error, such as thermal deformation between the FOG and the star sensor.

## 3.2 On-Board Processing

**Step 1** Data processing of three-axis angular rate of spacecraft body measured by gyroscope relative to reference system.

$$\begin{pmatrix} \omega_x\_\mathrm{JZ} \\ \omega_y\_\mathrm{JZ} \\ \omega_z\_\mathrm{JZ} \end{pmatrix} = \begin{pmatrix} \omega_{ix} \\ \omega_{iy} \\ \omega_{iz} \end{pmatrix} - \begin{pmatrix} B_x \\ B_y \\ B_z \end{pmatrix} - A_{b\mathrm{JZ}} \times \begin{pmatrix} \omega_{0x} \\ \omega_{0y} \\ \omega_{0z} \end{pmatrix} \tag{2}$$

Among that:

$(\omega_{ix}\ \omega_{iy}\ \omega_{iz})^T$ is the Gyroscope measurement of angular rate of spacecraft body relative to three-axis inertial system;

$\begin{pmatrix} B_x & B_y & B_z \end{pmatrix}^T$ is the constant drift of real-time estimation of three-axis angular rate on the spacecraft, and its initial value is $\begin{pmatrix} 0 & 0 & 0 \end{pmatrix}^T$; See “Step 2” for detailed calculation process;

$A_{bJZ}$ is the attitude transfer matrix from the reference coordinate system to the spacecraft system;

$\begin{pmatrix}\omega_{0x} & \omega_{0y} & \omega_{0z}\end{pmatrix}^T$ is the three-axis angular rate of the reference coordinate system relative to the inertial system, and its initial value is $\begin{pmatrix}0 & 0 & 0\end{pmatrix}^T$ ; See "Step 3" for detailed calculation process;

**Step 2** Calculate the constant drift of the three-axis gyroscope angular rate.

***Step 2.1*** In an integral period $T$ ($T$ *is* generally the orbit period), the sum of the three-axis angular rate of the gyroscope in this period is obtained by integration.

$$\begin{aligned} Bxx(k) &= Bxx(k-1) + \mathrm{ts}\cdot\omega_x\_\mathrm{JZ} \\ Byy(k) &= Byy(k-1) + \mathrm{ts}\cdot\omega_y\_\mathrm{JZ} \\ Bzz(k) &= Bzz(k-1) + \mathrm{ts}\cdot\omega_z\_\mathrm{JZ} \end{aligned} \tag{3}$$

Among that:

$ts$ is the time length of control cycles of the AOCS (attitude and orbit control subsystem);

$k$ is the sequence number of the control cycles in an integral period;

***Step 2.2*** Calculate the three-axis attitude angle difference in this period of time.

$$\begin{aligned} \Delta X &= X_{\left(\frac{T}{\mathrm{ts}}\right)} - X_{(0)} \\ \Delta Y &= Y_{t\left(\frac{T}{\mathrm{ts}}\right)} - Y_{(0)} \\ \Delta Z &= Z_{t\left(\frac{T}{\mathrm{ts}}\right)} - Z_{(0)} \end{aligned} \tag{4}$$

Among that:

$X_{(0)}$, $Y_{(0)}$, $Z_{(0)}$ is the first beat of an integral period ($k = 0$) to record the three-axis attitude angle;

$X_{\left(\frac{T}{ts}\right)}, Y_{\left(\frac{T}{ts}\right)}, Z_{\left(\frac{T}{ts}\right)}$ is the last beat of an integral period $\left(k = \frac{T}{ts}\right)$ recording thethree-axis attitude angle;

***Step 2.3*** Update gyroscope constant drift.

$$\begin{aligned} Bx &= Bx + K * (Bxx - \Delta X)/T \\ By &= By + K * (Byy - \Delta Y)/T \\ Bz &= Bz + K * (Bzz - \Delta Z)/T \end{aligned} \tag{5}$$

Among that, $K \in (0, 1]$ is the correction coefficient, $K \leq 0.01$ is recommended.

***Step 2.4*** Zero integral value.

$$\begin{aligned} Bxx &= 0 \\ Byy &= 0 \\ Bzz &= 0 \end{aligned} \tag{6}$$

In Step 2.1, the sum of the angular rate of the three-axis gyroscope is carried out at each beat; Step 2.2~2.4 is executed in the last beat of the algorithm.

In order to maintain the accuracy of gyroscope constant drift correction, the algorithm is used when the spacecraft is in stable working state. Once the spacecraft is in attitude maneuver or attitude instability, the calculation process is stopped immediately; at the same time, the integral value of three-axis gyroscope angular rate is reset: $Bxx(k) = 0$, $Byy(k) = 0$, $Bzz(k) = 0$, and the calculation results of the gyroscope constant drift $Bx$, $By$, $Bz$ are retained.

The integral period $T$ (is also the correlation time) takes the orbital period, for the following reasons: Firstly, the accuracy of FOG for spacecraft is getting higher, and the correlation time for estimating the RRW coefficient is getting longer; Secondly, to eliminate the influence of the orbit period error.

As for the calculation of gyroscope angular rate and the correction of constant drift, it can also be used as a general standard module, which is suitable for on-board processing mode of spacecraft equipped with star sensor and gyroscope during steady-state operation.

**Step 3** Angular rate calculation of reference frame relative to inertial frame.

***Step 3.1*** Calculation of quaternion variation of reference system.

$$\begin{bmatrix} \dot{q}_0 \\ \dot{q}_1 \\ \dot{q}_2 \\ \dot{q}_3 \end{bmatrix} = q_{iJZ(k-1)}^{-1} \otimes q_{iJZ(k)} \tag{7}$$

Among that:

$q_{iJZ(k)}$ is the quaternion from inertial system to reference system of current beat, which is the reference coordinate system of spacecraft attitude control;

$q_{iJZ(k-1)}$ is the quaternion from inertial system to reference system of last beat;

$\otimes$ is the quaternion multiplier;

***Step 3.2*** Calculation of angular rate of reference frame relative to inertial frame.

If $\dot{q}_0 > 0$, then:

$$\begin{pmatrix} \omega_{0x} \\ \omega_{0y} \\ \omega_{0z} \end{pmatrix} = 2 \begin{bmatrix} \dot{q}_1 \\ \dot{q}_2 \\ \dot{q}_3 \end{bmatrix} \tag{8}$$

Else if $\dot{q}_0 \leq 0$, then:

$$\begin{pmatrix} \omega_{0x} \\ \omega_{0y} \\ \omega_{0z} \end{pmatrix} = (-2) \cdot \begin{bmatrix} \dot{q}_1 \\ \dot{q}_2 \\ \dot{q}_3 \end{bmatrix} \tag{9}$$

**Step 4** Telemetry output of gyroscope integration angle of each beat.

$$\begin{aligned} x\_\text{download}(k) &= Bxx(k) - \left(X_{(k)} - X_{(0)}\right) \\ y\_\text{download}(k) &= Byy(k) - \left(Y_{(k)} - Y_{(0)}\right) \\ z\_\text{download}(k) &= Bzz(k) - \left(Z_{(k)} - Z_{(0)}\right) \end{aligned} \tag{10}$$

Among that, $X_{(0)}$, $Y_{(0)}$, $Z_{(0)}$ are the attitudes of the first beat of the integral period; $X_{(k)}$, $Y_{(k)}$, $Z_{(k)}$ are the attitudes of the current beat of the integral period.

It should be noted in Step 4 that: The attitude determination noise is white noise, which is a constant value independent of time; the larger the integration period, the smaller the proportion of attitude determination noise, and the smaller the impact on the gyroscope integration error.

Step 4 can be understood in this way: at the cost of discarding the lower correlation time gyroscope data, the gyroscope data on-orbit is deeply compressed, so the ground data processing and storage capacity can be greatly reduced.

## 3.3 *On-Ground Processing*

After being processed by onboard computer, the processing results are transmitted to the ground through telemetry, we can get the gyroscope integral angle with correlation time $T$, fot the data of the three-axis gyroscope are independent with each other. The same method can be used to process the data respectively.

**Step 5** Data acquisition.

If only the integral score of the last beat of each integration period $T$ is counted, we can get the gyroscope integration angle in long correlation time, so the data to be processed is greatly reduced, and the workload of ground data processing is correspondingly greatly reduced, and in this way, we can eliminate the influence of the orbit period error.

Also, if all the gyroscope integral angles are differential according to telemetry period $T\_$ telemetry, we can get the gyroscope integration angle in short correlation time ($T\_$ telemetry), so we can also get the parameters that with the shorter correlation time, such as ARW and BI coefficients, but in this way, we should accept the influence of the orbit period error.

Allan variance method is used for data processing [3], as follows:

**Step 6** Generated array.

Divide every $m$ ($m = 1,2,\ldots, M$, $M \leq L/4$) data in the sample space into a group, and get $J = [L/m]$ independent arrays, where operator $[]$ indicates rounding to 0.

Among that, $L$ is the number of data in Step 5.

In Step 6, it is recommended that the maximum of $M$ is no more than $L/4$, which is good for the estimation of coefficients, while in the Allan variance method [3], it is recommended that the maximum of $M$ is $<L/2$.

**Step 7** Average data and get new elements.

Take the average value of each group of original data (calculate the group average) to get the new elements:

$$\bar{\omega}_k\left(m\right) = \frac{1}{m}\sum_{m}^{i=1}\omega_{(k-1)m+i}, k = 1,2,\cdots,J \tag{11}$$

**Step 8** Variance calculation.

Allan variance with correlation time $\tau_m = mT$ is calculated as follows:

$$\sigma^2\left(\tau_m\right) = \frac{1}{2\left(K-1\right)}\sum_{K-1}^{i=1}\left(\bar{\omega}_{i+1}\left(m\right) - \bar{\omega}_i\left(m\right)\right)^2 - \frac{2}{m}\sigma^2{}_{xm} \tag{12}$$

Among that, $\sigma^2{}_{xm}$ is the measurement variance of the star sensor.

**Step 9** Calculation of fitting coefficients.

Taking the correlation time $\tau$ as the variable, the least square method is used to fit the square error:

$$\sigma^2\left(\tau\right) = A_{-1}\tau^{-1} + A_1\tau \tag{13}$$

Among that, $A_{-1}$, $A_1$ are the coefficients to be fitted, $A_{-1}$ is the fitting coefficient of ARW; $A_1$ is the fitting coefficient of RRW;

**Step 10** Calculation of ARW and RRW coefficients.

$$N = \frac{\sqrt{A_{-1}}}{60}\left(\text{deg}/h^{0.5}\right)$$
$$K = 60\sqrt{3A_1}\left(\text{deg}/h^{1.5}\right) \tag{14}$$

Among that, $N$ is the ARW coefficient, $K$ is the RRW coefficient.

It should be noted that, the coefficients to be estimated should be selected according to the correlation time, otherwise it is easy to get the illegal estimation coefficient.

## 4 On Orbit Data Processing and Analysis

In order to estimate both the ARW and RRW coefficients, we make the sample time of the data to be the telemetry period: $T\_\text{telemetry} = 96$ s, and the orbit period is $T = 5960$ s, the duration of data is 2 days, the integral angle curves (in orbit period) are as Fig. 1.

As we can see from Fig. 1, the number of integral angle curves is 28, and the angle always starts from 0.

The angular rate of the FOG is as Fig. 2.

Allan variance curve is as Fig. 3.

From Figs. 2 and 3, the sampling time of the data sample is 96 s, and the duration of all data is 17,280 s, the standard deviation of angular rate is about $1.7 \times 10^{-5}$°/s.

As can be seen from Fig. 3, when the correlation time is about 20,000 s, the variance is the smallest, when the time is <20,000 s, and the shorter the time is, the variance of ARW is dominant; when the time is longer than 20,000 s, and the longer the time is, the variance of RRW is dominant. So the correlation time of RRW is longer than 20,000 s.

According to Eq. (1), we can get the unreliability of RRW(with the correlation time 20,000 s) is about 34%.

The ARW and RRW coefficients are as Table 1.

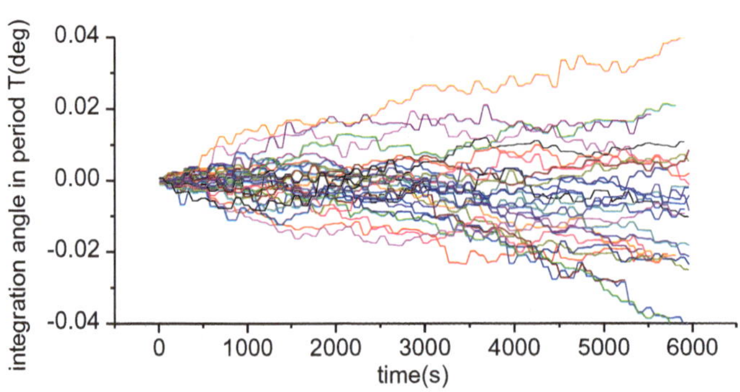


**Fig. 1** 28 integral angle curves in orbit period

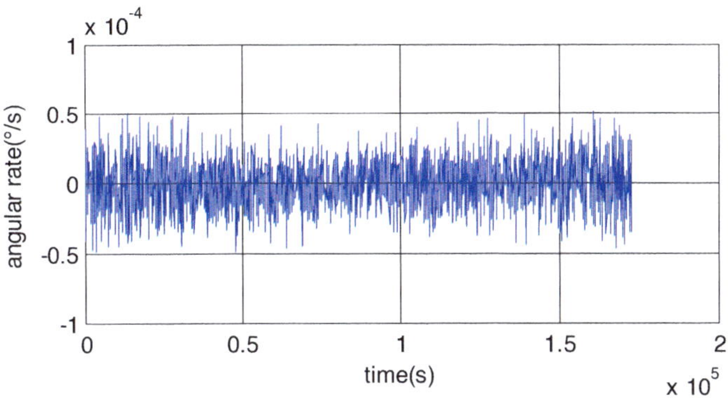


**Fig. 2** 28 integral angle curves in orbit period

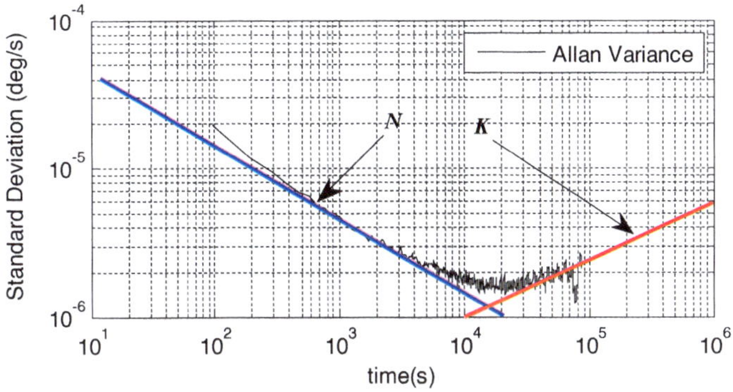


**Fig. 3** Allan variance curve

**Table 1** The ARW and RRW coefficients

| Coefficient | Value |
|---|---|
| ARW | $N = 0.0107^{\circ} / \sqrt{\mathrm{h}}$ |
| RRW | $K = 0.0027^{\circ} / \sqrt{\mathrm{h}}^{3}$ |

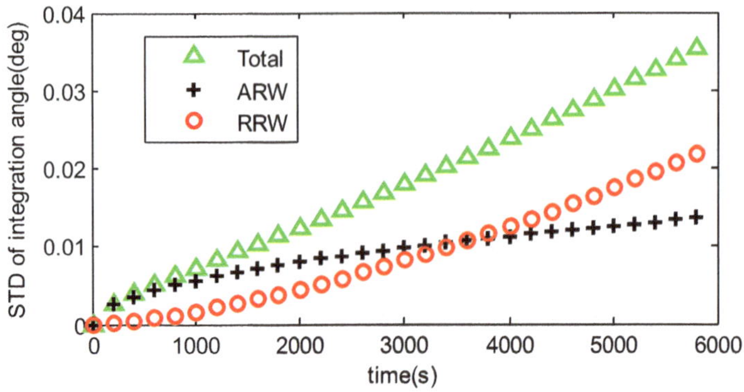


**Fig. 4** The relationship between the standard deviation of integral angle and time

With the parameters in Table 1, the relationship between the total standard deviation $\sigma_T$ of the integral angle and time $t$ can be calculated:

$$\begin{aligned} \sigma_T &= \sigma_N + \sigma_K \\ \sigma_N &= \frac{N}{60} \cdot t^{0.5} \\ \sigma_K &= \frac{N}{216{,}000} \cdot t^{1.5} \end{aligned} \tag{15}$$

Among that, $\sigma_N$ is standard deviation of the integral angle caused by ARW; $\sigma_K$ is standard deviation of the integral angle caused by RRW;

As we can see, the integration angle error caused by ARW is directly proportional to the 0.5th power of time, and the integration angle error caused by RRW is directly proportional to the 1.5th power of time.

The relationship between the standard deviation of integral angle and time is shown as Fig. 4.

Compared with Figs. 1 and 4, the calculated standard deviation (Fig. 4) is consistent with the real curves (Fig. 1), which shows the correctness of the algorithm.

## 5 Conclusion

The longer the correlation time is, the longer the duration of data for coefficient estimation is, though the data preprocessing on-board, the sample step length of integral angle is greatly increased, and data duration is correspondingly greatly increased on the premise that the total amount of data does not increase, so that RRW coefficient can be estimated more accurately.

In Allan variance coefficient estimation, the coefficients to be estimated should be selected according to the characteristics of variance curve, otherwise, illegal estimation values are likely to occur, resulting in the reduction of estimation accuracy.

Set the telemetry period (96 s) as the data sampling step, the total duration of data is 2 days, and the RRW coefficient with the correlation time of 20,000 s is estimated. If the integration period (orbit period) is set as the data sampling step, the duration of the data is several years, then the RR coefficient may be estimated.

## References


1. Jiankang, Z., Chao, C., Jianbin, Z.: A robust double-stage EKF Design for navigation system of micro UAVs based on 9-D MIMU/GPS. Flight Control Detect. **2**, 1–9 (2019)
2. Jianxiang, W., Lijun, C., Qian, W., Fufei, P., Zhenyi, C., Tingyun, W.: Preparation and spectral properties of bismuth and lead co-doped silica fiber using for fiber optic gyroscope. Flight Control Detect. **1**, 56–62 (2018)
3. IEEE Std 952–1997: IEEE Standard Specification Format Guide and Test Procedure for Single-Axis Interferometric Fiber Optic Gyroscopes
4. Haibo, Z., Jianye, L., Jizhou, L.: Research on IFOG random noise. Transducer Microsyst. Technol. **25**, 73–76 (2006)
5. Kai, X., Yongjun, L., Haibo, Z.: Modeling and simulation of fiber optic gyroscopes based on Allan variance method. Aerosp. Control Appl. **36**, 7–11 (2010)
6. Jue, W., Ruicai, J., Liwei, Q.: Realistic simulation method of inertial sensor random noise based on band pass filter. J. Chin. Inert. Technol. **26**, 112–117 (2018)
7. Baoliang, L., Yufen, D., Bo, Z.: Allan variance based testing for fiber optic gyroscope. Mod. Electron Tech. **35**, 126–127+133 (2012)
8. Sihao, Z., Mingquan, L., Zhenming, F.: MU error analysis based on a simplified Allan variance method. Transducer Microsyst. Technol. **29**, 12–18 (2010)

# A Target Recognition Algorithm in Remote Sensing Images

**Fei Jin, Xiangyun Liu, Zhi Liu, Jie Rui, and Kai Guan**

**Abstract** In the remote sensing image, the background is complex, which results in the low detection rate and slow speed of some target recognition algorithms. Therefore, an improved RFB Net model is used in this paper. The algorithm builds a feature pyramid network based on RFB Net model and adds more branches to the RFB module to improve the network recognition ability. The experimental results show that the improved algorithms have higher detection rate and better recognition effect.

**Keywords** RFB Net · Remote sensing image · Aircraft targets recognition · Deep learning

## 1 Introduction

With the development of remote sensing technology, the resolution of remote sensing images continues to increase, and the identification of targets in remote sensing images is of great significance in smart cities, transportation planning, military warfare, and so on. The traditional target recognition algorithms mainly use the artificially designed features for recognition [1], but these features are poorly targeted and the recognition effect is not good. Deep learning is a new direction in the field of machine learning in recent years. It acquires the local features of images through low-level filters, and then combines these local features into global features through high-level filters. The whole process does not require manual extraction and design features. Because of its strong learning ability and high recognition accuracy [2], deep learning has become a research hotspot in the field of target recognition.

At present, deep learning target detection algorithms are developing rapidly. These algorithms can be divided into two categories: two-stage algorithm and one-stage algorithm. Two-stage algorithms such as Faster R-CNN (Regions with CNN

---

F. Jin · X. Liu (✉) · Z. Liu · J. Rui · K. Guan
Information Engineering University, Zhengzhou, China



H. P. Urbach, Q. Yu (eds.), *6th International Symposium of Space Optical Instruments and Applications*, Space Technology Proceedings 7,
https://doi.org/10.1007/978-3-030-56488-9_12

features) [3] generate a series of candidate boxes as samples by the algorithm, and then classify the samples by convolutional neural network. The accuracy of two-stage algorithms is high, but because of the huge computational cost, the speed is slower than the one-stage algorithms. One-stage algorithms such as SSD (Single Shot MultiBox Detector) [4] do not generate candidate boxes, directly generate class probability and position coordinate values of objects. The one-stage algorithms is faster than two-stage algorithms, but the accuracy is lower.

RFB (Receptive Field Block) Net is a deep learning network model proposed by Liu [5] in 2017. This model is a deep learning target detection model based on SSD. It mainly introduces RFB into SSD. By simulating the receptive field of human vision, the feature extraction ability of the network is enhanced, which not only inherits the faster detection speed of SSD, but also has better accuracy. However, the environment of remote sensing image is complex and there are many disturbances. Most deep learning target detection algorithms are prone to have the problem of miss detection and false detection in target recognition of remote sensing image. The same is true for RFB Net. In this paper, an improved version of RFB-E is used for target recognition.

## 2 Method

### 2.1 *Principle of SSD*

SSD is based on VGG-16 Net [6]. After VGG-16 Net, feature extraction layer with decreasing size is added to generate predictive values of multi-scale detection. Its structure is shown in Fig. 1.

Given the input image and the truth label, the image is transmitted through a series of convolution layers of different scales in SSD, and different feature mappings are generated on different scales. For each location in each feature mapping, the default boundary box is evaluated by using a $3 \times 3$ convolution filter.

### 2.2 *RFB Module*

Some neurological studies have shown that in the human visual cortex, the size of pRF increases with the increase of eccentricity in retinal localization maps, which helps to highlight the importance of areas close to the center. Inspired by this, Liu proposed RFB to simulate human visual perception field, so as to enhance the deep features learned by lightweight CNN model. The internal structure of RFB is mainly divided into two parts: multi-branch convolutional layer and atrous convolutional layer. Therefore, RFB adopts a multi-branch structure, and adopts $1 \times 1$, $3 \times 3$, and $5 \times 5$ convolution kernels in each branch. The bottleneck structure adopted in each

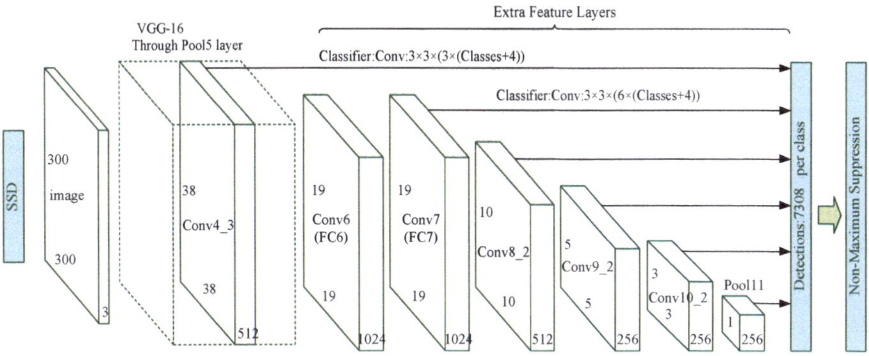


**Fig. 1** Structure of SSD

branch can effectively reduce training parameters and the complexity of network training. The structure of RFB and RFB-S is shown in Fig. 2.

1. Multi-branch convolutional layer.
   According to the definition of receptive field in CNN, different size convolution kernels are better than the same size convolution kernels for feature extraction. In addition, RFB module also uses the shortcut structure in ResNet. This is because the depth of the network has a great impact on the recognition effect of the network. Simply increasing the depth of the network sometimes does not improve the recognition ability of the network. Because of the divergence of the gradient, it may even be bad to the network model, the shortcut structure can solve this problem very well. RFB-S is used to simulate the smaller sensory field in shallow retina. Compared with RFB, RFB-S uses $3 \times 3$ convolution layer instead of $5 \times 5$ convolution layer. In addition, it uses $1 \times 3$ and $3 \times 1$ convolution layer instead of $3 \times 3$ convolution layer, which has more branches and smaller convolution nucleus size and reduces computation.
2. Atrous convolution.
   The atrous convolution theory was born in the field of image segmentation, which is essentially a generalization of general convolution operations. Atrous convolution adds a parameter $r$ to the original convolution, which can expand the convolution core to the scale constrained by expansion coefficient and fill the unoccupied area of the original convolution core with 0. The parameter $r$ is rate, which represents the expansion coefficient of the convolution core. So the obtained effective convolution kernel has a height of $f_h + (f_h - 1) \cdot (r - 1)$ and a width of $f_w + (f_w - 1) \cdot (r - 1)$, $f_h$ represents the height of the original convolution kernel, and $f_w$ represents the width of the original convolution kernel. Specifically, the convolution kernel performs multiplication with adjacent pixel points of the feature map when performing general convolution operations, and the atrous convolution operation allows the convolution kernel to be multiplied with pixels of a fixed interval $r$, so as to increase the receptive field without adding extra

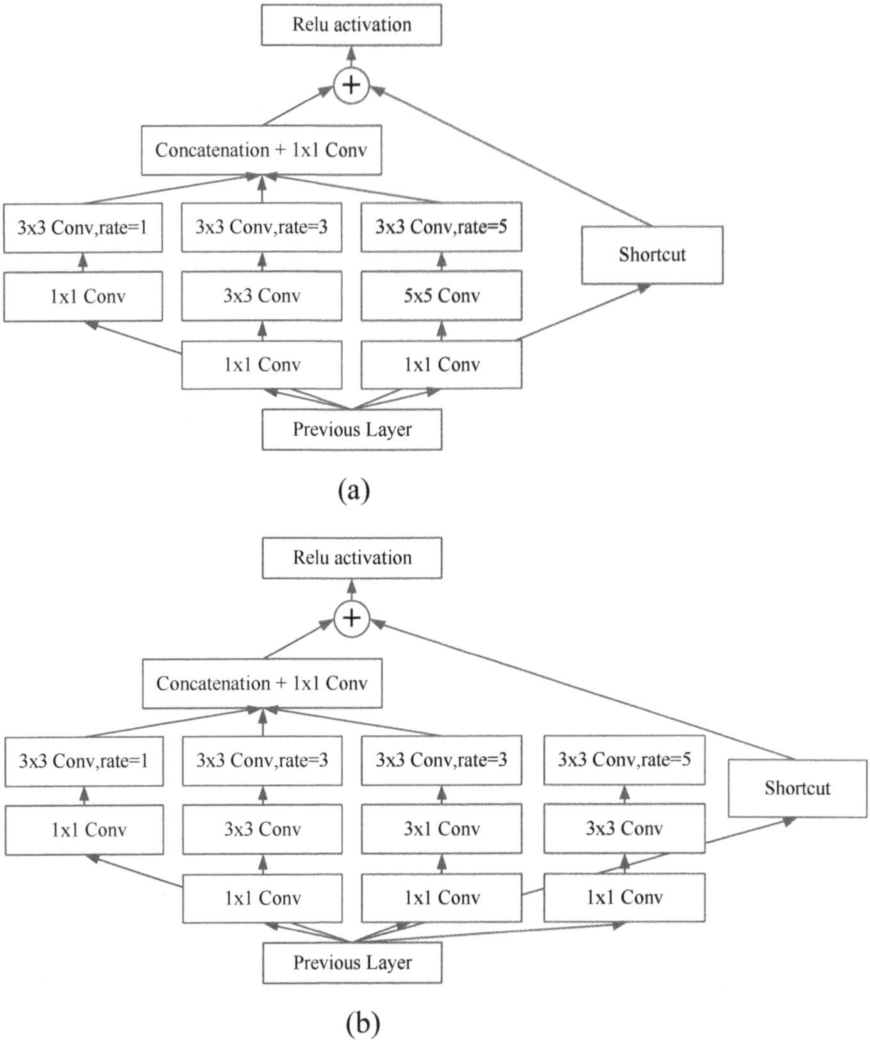


**Fig. 2** Structure of (**a**) RFB and (**b**) RFB-S

calculations. For high resolution images, there are more redundant information between adjacent pixels, so atrous convolution can be used for optimization. The basic purpose of atrous convolution is to provide a larger receptive field without adding parameters. This design has achieved good results in both semantic segmentation and target detection. RFB uses atrous convolution layer to simulate the centrifugal effect of pRF in human visual cortex.

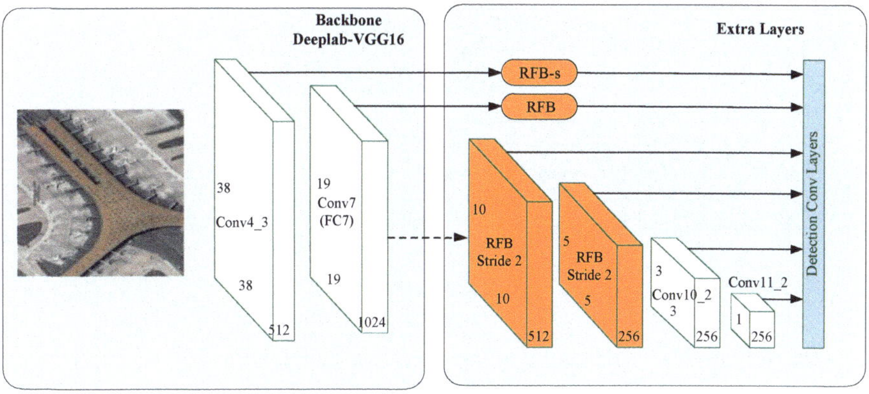


**Fig. 3** Structure of RFB Net

## 2.3 *RFB Net*

RFB Net is based on SSD and mainly adds RFB module to SSD. The structure of RFB Net is shown in Fig. 3.

In the SSD, a series of scale-decreasing convolution layers are connected behind the basic backbone network to generate a series of feature maps with decreasing spatial resolution and increasing receptive fields. RFB Net retains this structure, but two layers of extra layers in the SSD is replaced with RFB modules. Then, considering the different proportion of pRF in different visual atlases, RFB-s is used to simulate the shallow pRF in human retina, which is placed behind conv4_3 convolution layer.

## 2.4 *RFBE Net*

RFBE is an improved version of RFB. Compared with RFB, RFBE takes two efficient updates: (1) to up-sample the conv7_fc feature maps and concat it with the conv4_3 before applying the RFB module, sharing a similar strategy as in FPN (Feature Pyramid Networks, FPN). (2) Add a branch with a $7 \times 7$ kernel in all RFB layers. It is more accurate than RFB. So this paper replaces RFB and RFB-S in RFB Net with RFBE and RFBE-S. The structure of RFBE and RFBE-S is shown in Fig. 4.

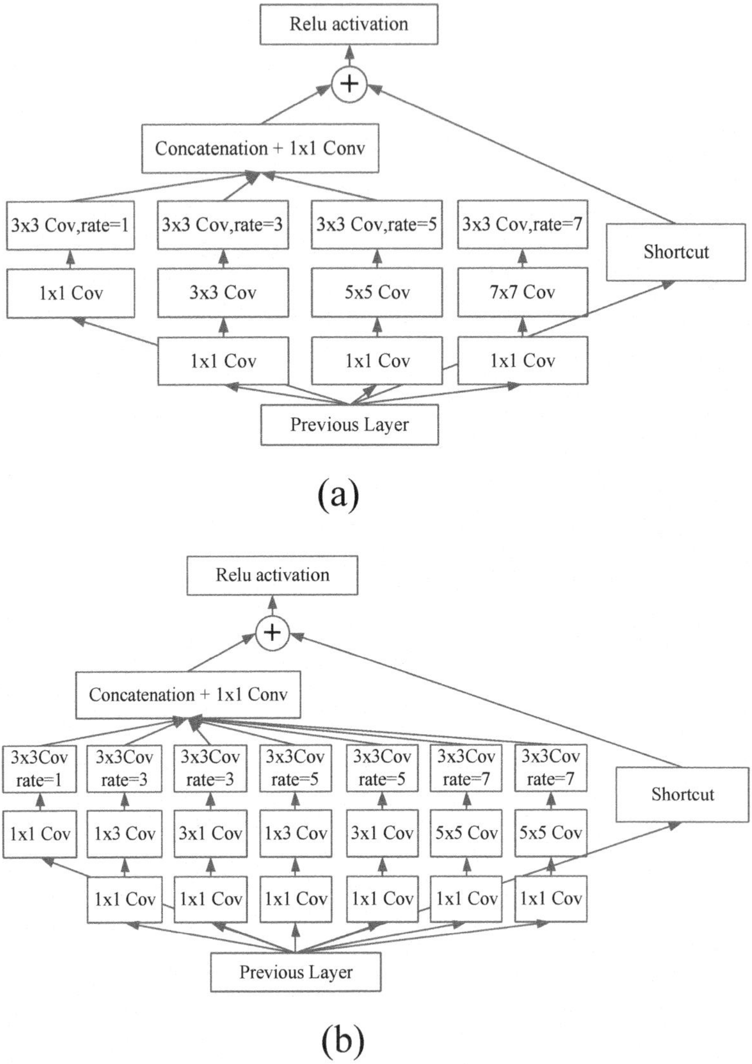


**Fig. 4** Structure of (**a**) RFBE and (**b**) RFBE-S

# 3 Experiments and Results

## *3.1 Experimental Environment and Data*

Ubuntu 16.04 operating system is used in this experiment. The GPU is NVIDIA GeForce GTX 950, 8GB of memory and Pytorch 0.4.0 is chosen as the framework of deep learning. The training dataset is UCAS-AOD [7], contains 1000 remote sensing images from the Google Earth satellite map. The content is the civil airport

**Fig. 5** Data example

in Europe, the USA, China, and other places, the image resolution is about 0.6~1.19 m, and the aircraft type is various types of civil aircraft. The training dataset example is shown in the Fig. 5.

## 3.2 *Detection Result*

SSD, RFB Net, and RFBE Net are used to train the dataset. The number of images used for training and verification is 1000, the ratio is 1:1, and the number of images used for testing is 55. The test data do not participate in training. Some test results of RFBE Net are shown in Fig. 6.

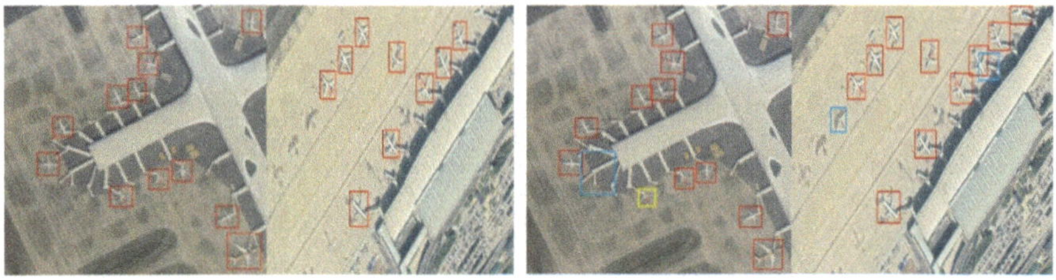

(a) (b)

**Fig. 6** Detection results. (**a**) Detection results of RFBE Net, (**b**) Detection results of RFB Net

As can be seen from Fig. 6a, the RFBE Net can detect aircraft targets very well. In addition, in order to compare the recognition effect of RFBE Net and RFB Net, test results of RFB Net are shown in Fig. 6b.

As can be seen from Fig. 6, when the scene is complex, RFB Net will be easy to have missed detection (yellow box) and false detection (blue box). Because compared with other application scenarios of target detection algorithm, the background of remote sensing image is complex and the algorithm is difficult to detect. In addition, RFB Net is easy to misdetect the boarding bridge as an aircraft target, because the boarding bridge is similar to the aircraft target, and it is easy to label the boarding bridge adjacent to the aircraft target when labeling. It is also one of the main reasons for false detection in aircraft target recognition algorithm. Based on the original RFB Net, the RFBE Net builds a feature pyramid network, and adds more branches to the RFB module, which improves the recognition ability of the network and effectively reduces the phenomenon of missed detection and false detection.

## 3.3 Accuracy Evaluation

The main indexes to evaluate the performance of aircraft target recognition algorithms are detection rate (DR) and false alarm rate (FA). The detection rate of aircraft targets refers to the proportion of the number of aircraft targets detected by the algorithm to the number of all aircraft targets. Its formula is

$$\mathrm{DR} = \frac{\sum_{i=1}^{n} m_i}{\sum_{i=1}^{n} N_i} \tag{1}$$

In the formula, DR represents the aircraft detection rate, $m_i$ represents the number of aircraft targets in the detected $i$-th image, $N_i$ is the total number of aircraft targets in the $i$-th image, and $n$ is the total number of experimental images. False alarm rate is the proportion of non-aircraft targets in the samples identified as aircraft targets. The formula is

$$FA = \frac{FP}{TP + FP} \tag{2}$$

In the formula, FA is the false alarm rate, FP is the number of judging a target that is not an aircraft as an aircraft target. TP is the number of aircraft targets that are correctly recognized. Of the 55 remote sensing images used for testing, 417 aircraft targets were included. The performance comparison between the RFBE Net and other algorithms is shown in Table 1.

It can be seen from Table 1 that under the same training strategy, the detection rate, false alarm rate, and detection speed of aircraft target in test set are better than SSD by using RFB Net. This is because RFB Net has improved SSD and embedded RFB module in original SSD network, which not only improves the detection accuracy, but also improves the detection speed. RFBE Net detects 404 aircraft targets in the test set, nine false detectors, the detection rate is 96.8%, false alarm rate is 2.17%. Although the detection speed is slightly slower than that of RFB Net due to the complexity of the network, it is faster than that of SSD network.

In addition, in order to verify the generalization of the RFBE Net, the oil tank and playground in RSOD-Dataset [8] are tested. In RSOD-Dataset, there are 1586 oil tanks in 165 images and 191 playgrounds in 189 images. 88 images of the oil tank and 112 images of the playground are used for training, 77 images of the oil tank and 77 images of the playground are used for testing. There are 731 oil tanks and 79 playgrounds in test set. Faster R-CNN, YOLO v2 [9], and SSD are selected for comparative experiment. The performance comparison between the RFBE Net and other algorithms is shown in Table 2.

It can be seen from Table 2 that under the same training strategy, the detection rate and false alarm rate of oil tank and playground in test set are better than Faster R-CNN, YOLO v2, and SSD by using RFBE Net. RFBE net not only performs well in aircraft target detection, but also achieves good results in oil tank and playground target detection. The detection results of oil tank and playground are shown in Fig. 7.

**Table 1** Performance comparison of different recognition algorithms for aircraft

| Method | DR/(%) | FA/(%) | Detection time per image/(s) |
|---|---|---|---|
| SSD | 94.00 | 5.08 | 0.131 |
| RFB Net | 95.92 | 4.07 | 0.092 |
| RFBE Net | 96.88 | 2.17 | 0.118 |

**Table 2** Performance comparison of different recognition algorithms for oil tank and playground

| | Oil tank | | Playground | |
|---|---|---|---|---|
| Method | DR/(%) | FA/(%) | DR/(%) | FA/(%) |
| Faster R-CNN | 87.70 | 3.60 | 86.08 | 17.07 |
| YOLO v2 | 93.17 | 4.48 | 81.01 | 11.54 |
| SSD | 96.86 | 3.14 | 93.67 | 1.33 |
| RFBE Net | 97.13 | 2.34 | 97.47 | 0 |

**Fig. 7** Detection results of oil tank and playground

# 4 Conclusion

Aiming at the problem that the background of remote sensing image is complex and the interference is too much, which leads to high miss detection rate and false detection rate in target recognition, this paper uses the RFBE Net model, which introduces the feature pyramid structure in RFB Net and adds more branches to the RFB module, the recognition ability of RFB Net is improved. The experimental results show that the RFBE Net has higher detection rate and better recognition effect.

# References


1. Li, W., Xiang, S., Wang, H., Pan, C.: Robust airplane detection in satellite images. In: 18th IEEE International Conference on Image Processing, pp. 11–14. IEEE (2011)
2. Krizhevsky, A., Sutskever, I., Hinton, G.E.: Image net classification with deep convolutional neural networks. In: International Conference on Neural Information Processing Systems. Springer, Berlin (2012)
3. Ren, S., He, K., Girshick, R., Sun, J.: Faster R-CNN: towards real-time object detection with region proposal networks. IEEE Trans. Pattern Anal. Mach. Intell. **39**(6), 1137–1149 (2017)
4. Liu, W., Anguelov, D., Erhan, D., et al.: SSD: single shot multibox detector. In: Proceedings of the 2016 European Conference on Computer Vision. Springer, Berlin (2016)
5. Liu, S., Huang, D., Wang, Y.: Receptive field block net for accurate and fast object detection. In: Proceedings of European Conference on Computer Vision, pp. 404–419. Springer, Berlin (2018)
6. Simonyan, K., Zisserman, A.: Very deep convolutional networks for large-scale image recognition. Comput. Sci. arXiv:1409.1556 (2014)
7. Zhu, H., Chen, X., Dai, W., et al.: Orientation robust object detection in aerial images using deep convolutional neural network. In: IEEE International Conference on Image Processing. IEEE (2015)
8. Long, Y., Gong, Y., Xiao, Z., Liu, Q.: Accurate object localization in remote sensing images based on convolutional neural networks. IEEE Trans. Geosci. Remote Sens. **55**(5), 2486–2498
9. Redmon, J., Farhadi, A.: YOLO9000: better, faster, stronger. In: IEEE Conference on Computer Vision and Pattern Recognition, pp. 6517–6525. IEEE (2017)

# AIR-Portal: Service for Urban Air Quality Monitoring

**A. F. Loenen, V. Huijnen, J. Douros, L. Breebaart, L. Miguens, and K. G. Biserkov**

**Abstract** Currently more people live in urban areas than ever before, and since 2007 the global urban population exceeds the global rural population and is still growing. These growing urban areas are increasingly vulnerable to air pollution and the compounding issues that result from it. Given that air pollution risks have a negative impact on local and national economies, ecosystems, and people living in urban environments, cities around the world are identifying air pollution as one of the most urgent environmental challenges.

The increasing urbanization and its associated intense traffic and industry in combination with long-range transport of pollution lead to air quality problems of varying nature. Part of the air pollution problem can be caused by local sources, such as traffic emissions during rush hours and coal or wood burning for heating or cooking. However, much of the problem must (also) be considered on a larger scale, such as with regional and seasonal variation in pollution levels, as well as large-scale episodes associated with e.g. livestock emissions, dust outbreaks, landscape fires, or volcanic eruptions. It is therefore clear that any system or service that wishes to address this urgent problem needs to account for large-scale effects, yet provide information on a useful urban scale.

AIR-Portal is an Air Quality Dashboard for urban areas, which uses a custom-designed, local Air Quality Model that combines remote sensing, land use, traffic, and local monitoring data into air quality forecasts on a city scale at a usable level of 100 × 100 m resolution throughout target cities. The forecasts can be used in decision-making processes for the cities involved and to inform the public about air quality on a local scale. The combination of large-scale and local effects makes AIR-Portal a unique service, able to address urban air quality issues on a large scale.

**Keywords** ISSOIA 2019 · Air quality · Earth observation

A. F. Loenen (✉) · L. Breebaart · L. Miguens · K. G. Biserkov
Science & Technology Corporation, Delft, The Netherlands
e-mail: edo.loenen@stcorp.nl

V. Huijnen · J. Douros
KNMI, De Bilt, The Netherlands



H. P. Urbach, Q. Yu (eds.), *6th International Symposium of Space Optical Instruments and Applications*, Space Technology Proceedings 7,
https://doi.org/10.1007/978-3-030-56488-9_13

# 1 Air Quality: An Urban Challenge on Many Scales

Currently more people live in urban areas than ever before, and since 2007 the global urban population exceeds the global rural population and is still growing. These growing urban areas are increasingly vulnerable to air pollution and the compounding issues that result from it. According to the WHO,[1] more than 80% of people living in urban areas that actually monitor air pollution are exposed to air quality levels that exceed the World Health Organization (WHO) limits. It is easy to hypothesize that non-monitored urban areas suffer at least as much. Given that air pollution risks have a negative impact on local and national economies, ecosystems, and people living in urban environments, cities around the world are identifying air pollution as one of the most urgent environmental challenges.

The increasing urbanization and its associated intense traffic and industry in combination with long-range transport of pollution lead to air quality problems of varying nature. Part of the air pollution problem can be caused by local sources, such as traffic emissions during rush hours and coal or wood burning for heating or cooking. However, much of the problem must also be considered due to larger scale effects such as for instance regional industrial and livestock emissions, of seasonally varying magnitude, as well as large-scale episodes associated with e.g. dust outbreaks, landscape fires, or volcanic eruptions.

*It is therefore clear that any system or service that wishes to address this urgent problem needs to account for large-scale effects, yet provide information on a useful urban scale.*

Given the challenges presented above, S[&]T, building on expertize from KNMI, has developed an Air Quality Dashboard for urban areas, called AIR-Portal.[2] AIR-Portal uses a custom-designed, local Air Quality Model that combines remote sensing, land use, traffic, and local monitoring data into air quality forecasts on a city scale at a usable level of 100 × 100 m resolution throughout target cities. The forecasts can be used in decision-making processes for the cities involved and to inform the public about air quality on a local scale.

*The combination of large-scale and local effects makes AIR-Portal a unique service, able to address urban air quality issues on a large scale.*

# 2 Business Opportunity

## 2.1 Service Description and Rationale

Citizens, health institutes, governments, and industry have little to no accurate means to monitor exactly what air pollution they are exposed to on a local scale and to access this data in an accessible (visual) manner. The opportunity is to build a

[1] http://www.who.int/phe/health_topics/outdoorair/databases/cities/en/

[2] https://air-portal.nl

system and service that provides historic, current, and forecast data and tools to visualize local air quality over the whole of Europe. An accurate monitoring and analysis tool will help to make decisions and to minimize exposure to air pollution and all of its negative effects—from the individual to the government level.

AIR-Portal provides a reliable, 24/7 operational service for displaying and analysing current air quality ($NO_2$, PMx, $O_3$, and Air Quality Index), as well as providing historical data and 3-day forecasts. AIR-Portal does this by combined processing of geo-information, available sensors, satellite data, weather information, and forecasting models. This has been proven to work for a spatial resolution of 100 m.

The targeted users can be found primarily in the market segments of large and medium-sized municipalities, local governments, and in third party service providers. For all segments, monitoring of air quality is necessary to decide on mitigating actions and to experience the effects of these actions. The data portal provides access to historical, current, and forecasted data. Historical data delivers trends through time and can be compared with current measurements. Combinations can be made with other data trends and long-term results of policy implementations can be demonstrated. Actual data visualizes the direct environment and its potential health effects. Actual and predicted data can be used as decision and planning support on different levels.

*Planning of tactical (policy) actions needs precise data, knowledge of situations, and traceability of pollution. With AIR-Portal, local effects can be compared to more global effects and over larger regions. For less polluted regions often no direct sensor measurements are available to begin with, so AIR-portal represents a unique source of AQ/pollution information.*

## 2.2 Value Chain

Figure 1 shows the AIR-Portal value chain, going from space assets on the left to the end-users on the right. In this value chain, the key partners described in the previous section are included. The figure shows the position of AIR-Portal the middle of the value chain. On the upstream side there are the space asset operators, who generate the raw data, which are processed and distributed by the data providers. Amongst the data providers are also ground based data sources. AIR-Portal combines these data together with the knowledge provided by the R&D partners into data and information products. They are sold either directly to the end-user, or via an information reseller. As can be seen, AIR-Portal has a crucial position in the value chain, bridging the gap between data and knowledge providers, and those parties that are looking for information.

## 2.3 Market and Competitive Landscape

AIR-Portal is not the only provider of information/data on air quality. The competition can be divided into four categories: National institutions, (international) commercial providers, open data platforms, and city specific platforms. The main

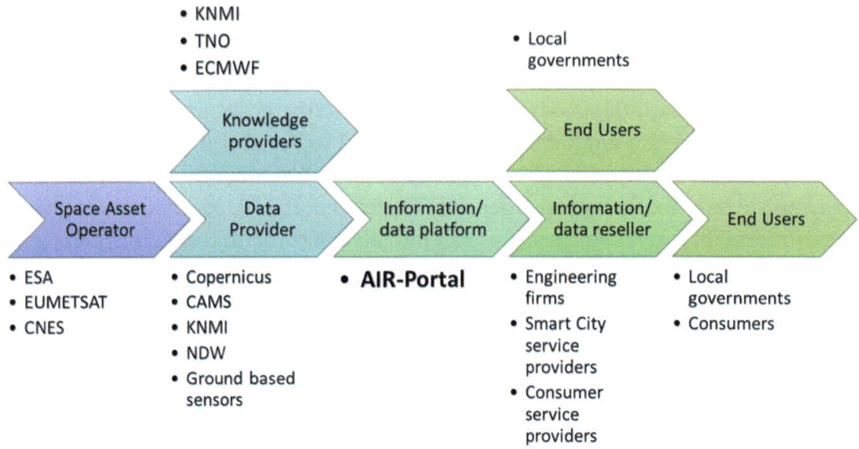


**Fig. 1** AIR-Portal value chain

advantage that many of these providers have over AIR-Portal is that they are government-funded and are therefore free to use. However, many of them are limited in the kind of data they use (often only local measurements), and in the amount of customization they offer (only available in a limited area, or a generic model applied everywhere).

## *2.4 Unique Selling Points of AIR-Portal*

The main strength of our approach in comparison to more traditional approaches for regional to local air quality assessment based on chemistry transport models (CTMs) is that there is no reliance on high-resolution emission inventories which are very difficult and expensive to compile and equally difficult and expensive to keep up to date. Although this is an advantage in terms of effort it takes to operate the AIR-Portal service, merely generating high-resolution AQ data is not sufficient to set AIR-Portal apart from the other data providers in the market. These data rather form the foundation for AIR-Portal's two USPs (see also Fig. 2):

1. *Customization using local information*: AIR-Portal users are those looking for data that is of high quality and tuned to the local situation. Rather than having one generic solution that will be applied for large geographic areas such as done by many AQ data providers. AIR-Portal can be tailored for each client by adding local data sources and insights to the model. Such customizations are local knowledge of the current/future land use situation and AQ in situ sensor data. This will lead to a level of accuracy that can never be reached with a generic method, since every situation is unique.
2. *Decision support tools*: Rather than just providing an AQ *database*, AIR-Portal will provide additional tools to generate information for AQ *decision* support.

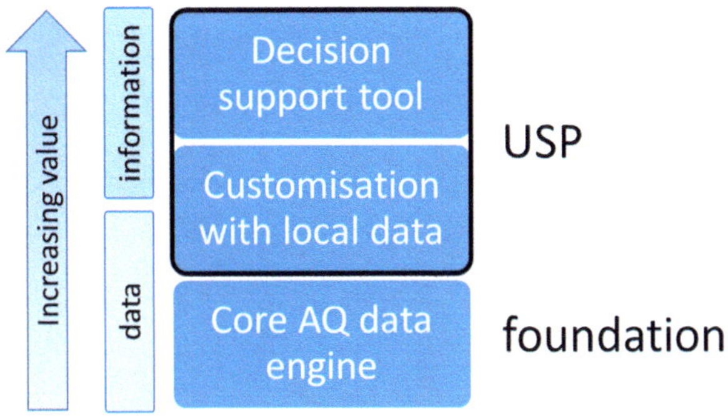


**Fig. 2** AIR-Portal USPs

The meetings held with several policy makers of small to large cities showed that decision support tools are a substantial value-add for these stakeholders.

## 3 Technical Solution

AIR-Portal is a software system for creating and providing access to new air quality-related data products to customers. The main functions of the AIR-Portal system are:

1. To provide near-future forecasts and analysis data products concerning pollutant species concentrations at a high (100 × 100 m) resolution grid;
2. To provide new products such as an Air Quality Index (AQI) and other derived information of interest;
3. To provide a user interface for exploring and interpreting these data results.

### *3.1 The AIR-Portal Processing Chain*

Figure 3 shows the overall system architecture of the AIR-Portal processing chain:

### *3.2 Ingestion*

The *Ingestion* component consists of a command-line tool that is responsible for daily retrieval of external input data and preparing that data so it can subsequently be used by the *Processing* component.

*Ingestion* maintains a local archive with the unchanged, retrieved sources. It applies selection filters (because raw input sources may often contain much more

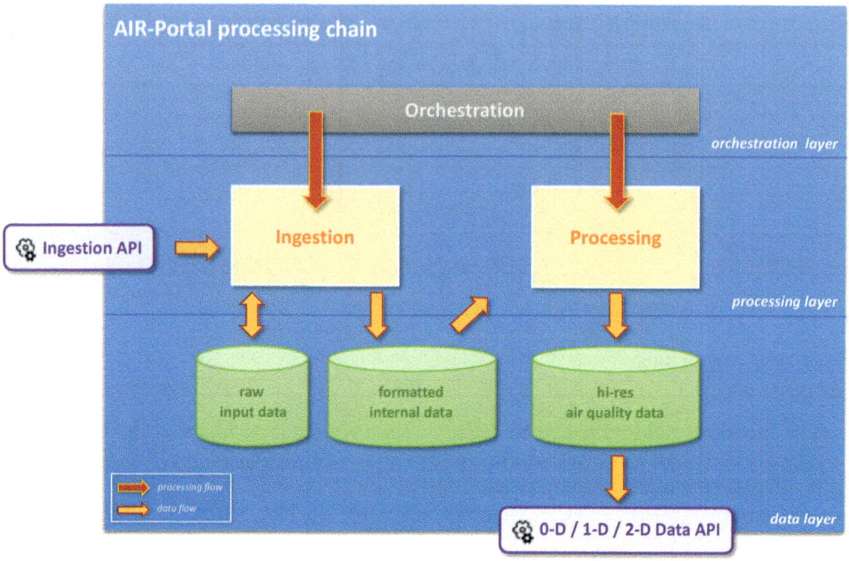


**Fig. 3** The AIR-Portal processing chain architecture

data than what is needed for AIR-Portal processing) and performs various conversions and harmonizations, leading to a representation of the different types of data in uniform, consistent formats that can be further processed without requiring knowledge of the original provenance. These harmonized files are also stored in a database.

The AIR-Portal pilot system requires the ingestion of following types of input data:

- Background pollution analyses and forecasts
- Meteorology information
- Land use information
- Ground sensor data

*Ingestion* is a component that is horizontally scalable both at the process (e.g. starting parallel runs for different geographical regions) as well as at thread level (within one *Ingestion* process, multiple input products can be downloaded in parallel).

## 3.3 Processing

The *Processing* component consists of a command-line tool that is responsible for applying the air quality scaling models to create the high-resolution pollution concentrations and Air Quality Index products.

*Processing* is fully parameterized with respect to concepts such as the resolution of the input data and can, therefore, work on data from any available source, provided it has been formatted by the *Ingestion* module first. The only limitation is that the resolution of the land use product must be higher than any of the other sources (background pollution and meteorology components).

Like *Ingestion*, the *Processing* component is scalable at both the process level (e.g. starting parallel runs for different geographical regions) and at the thread level (within one process, different time steps can be calculated in parallel).

## 3.4 Orchestration

The *Orchestration* component consists of a command-line script that is responsible for scheduling and running the *Ingest* and *Processing* components on a daily basis at the appropriate times and in the appropriate order.

*Orchestration* is also responsible for managing and providing configuration files and invocation parameters to the *Ingestion* and *Processing* components. AIR-Portal contains extensive logging support for monitoring and assessing the status of the processing chain steps.

## 3.5 AIR-Portal Web Application

In order for the users to easily view the AIR-Portal data, a progressive web application was developed, allowing it to support a wide variety of platforms, including mobile devices (see Fig. 4).

## 3.6 Air Quality Scaling Model

Air quality conditions can change significantly on very local scales, due to pollution sources such as traffic emissions during rush hours, and wood burning for heating. At the same time, background conditions can vary significantly due to pollution transport from larger distances, such as from industrial areas or airports. Also occasionally large-scale events such as dust outbreaks or wildfires can lead to significantly enhanced pollution. Therefore, to be able to describe air pollution on a very local ($<100$ m) scale, this requires a configuration where many data sources, describing pollution levels on various temporal and spatial scales, or proxies of pollution sources, are combined in a tuned manner. The system needs to be configured such that it automatically corrects for model biases based on the most recent observations of pollution levels.

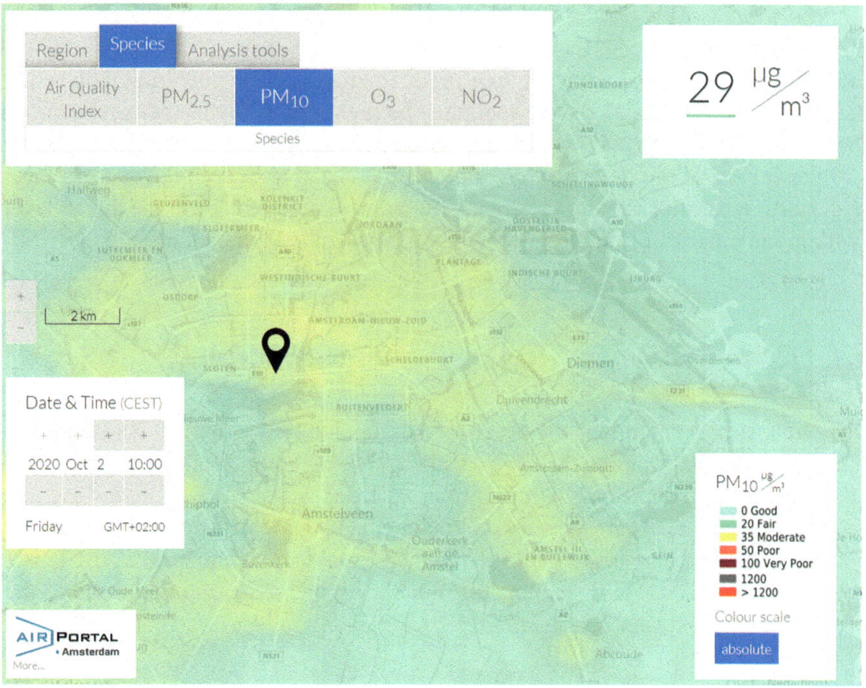


**Fig. 4** AIR-Portal progressive web application

In practice, this implies that the core of the global AIR-Portal methodology relies on a dynamic component which utilizes CAMS-regional atmospheric composition as well as regional meteorological model forecasts, and a local, static component which includes the land use information, as well as local observations to calibrate and validate the system. The outline of the method is indicated in Fig. 6.

After collection of the necessary input data, the first step is the application of a land use regression scheme to derive a first estimate for the local-scale air quality forecast. The basic concept is that the global atmospheric composition fields of CAMS-regional are downscaled towards resolutions of $100 \times 100$ m$^2$ using geo-referenced data for the land use types (vegetation, water bodies, various types of urban fabric, etc.). The next step in the procedure is the advection of pollutants by the wind, as well as a correction for the boundary layer height. Finally, calibration of the model aims at optimizing the model behaviour by operationally employing an automated routine which has access to the historical archive of model runs, as well as available measurements and will re-evaluate the land use scaling factors based on a statistical evaluation of past forecasts against available measurements.

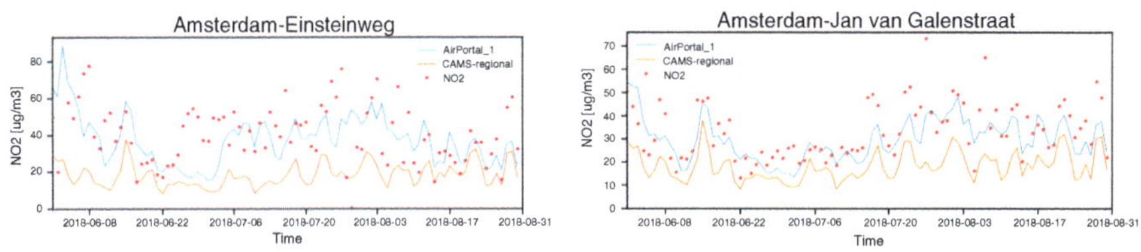


**Fig. 5** Time-series comparisons of daily mean values between CAMS-regional, AIR-Portal, and measurements at two stations in the Amsterdam area for the summer validation period. Both measuring stations are characterized as "traffic"

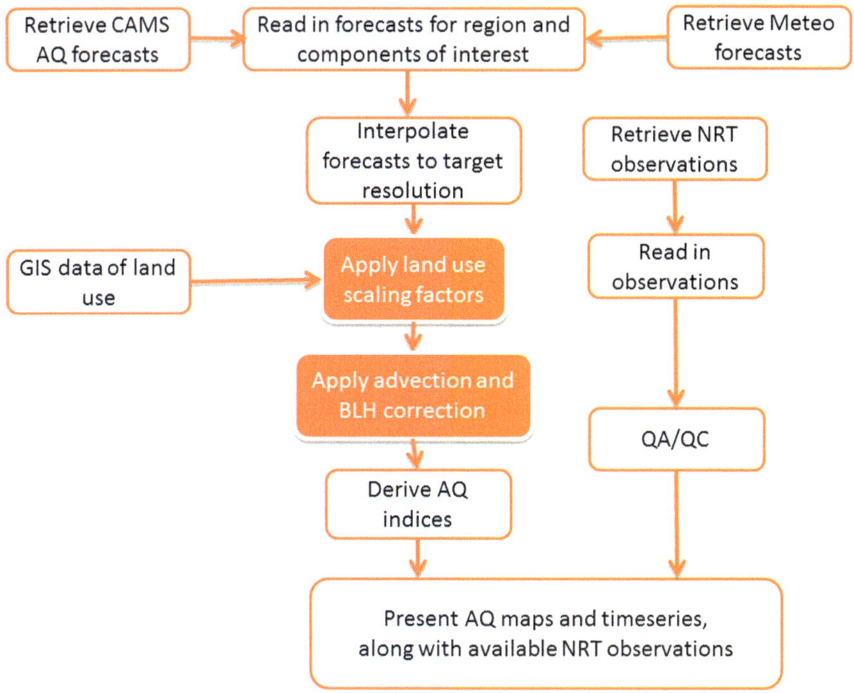


**Fig. 6** Outline of the structure of the AIR-Portal modelling system

## 4 Validation

During the pilot period that followed AIR-Portal's development, the data quality of the system was validated. Comparisons of the AIR-Portal approach with measured values provide a strong case for the added value of the approach for downscaling coarse scale models for the urban/local-scale environment.

The main added value of AIR-Portal is the downscaling step: AIR-Portal data has a resolution of 100 m, compared to 5–10 km for reference CAMS data. This in particular impacts pollutants with a strong primary component such as NO2, which was reflected in most evaluation metrics (see Fig. 5 for an example).

Within urban environments, ozone is often slightly underestimated due to an over-estimate of the "titration effect". The strongest benefits of using AIR-Portal as a means of downscaling and refining coarser resolution models are confirmed by the performance of the system near strong emission sources (e.g. in the vicinity of the city road network) as established by the evaluation at traffic stations. For the urban background the forecast quality from the CAMS system has been shown generally very good, and the additional benefits of the downscaling approach are more limited. Still, for these cases a substantial improvement is seen generally for the hit ratios, although together with a small deterioration of false alarm ratios for the key pollutants. In summary, the validation activities show that the AIR-Portal makes optimal use of a combination of CAMS background air pollution evaluations, and its specific downscaling capabilities, resulting in an overall well-characterized, and therefore reliable product for providing timely air quality information (Fig. 6).

# Comparative Analysis of Fire Detection Algorithms in North China

**Chang Guo, Ke Liu, Guohong Li, Yuanping Liu, and Yujing Wang**

**Abstract** Forest fires are sudden and destructive natural disasters, which have irreversible impacts on ecological environment. Therefore, it is of great significance to master forest fires in a timely manner. To analysis the accuracy of absolute fire hotspots detection algorithm, MODIS fire detection algorithm and fire hotspots detection algorithm based on variance between class three forest fires in north China since 2019 were selected. MODIS data was used to detect forest fires. Take the medium and spatial resolution data as the comparison standard, the accuracy comparison of algorithms were carried out. The results indicated that the absolute fire hotspots detection algorithm was better than the other two algorithms. In general, combined with the absolute fire hotspots detection algorithm, data from satellite ground receiving station can apply to real-time supervision of fire hotspots.

**Keywords** Forest fires · Absolute fire hotspots detection algorithm · MODIS fire detection algorithm · Fire hotspots detection algorithm based on variance between class

---

C. Guo · Y. Wang
North China Institute of Aerospace Engineering, Langfang, China

K. Liu · G. Li (✉) · Y. Liu
North China Institute of Aerospace Engineering, Langfang, China

National and Regional Joint Engineering Research Center for Aerospace Remote Sensing Application Technology, North China Institute of Aerospace Engineering, Langfang, China

Hebei Aerospace Remote Sensing Information Engineering Technology Research Center, North China Institute of Aerospace Engineering, Langfang, China



H. P. Urbach, Q. Yu (eds.), *6th International Symposium of Space Optical Instruments and Applications*, Space Technology Proceedings 7,
https://doi.org/10.1007/978-3-030-56488-9_14

## 1 Introduction

Forests are the main body of land ecosystems. The fierce forest fires threaten the ecological environment and biodiversity. It is important to accurately and timely grasp the fires to reduce disaster losses and protect forest resources [1]. The traditional way of monitoring forest fires not only wastes resources, but also cannot provide timely information. Satellite remote sensing technology is widely used in monitoring forest fires due to its macroscopic large-scale and rapid observation of the earth. The technology to monitoring fire hotspots is achieved by detecting the brightness temperature changes before and after the fires [2, 3]. At present, the detection algorithms of forest fires mainly include fixed threshold method [4], absolute fire hotspots detection algorithm [5], contextual algorithm [6], MODIS fire hotspots detection algorithm [7], fire detection algorithm based on variance between-class algorithm [8], and so on.

Flannigan and Vonder Haar [4] proposed a fixed threshold model based on the North American forest area. Xiaocheng Zhou [9] used the absolute fire hotspots detection algorithm and improved algorithm to verify and analyze nine forest fires in China. Xin Hang [10] based on contextual algorithm detected the fire hotspots of Jiangsu Province from June 2010 to June 2015 in China, combined with satellite monitoring data and meteorological observation data, analyzing the temporal and spatial variation characteristics of AOD and fire hotspots. Wanting Wang [11] used the MODIS fire detection algorithm to analyze the fires in southeastern USA, and proposed a fire hotspots monitoring method for low-intensity fire areas. Yang Chen et al. [12] modified the threshold of MODIS fire detection algorithm to detect Amazon forest fires and reduced the false detection rate. Xia Xiao [8] detected fire hotspots from Fujian and Heilongjiang provinces in China, by using the fire hotspots detection algorithm based on variance between class. In general, the fixed threshold method is based on the regional and seasonal experience thresholds, simple and efficient, but the algorithm is less adaptable. Researchers have proposed an absolute fire hotspots detection algorithm based on the fixed threshold method to improve the accuracy in different regions and seasons. The concept of background pixels is introduced based on the contextual algorithm, and the background pixel information around the fire hotspots pixel is fully considered. The MODIS fire detection algorithm evolves on the basis of contextual algorithm.

The TERRA and AQUA satellite are equipped with MODIS sensor, which have taken the needs of fire monitoring into account. MODIS have higher accuracy and shorter return time than meteorological satellites. It can analyze fire hotspots in a more efficient way, which has special meanings and application prospects in monitoring forest fires. In this paper, based on the MODIS data received by the satellite receiving station, the fire hotspots algorithms were adopted to detect three forest fires in north China in 2019, which include the absolute fire hotspots detection algorithm, the MODIS fire detection algorithm, and the fire hotspots detection algorithm based on the variance between classes. Finally, the medium and high spatial resolution data were used to verify the accuracy of three algorithms. The results of the three algorithms were compared and analyzed.

## 2 Algorithm Introduction

Fixed threshold method, absolute fire hotspots detection algorithm, contextual algorithm, MODIS fire detection algorithm, and fire detection algorithm based on variance between class are universal algorithms to monitor fires. Absolute fire hotspots detection algorithm and MODIS fire detection are developed based on fixed threshold methods. The principle of these three algorithms are as follows. Each algorithm uses MATLAB to implement its functions.

### 2.1 *Absolute Fire Hotspots Detection Algorithm*

Based on the fixed threshold method, the algorithm mainly includes four parts: cloud detection, potential fire hotspots pixels, and background pixels detection, identify fire hotspots and remove misdetected hotspots.

#### 2.1.1 Cloud Detection

$$\rho_{0.66} > 0.2 \tag{1}$$

If the pixel meets the condition (1), it will be judged as the cloud pixel. $\rho_{0.66}$ represents the reflectance of 0.66 μm.

#### 2.1.2 Potential Fire Pixels and Background Pixels Detection

$$\Delta T_{411} = T_4 - T_{11} \geq 20\ \text{K} \,\&\, T_4 > 320\ \text{K} \tag{2}$$

The pixel, which satisfies the condition (2) is the potential fire pixels, otherwise it will be judged as the background pixel. $T_4$ corresponds to the brightness temperature of wavelength of 4 μm. Similarly, $T_{11}$ corresponds to the wavelength of 11 μm. $\Delta T_{411}$ is the brightness temperature difference between the wavelength 4 μm and 11 μm.

#### 2.1.3 Identify Fire Hotspots

$$T_4 < 315\ \text{K} \,\text{or}\, T_{411} < 5\ \text{K} \tag{3}$$

$$T_4 > 360\ \text{K} \tag{4}$$

$$T_4 > 320\ \text{K} \,\&\, T_{411} > 20\ \text{K} \tag{5}$$

$$\left(T_4 > T_{4b} + 4 * \Delta T_{4b}\right) \& \left(T_{411} > T_{411b} + 4 * \Delta T_{411b}\right) \tag{6}$$

If the land pixel satisfies the condition (3), then it will be classified as the non-fire hotspots pixel. And if the rest pixels satisfy the conditions (4) or (5) or (6), it will be determined as the fire hotspots. Among them, $T_{4b}$ and $\Delta T_{4b}$ are the mean value and variance value of 4 μm wavelength of the background pixels, respectively. Similarly, $T_{411b}$ and $\Delta T_{411b}$ are difference between the 4 μm and the 11 μm wavelength. If the variance of background pixels is <2, it will be calculated as 2.

#### 2.1.4 Remove Misdetected Hotspots

During the daytime, if the reflectance of 0.64 μm and 0.86 μm are more than 0.3, this pixel will take as the flare. The final fire hotspots should exclude these false detection (Fig. 1) [5].

### 2.2 *MODIS Fire Detection Algorithm*

The MODIS fire detection algorithm is also divided into four parts: cloud pixel detection, land pixel classification, fire hotspots classification, and eliminate the false detection. The algorithm is as follows:

#### 2.2.1 Cloud Pixel Detection

$$\text{Daytime}: \left(\left(\rho_{0.65} + \rho_{0.86}\right) > 0.9 \cup \left(T_{12} < 265\ \text{K}\right)\right) \cup \left(\left(\rho_{0.65} + \rho_{0.86}\right) > 0.7 \cap \left(T_{12} < 285\ \text{K}\right)\right) \tag{7}$$

$$\text{Nighttime}: T_{12} < 265\ \text{K} \tag{8}$$

If the pixels satisfy the conditions (7) or (8), respectively, this pixel will be judged as the cloud. In this formula, $\rho_{0.65}$ and $\rho_{0.86}$ correspond to the reflectance of band 1 and band 2, respectively. $T_{12}$ corresponds to the brightness temperature of wavelength 12 μm.

#### 2.2.2 Land Pixel Classification

$$\text{Daytime}: \left(T_4 > 310\ \text{K}\right) \cap \left(T_{411} > 10\right) \cap \left(\rho_{0.86} < 0.3\right) \tag{9}$$

$$\text{Nighttime}: \left(T_4 > 305\ \text{K}\right) \cap \left(T_4 > 10\ \text{K}\right) \tag{10}$$

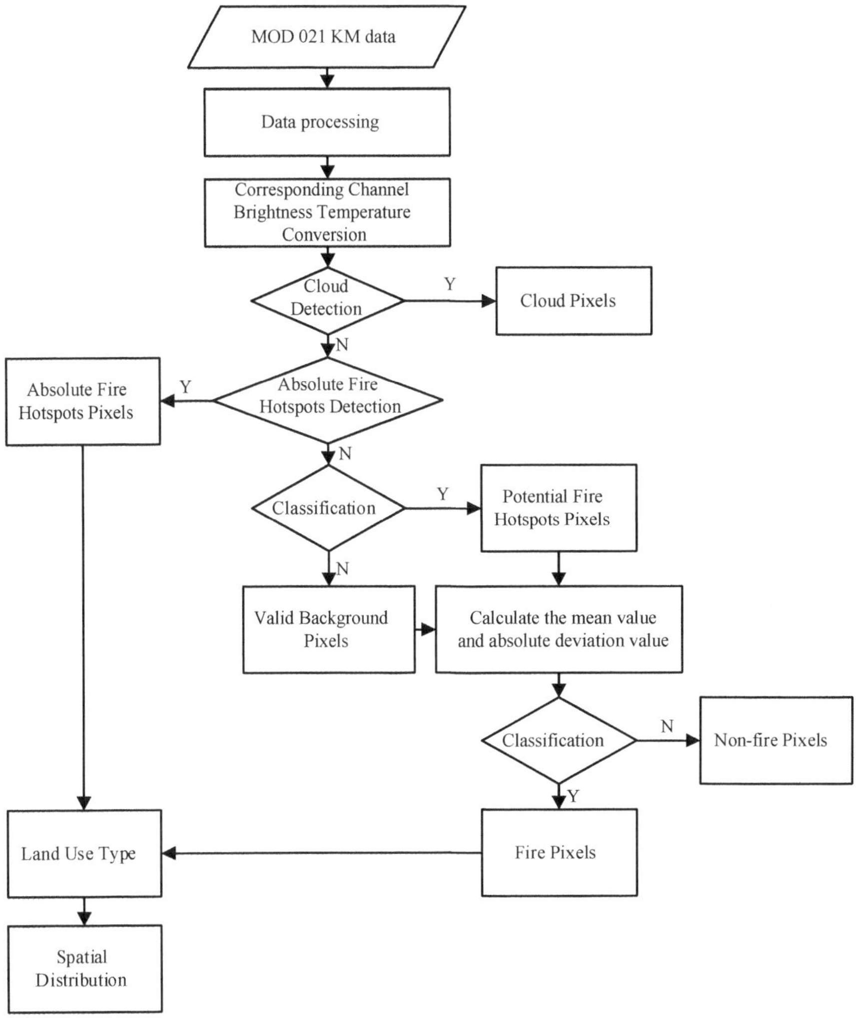


**Fig. 1** The technology route of absolute fire hotspots detection algorithm

Absolute fire hotspots classification:

$$\text{Daytime}: T_4 > 360\ \text{K};\ \text{Nighttime}: T_4 > 320\ \text{K} \tag{11}$$

Potential fire hotspots classification:

$$\text{Daytime}: (T_4 > 325\ \text{K}) \cap (T_{411} > 20\ \text{K});\ \text{Nighttime}: (T_4 > 310\ \text{K}) \cap (T_{411} > 10\ \text{K}) \tag{12}$$

If the land pixel satisfies the condition (9) or (10), the land pixel will be judged as the suspect fire hotspots pixel, otherwise it will be the non-fire land pixel. When the suspect fire hotspots pixel meets the condition (11), it will be judged as the absolute fire hotspots pixel and will not participate in subsequent operations. If the suspect fire hotspots pixel satisfies the condition (12), the pixel will be judged as the potential fire hotspots pixel, otherwise it is the valid background pixel. In the formula, $T_4$ is the brightness temperature of 4 μm wavelength, and $T_{411}$ is the difference between the wavelength of 4 μm and 11 μm.

### 2.2.3 Fire Hotspots Classification

In the process of detection, the window size $n \times n$ is expanded from 3 × 3 cycles to 21 × 21. Centering the potential fire pixel, the algorithm analyzes the brightness temperature difference between the potential fire pixel and its surrounding valid background pixels to determine whether the pixel is a real fire pixel or not. When the number of valid background pixels $N$ in the window satisfies $(N \geq n \times n \times 25\%) \cap (N \geq 8)$, the window is no longer expanded, and then the statistics of valid background pixel in window are calculated. If the potential fire hotspots pixels satisfy the condition (7)–(11), the pixels will be classified into the fire pixel. Otherwise, the potential fire pixels will be classified into the non-fire hotspots pixels, and then determine next potential fire hotspots pixel. If the number of valid background pixels does not satisfy the condition in a certain size window, the window will expand and repeat the above processes until the conditions are satisfied. If the window size expanded to 21 × 21 and the requirements are still not satisfied, the potential fire hotspots pixel will be classified as an indeterminate pixel.

Discriminant condition: Daytime : (13) ∩ (14) ∩ (15) ∩ ((16) ∪ (17); Nighttim e : (13) ∩ (14) ∩ (15)

$$T_4 > \overline{T_4} + 3 * \overline{\delta_4} \tag{13}$$

$$T_{411} > \overline{T_{411}} + 6 \tag{14}$$

$$T_{411} > \overline{T_{411}} + 3.5 * \overline{\delta_{T411}} \tag{15}$$

$$T_{11} > \overline{T_{11}} + \overline{\delta_{11}} - 4 \tag{16}$$

$$\delta_4^{'} > 5 \tag{17}$$

In the Eqs. (13)–(17), $\overline{T_4}$ and $\overline{T_{11}}$ are the average brightness temperature of valid background pixels of 4 μm and 11 μm. $\delta_4$ and $\delta_{11}$ are the average absolute deviation of the valid background pixels of 4 μm and 11 μm. $T_{411}$ is the average brightness temperature of valid background pixels of difference between 4 μm and 11 μm wavelength. $\delta_{411}$ is the average absolute deviation of the bright temperature of the valid background pixels of difference between 4 μm wavelength and 11 μm wavelength. $\delta_4^{'}$ is the average absolute deviation of bright temperature of the background fire pixel of 4 μm.

### 2.2.4 Eliminate the False Detection

Strong reflection on the surface, junctions of water and land, and desert edges may cause false detection. These false detection pixels can be further eliminated by other auxiliary band information (Fig. 2) [7].

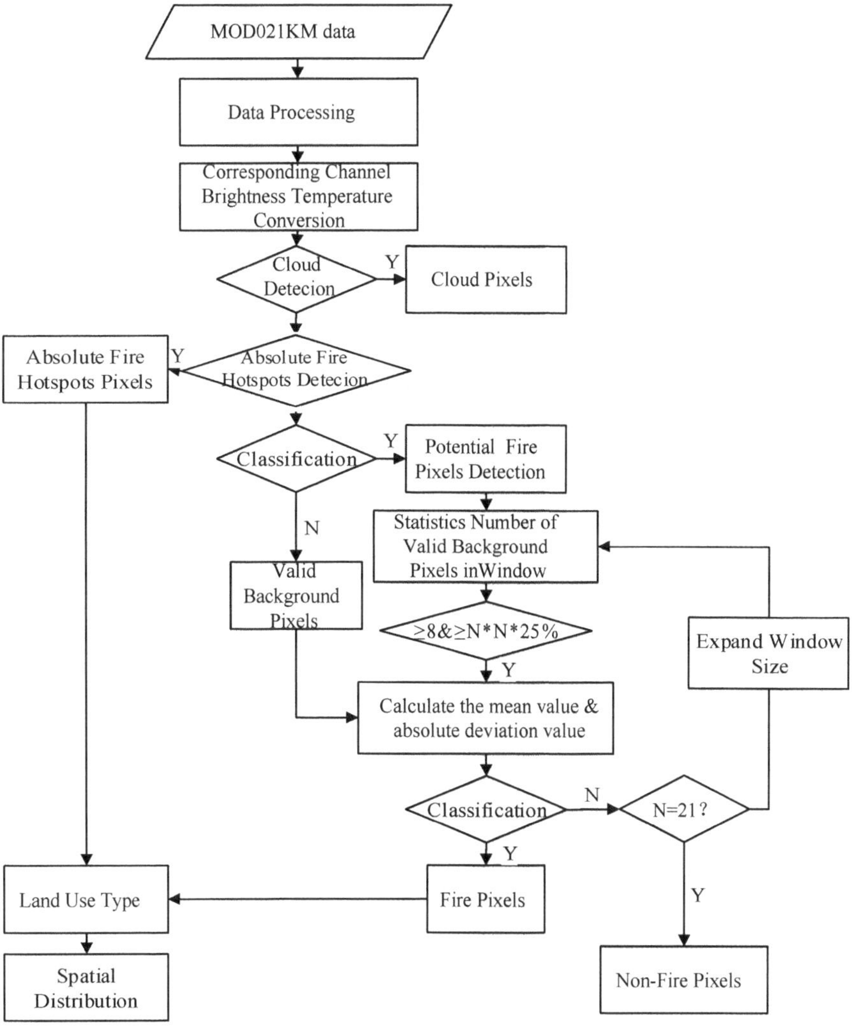


**Fig. 2** The technology route of MODIS fire detection algorithm

## 2.3 Fire Detection Algorithm Based on Variance Between Class

The principle of the fire detection algorithm based on the variance between class is to distinguish the variance between the fire hotspots pixels and the valid background pixels. At the same time, it can distinguish general fire hotspots pixels from the smolder fire hotspots pixels. The algorithm is mainly divided into four parts: cloud, water, and smoke plume classification, potential fire hotspots pixel discrimination, fire hotspots detection, and elimination of false fire hotspots. The content of the algorithm is as follows:

### 2.3.1 Cloud, Water, and Smoke Plume Classification

When the pixels in the image satisfy the conditions (18)–(20) respectively, the pixels are classified into different types. When recognized as cloud or water, the pixels will not perform subsequent operations.

Cloud pixels detection:

$$\text{Daytime}: \left(\left(\rho_{0.65}+\rho_{0.86}\right)>0.9 \cup \left(T_{12}<265\ \text{K}\right)\right) \cup \left(\left(\rho_{0.65}+\rho_{0.86}\right)>0.7 \cap \left(T_{12}<285\ \text{K}\right)\right) \tag{18}$$

$$\text{Nighttime}: T_{12}<265\ \text{K} \tag{19}$$

Water pixels detection:

$$R_2<0.15 \cap R_7<0.05 \cap \left(\left(R_2-R_1\right)/\left(R_2+R_1\right)\right)<0 \tag{20}$$

Smoke plume detection:

$$0.5 \geq \left(R_8-R_9\right)/\left(R_8+R_9\right) \geq 0.15 \tag{21}$$

$$\left(R_9-R_7\right)/\left(R_9+R_7\right) \geq 0.30 \tag{22}$$

$$\left(R_8-R_3\right)/\left(R_8+R_3\right) \leq 0.09 \tag{23}$$

$$R_8 \geq 0.09 \tag{24}$$

$\rho_{0.65}$ and $\rho_{0.86}$, respectively, correspond to the channel reflectance of band 1 and band 2, $T_{12}$ corresponds to the brightness temperature of 12 μm, and $R$ represents the reflectance of the corresponding band.

### 2.3.2 Potential Fire Hotspots Pixel Discrimination

$$T_4 > 305\ \text{K} \cap T_{411} > 10\ \text{K} \cap R_{16} < 0.3 \tag{25}$$

According to condition (25), the land pixels will be classified into potential fire hotspots pixels and valid background pixels. In the formula, $T_4$ is the brightness temperature of 4 μm, $T_{411}$ is the difference between the wavelength of 4 μm and 11 μm.

### 2.3.3 Fire Hotspots Detection

Traversing the potential fire hotspots pixels, the pixels meet the condition (28) will be judged as the absolute fire pixel. If it satisfies the condition, the pixels will be judged as the absolute fire pixel. Otherwise, taking this potential fire hotspots pixel as the center pixel, the algorithm will calculate the variance between potential fire hotspots pixel and the valid background pixel in the 21 × 21 window. When the potential fire hotspots pixel is not a smoke pixel and the condition (27) is meet, the potential fire hotspots pixel is classified as a general fire hotspots pixel. When the potential fire hotspots pixel is a smoke pixel and the condition (29) is satisfied, then the fire hotspots pixel is classified as a smolder fire hotspots pixel. Otherwise, this potential fire hotspots pixel is judged as a non-fire hotspots pixel.

The formula of variance between class $\sigma^2$ is as follows:

$$\sigma^2(t) = \omega_0(\mu_0 - \mu)^2 + \omega_1(\mu_1 - \mu)^2 = \omega_0\omega_1(\mu_0 - \mu_1)^2 \tag{26}$$

$$\sigma^2 T_4 > 8 \cap \sigma^2 T_{411} > 6 \tag{27}$$

$$T_4 > 360\ \text{K} \tag{28}$$

$$\sigma^2 T_4 > 4 \cap \sigma^2 T_{41} > 3 \tag{29}$$

If n is the number of valid background pixels in window 21 × 21, then $\omega_1$ is the probability of the potential fire hotspots in total pixels. $\omega_0$ is the probability of the valid background pixels. $\mu_1$ is the average value of the potential fire hotspots pixels. Similarly, $\mu_0$ is the average values of the valid background pixels. $\sigma^2$ is the variance between-class value between the fire hotspots and the valid background pixel of corresponding wavelength.

### 2.3.4 Elimination of False Fire Hotspots

The misdetected fire pixels satisfy condition $R_{0.65} > 0.3 \cap R_{0.86} > 0.3$ (Fig. 3) [8].

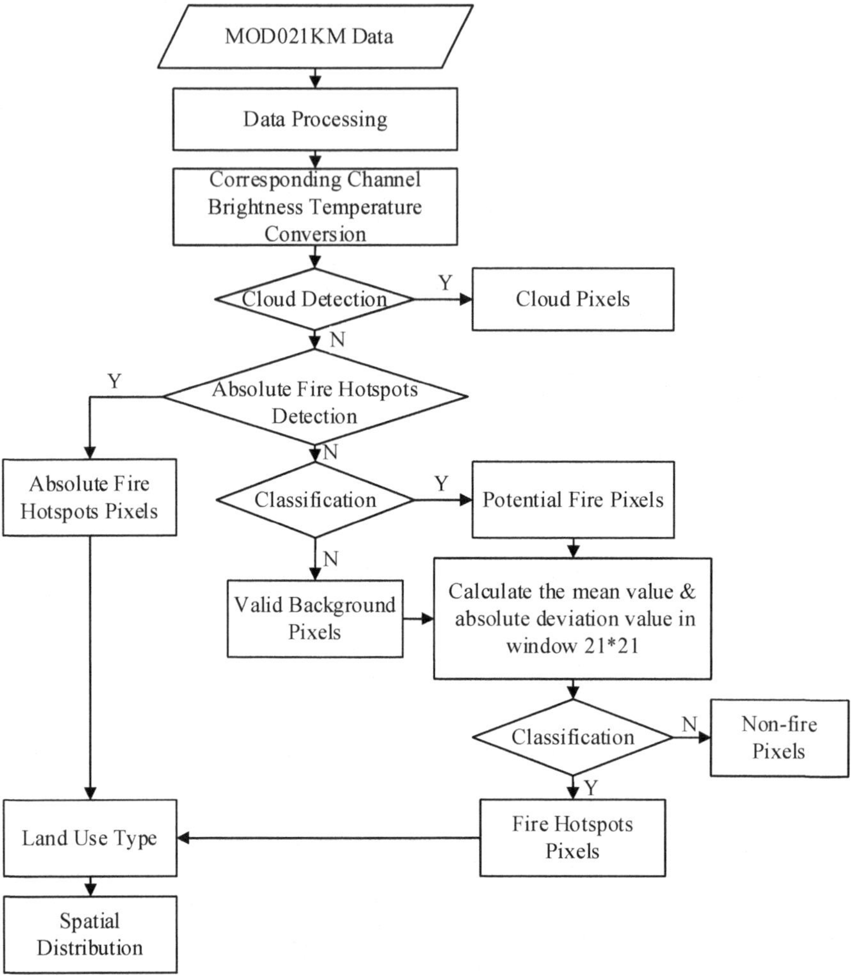


**Fig. 3** The technical route of fire hotspots detection algorithm based on variance between class

# 3 Comparative Analysis of Fire Hotspots Extarction Algorithm

## *3.1 Data Source and Data Processing*

The MODIS data received by the ground satellite receiving station in real time was taken as the data source, which covers the range from visible to far-infrared. Among them, band 21, band 22, band 31, and band 32 are sensitive to temperature changes, as we can see from Table 1.

**Table 1** Band distribution characteristics of MODIS data

| Band | Bandwidth (nm) | Bandwidth (μm) | Required SNR[3] | Primary use | Resolution (m) |
|---|---|---|---|---|---|
| 1 | 620–670 | | 128 | Land/cloud/aerosols boundaries | 250 |
| 2 | 841–876 | | 201 | | |
| 3 | 459–479 | | 243 | Land/cloud/aerosols properties | 500 |
| 4 | 545–565 | | 228 | | |
| 5 | 1230–1250 | | 74 | | |
| 6 | 1628–1652 | | 275 | | |
| 7 | 2105–2155 | | 110 | | |
| 8 | 405–420 | | 880 | Ocean color/phytoplankton/ biogeochemistry | 1000 |
| 9 | 438–448 | | 838 | | |
| 10 | 483–493 | | 802 | | |
| 11 | 526–536 | | 754 | | |
| 12 | 546–556 | | 750 | | |
| 13 | 662–672 | | 910 | | |
| 14 | 673–683 | | 1087 | | |
| 15 | 743–753 | | 586 | | |
| 16 | 862–877 | | 516 | | |
| 17 | 890–920 | | 167 | Atmospheric water vapor | 1000 |
| 18 | 931–941 | | 57 | | |
| 19 | 915–965 | | 250 | | |
| 20 | | 3.660–3.840 | 0.05 | Surface/cloud temperature | |
| 21 | | 3.929–3.989 | 2.00 | | |
| 22 | | 3.929–3.989 | 0.07 | | |
| 23 | | 4.020–4.080 | 0.07 | | |
| 24 | | 4.433–4.498 | 0.25 | Atmospheric temperature | |
| 25 | | 4.482–4.549 | 0.25 | | |
| 26 | | 1.360–1.390 | 150 | Cirrus clouds Water vapor | |
| 27 | | 6.535–6.895 | 0.25 | | |
| 28 | | 7.175–7.475 | 0.25 | | |
| 29 | | 8.400–8.700 | 0.25 | Cloud properties | |
| 30 | | 9.580–9.880 | 0.25 | Ozone | |
| 31 | | 10.780–11.280 | 0.05 | Surface/cloud temperature | |
| 32 | | 11.770–12.270 | 0.05 | | |
| 33 | | 13.185–13.485 | 0.25 | Cloud top altitude | |
| 34 | | 13.485–13.785 | 0.25 | | |
| 35 | | 13.785–14.085 | 0.25 | | |
| 36 | | 14.085–14.385 | 0.35 | | |

**Table 2** Data information of MODIS and verify image

| Num | Place of fire | MODIS data | Verify data |
|---|---|---|---|
| 1 | Boli County, Heilongjiang Province | MOD021KM. A2019078.0210.006.2019078130619.hdf MOD021KM. A2019078.0215.006.2019078130654.hdf | Sentinel-2B |
| 2 | Qinyuan County, Shanxi Province | MOD021KM. A2019091.0320.061.2019091132950.hdf | Sentinel-2A |
| 3 | Pingquan County, Hebei Province | MOD021KM. A2019126.0215.061.2019126132037.hdf MOD021KM. A2019126.0350.061.2019126132011.hdf | GF-6 |

Through collecting forest fires in 2019, the data of Table 2 were determined as data sources for fire detection. We used the band 21, band 22, band 31, and band 32 of MOD021KM data to detect fires. Through performing radiation calibration, geometric correction, and brightness temperature conversion, the value of visible band and the infrared band can be transformed into the reflectivity data and the brightness temperature data. The technology route of fire detection is shown in Fig. 4.

Medium and high spatial resolution data can be used to verify the accuracy of the algorithms. And the transit time of different satellites were closed. Finally, we collected the data in Table 2. The fires obtained by the three algorithms were overlapped with medium and high spatial resolution data to analyze the number and spatial distribution of fire hotspots. The accuracy evaluation and model optimization of algorithms were carried out from the detection rate of fire events, the accuracy rate of fire detection, the rate of omission, and the rate of false detection.

## 3.2 *Evaluation of Experimental Results Accuracy*

Taken the spatial location of three events as the center, three algorithms were used to detect fires. The experimental results are shown in Table 3. By analyzing the number and spatial distribution of different fire detection algorithms, the accuracy evaluation and model optimization were carried out from the detection rate of fire event, the accuracy rate of fire hotspots detection, the rate of omission, and the rate of false detection. We use the absolute fire hotspots detection algorithm, MODIS fire hotspots detection algorithm, and the fire hotspots detection algorithm based on the variance between classes to detect the fires.

The detection rate of fire event is designed to describe the probability of detecting a forest fire, not to describe the missed or misdetected fire hotspots in the certain fire.

It can be seen from Table 3 that MODIS fire hotspots detection algorithm and fire hotspots detection algorithm based on the variance between classes both have missed detecting forest fire events. The rate of detecting fire events of absolute fire

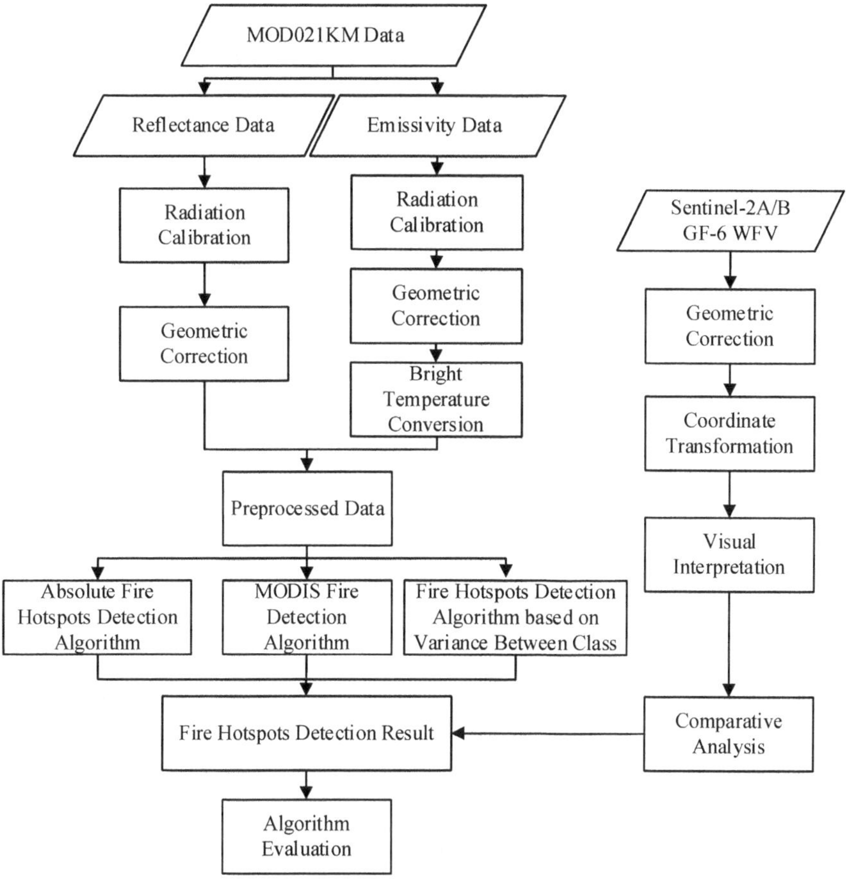


**Fig. 4** The technology route of fire hotspots detection

**Table 3** Statistics of fire hotspots of each algorithm

| Num | Phase | Absolute fire hotspots detection algorithm | MODIS fire hotspots detection algorithm | Fire hotspots detection algorithm based on variance between classes |
|---|---|---|---|---|
| 1 | 20,190,319 | 10 | 0 | 21 |
| 2 | 20,190,401 | 4 | 0 | 0 |
| 3 | 20,190,506 | 16 | 3 | 48 |

hotspots detection algorithm was 100%. The rate of MODIS fire hotspots detection algorithm and fire hotspots detection algorithm based on the variance between classes were 33.33% and 66.67%, respectively.

Only based on the quantitative analysis of the experimental results, we cannot get the scientific conclusions about the accuracy of the algorithms. In order to further evaluate the accuracy of fire algorithms, the interpretation of medium and high spatial resolution data was used as the evaluation standard, at the same time the accuracy rate of fire hotspots detection, the rate of omission, and the rate of false detection were analyzed, respectively. The experimental results are described below:

### 3.2.1 Case1: March 19th, 2019

A forest fire occurred in Boli County, Heilongjiang Province on March 19th. Using the same period image data, Sentinel-2B MSI image were interpreted ten fire hotspots pixels (Fig. 5). The absolute fire hotspots detection algorithm extracted ten pixels, correctly detected six pixels, missed four pixels, and misdetected four pixels. The correct rate of absolute fire hotspots detection algorithm was 60% and the omission rate was 40%. While the false detection rate was 40%. MODIS fire detection algorithm failed to detect all the fire pixels. The algorithm based on the variance between classes missed three pixels. At the same time, there were 12 misdetected pixels. The missed detection rate was 30%, and the false detection rate was 63.16%. The red points in picture were the result of visual interpretation.

### 3.2.2 Case2: April 1th, 2019

We used Sentinel-2A data to visually interpret the forest fires in Qinyuan county on April 1th. The results of three algorithms are shown in Fig. 6. The absolute fire hotspots detection algorithm had two false detection pixels and one pixel omission. The correct rate of the algorithm was 66.67%, which was the same with the false detection rate. The omission rate was 33.33%. The MODIS fire hotspots detection algorithm and the detection algorithm based on the variance between classes did not detect this fire event successfully. The red points in picture were the result of visual interpretation.

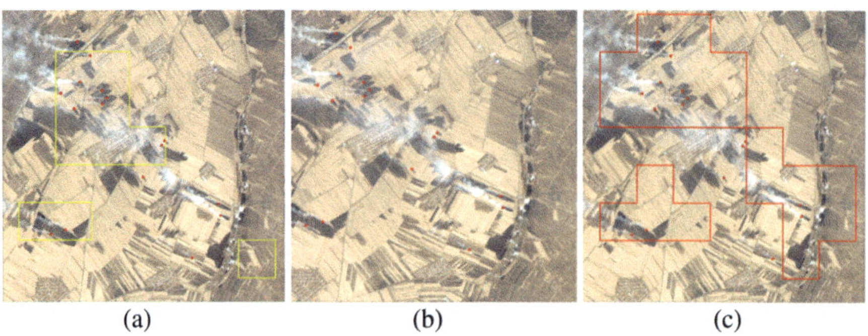


**Fig. 5** The spatial distribution of fire hotspots in Boli County on March 19th. (**a**) Absolute fire hotspots detection algorithm, (**b**) MODIS fire detection algorithm, (**c**) Algorithm based on variance between class

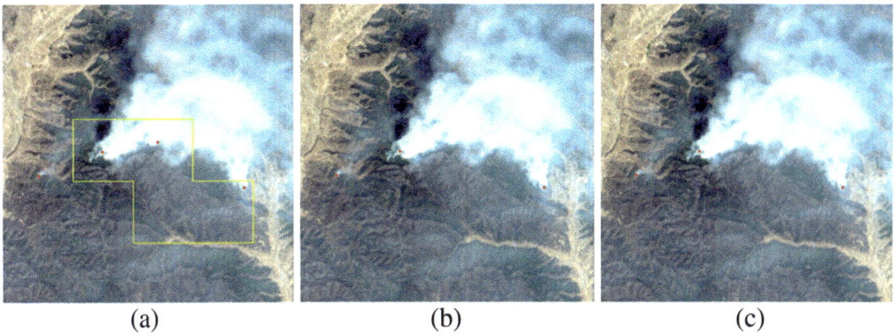

Fig. 6 The spatial distribution of fire hotspots in Qinyuan County, Shanxi Province on April 1th. (a) Absolute fire hotspots detection algorithm, (b) MODIS fire detection algorithm, (c) Algorithm based on variance between class

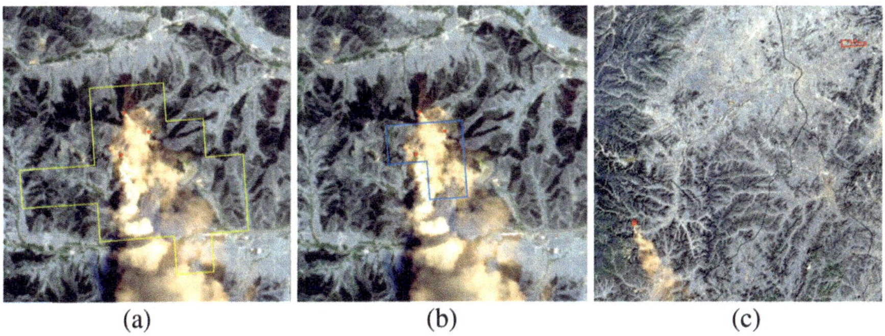

Fig. 7 The spatial distribution of fire hotspots in Pingquan County, Hebei Province on May 6th. (a) Absolute fire hotspots detection algorithm, (b) MODIS fire detection algorithm, (c) Algorithm based on variance between class

### 3.2.3 Case3: May 6th, 2019

The GF-6 WFV image was used to visually interpret the forest fire in Pingquan county on May 6th. The absolute fire hotspots detection algorithm could detect all fire hotspots pixels, but there existed a certain number of false detection of non-fire hotspots pixels. Sixteen fire hotspots pixels were detected by this algorithm, which had misdetected 13 pixels. MODIS fire hotspots detection algorithm detected three pixels. There was one pixel omission detection and one pixel false detection. The algorithm based on the variance between class detected 48 fire pixels, including three real pixels. As we can see from the upper right corner of Fig. 7. The red points in picture were the result of visual interpretation.

Based on the data of Sentinel-2A/B and GF-6 satellite, the correct rate, omission rate, and false detection rate of three algorithms were compared and analyzed, as shown in Table 4. A–C stands for absolute fire hotspots detection algorithm, MODIS

**Table 4** Verification of fire hotspots detection results

| Phase | Category | Algorithm results | Actual amount | Missed/ misdetected | Correct rate | Omission rate | False detection rate |
|---|---|---|---|---|---|---|---|
| 2019.3.19 | A | 10 | 10 | 4/4 | 60% | 40% | 40% |
| | B | 0 | | 10/0 | 0 | 100% | 0 |
| | C | 21 | | 3/12 | 70% | 30% | 63.16% |
| 2019.4.1 | A | 4 | 3 | 1/2 | 66.67% | 33.33% | 50% |
| | B | 0 | | 3/0 | 0 | 100% | 0 |
| | C | 0 | | 3/0 | 0 | 100% | 0 |
| 2019.5.6 | A | 16 | 3 | 0/13 | 100% | 0 | 81.25% |
| | B | 3 | | 1/1 | 66.67% | 33.33% | 33.33% |
| | C | 48 | | 0/45 | 100% | 0 | 93.75% |

fire hotspots detection algorithm, and the algorithm based on variance between class, respectively.

The above comparison analysis showed that the absolute fire hotspots detection algorithm had detected all the forest fire events, the detect rate of fire events was high among other algorithms. The MODIS fire hotspots detection algorithm only detected one forest fire. Although the algorithm had the highest correct rate and the lowest false detection rate in already detected events. This algorithm had a large rate of fire events omission, which could not meet the actual fire monitoring need. In the three phases of fire events, the correct rate of the algorithm based on variance between class was relatively low, but the false detection rate was high, especially in the large number of false detection pixels on May 6th.

Although the MODIS fire hotspots detection algorithm reduced the false detection rate of the absolute fire detection method, there also existed high omission rate. The threshold of the MODIS algorithm was designed on worldwide scale. In the process of detecting fires, if the algorithm sets a higher threshold for identifying, some fire hotspots of lower temperature will be classified into non-fire hotspots land pixels, which forming omission of fire pixels. When distinguishing potential fire pixels and valid background pixels, the pixels with low brightness temperature may be classified into the background pixels, which reduces the number of potential fire pixels, and then causes fire hotspots omission.

It can be seen from the result, this algorithm existed large misdetected pixels and omission pixels. According to the analysis, different thresholds can effect the detection result. If the threshold is set too high, the fire pixels will be missed. Otherwise, the fire hotspots pixels will be misdetected.

## 4 Conclusion

In order to monitor forest fires accurately in north China, the MODIS data received from the ground satellite receiving station was used as the data source. The absolute fire hotspots detection algorithm, MODIS fire detection algorithm, and fire hotspots detection based on variance between class were used to detection fires. The image of Sentinel-2A/B and GF-6 WFV were used as the standard by visual interpretation. The results showed that the detection rate of absolute fire hotspots detection algorithm was highest, which was 100%. Considering the correct rate, omission rate, and false detection rate, the algorithm is better than other two algorithms. The omission rate of MODIS fire hotspots detection algorithm was obvious. The threshold of this algorithm was not suitable for the study area, which resulting in the high omission rate. The algorithm based on the variance between classes uses different thresholds to distinguish the potential fire hotspots pixels, which can be classified into the smoldering fire pixels and the general fire pixels. However, the improved algorithm also has higher misdetected rate,which need modify in the future. Finally, the above three algorithms inevitably have the problem of weak applicability of fixed threshold, resulting in the omission rate or the false detection rate. In the future, we will focus on how to scientifically set threshold to meet the differences in various environment, and more experiments will be done.

## References


1. Qingyun, X., Weiwei, G., Xie, T., Liu, R.: CropStraw fire remote sensing monitoring and its algorithm implementation. Remote Sens. Technol. Appl. **32**, 728–733 (2017)
2. Hantson, S., Padilla, M., Corti, D., Chuvieco, E.: Strengths and weaknesses of MODIS hotspots to characterize global fire occurrence. Remote Sens. Environ. **131**, 152–159 (2013)
3. Li, J., Xingfa, G., Tao, Y.: Detection of Australian southeast forest fire using HJ satellite. J. Beijing Univ. Aeronaut. Astronaut. **36**, 1221–1224 (2010)
4. Vonder Haar, T.H., Flannigan, M.D.: Forest fire monitoring using NOAA satellite AVHRR. Can. J. For. Res. **16**, 975–982 (1986)
5. Kaufman, Y.J., Justice, C.O., Flynn, L.P., Kendall, J.D., Prins, E.M., Giglio, L., et al.: Potential global fire monitoring from EOS-MODIS. J. Geophys. Res. **103**, 32215–32238 (1998)
6. Shimabukuro, Y.E., Smith, J.A.: The least-squares mixing models to generate fraction images derived from remote sensing multispectral data. IEEE Trans. Geosci. Remote Sens. **29**, 16–20 (1991)
7. Giglio, L., Descloitres, J., Justice, C.O., Kaufman, Y.J.: An enhanced contextual fire detection algorithm for MODIS. Remote Sens. Environ. **87**, 273–282 (2003)
8. Xiao, X., Song, W., Wang, Y., Ran, T., Liu, S., Zhang, Y.: An improved method for forest fire spot detection based on variance between-class. Spectrosc. Spectr. Anal. **30**, 2065–2068 (2010)
9. Zhou, X., Wang, X.: Validate and improvement on arithmetic of identifying forest fire based on EOS-MODIS data. Remote Sens. Technol. Appl. **21**, 206–211 (2006)
10. Hang, X., Li, Y., Zhang, M., Xie, X., Ren, Y., Zhang, L.: Study on the spatial and temporal distribution of AOD in Jiangsu province influenced by straw burning based on satellite remote sensing. Ecol. Environ. **26**, 111–118 (2017)
11. Wang, W., John, J.Q., Hao, X., Liu, Y., Sommers, W.T.: An improved algorithm for small and cool fire detection using MODIS data: a preliminary study in the southeastern United States. Remote Sens. Environ. **108**, 163–170 (2017)
12. Chen, Y., Velicogna, I., Famiglietti, J.S., Randerson, J.T.: Satellite observations of terrestrial water storage provide early warning information about drought and fire season severity in the Amazon. J. Geophys. Res. Biogeo. **118**, 495–504 (2013)

# Achievements and Prospects of Space Optical Remote Sensing Camera Technology

**Guo Chongling and Yang Peng**

**Abstract** In this paper, the development of optical remote sensor in China is reviewed briefly. The following technical achievements are reviewed, included: the collaborative design ability of the integration of satellite and earth, imaging stability, quantitative detection, optical system development capability, and the integration design of imaging electronics. Finally, the future development trend of optical remote sensing system and technology is analyzed.

**Keywords** Space optical remote sensing camera · The development trend

## 1 Foreword

The development of China's space optical remote sensor began in November 1967, and it has gone through more than 50 years. After long-term hard work and independent innovation, China's space optical remote sensor technology has made a series of major breakthroughs, established a complete first-class space optical remote sensor design, manufacturing and test system, and entered a new development format. As the main force of China's space optical remote sensing industry, China Academy of space technology has presided over and participated in a number of national major scientific and technological projects related to remote sensing, greatly promoting the progress of remote sensing theory, technology, and application.

Through 50 years of construction and development, we can see that multispectral detection technology, infrared detection technology, and spectral detection technology have made considerable progress [1]. Satellite systems such as land observation satellite, environment and disaster monitoring and prediction satellite, ocean satellite, meteorological satellite, and high-resolution earth observation satellite have been formed, and a series of space optical remote sensor products have been devel-

G. Chongling (✉) · Y. Peng
Beijing Key Laboratory of Advanced Optical Remote Sensing Technology, Beijing Institute of Space Mechanics & Electricity, Beijing, China



H. P. Urbach, Q. Yu (eds.), *6th International Symposium of Space Optical Instruments and Applications*, Space Technology Proceedings 7,
https://doi.org/10.1007/978-3-030-56488-9_15

oped. The products are widely used in the related fields of national economic construction, constantly meet the needs of national security and people's life, and also make important contributions to human exploration of the mysteries of space and better protection of the earth home.

## 2 Development of Optical Remote Sensing Camera

So far, China has successfully developed and launched a variety of space optical remote sensing products. From film camera to all kinds of transmission cameras in high and low orbit, from the camera for earth observation to the camera for deep space exploration, we have made important breakthroughs in filling the gaps.

In the 20 years since 1967, we have developed two generations of film type earth observation cameras and mapping cameras. These camera systems include ground camera and star camera. The ground camera takes pictures of the ground objects and records the information on the film; the satellite camera takes pictures of the stars at the same time, records the attitude of the satellite photography, and determines the position of the ground objects. The photographic film of the camera is wound to the recycling box, and then returns to the ground together with the return cabin, and is used by the user after subsequent processing. This type of product realizes the earth observation of 2 m resolution and mapping of 1:100000 scale map.

In 1999, two cameras of CBERS-1 satellite [2] were used in orbit, which realized the leap from film camera to transmission camera and the expansion from visible panchromatic camera to infrared camera. The satellite remote sensing system covers 11 spectral segments and 4 resolutions. The CCD camera can realize 3-day repeated observation of key ground objects through side swing mirror, which solves the problems of multispectral segments, high resolution, and short observation period. In 2002, the Ocean-1 coastal zone imager was successfully launched to expand from land observation to ocean observation. The Ocean-1 coastal zone imager adopts a short focal length and large field of view optical system, and a CCD detector with a pixel size of 13 μm. The resolution of the ground pixel is 250 m and the width is 500 km. In order to meet the needs of multicamera combination imaging, multiview axis parallel technology and image height measurement technology are used to achieve high-precision multicamera full field registration imaging.

Stereo mapping camera is always the leader of many remote sensing cameras because of its high-precision requirements of internal orientation elements, stability, and time synchronization. The ZY-3 three linear array mapping camera launched in 2012 [3] is the first civil high-resolution stereo mapping camera system independently designed and launched by China. It can provide panchromatic images with a width of more than 51 km and a resolution of 2.1 m and multispectral images with a resolution of 5.8 m. It is mainly used for the production of 1:50000 basic geological information products, and the repair and update of 1:25000 and larger scale topographic maps. The application of high image quality and high stability lens technology and high-precision time scale technology has greatly improved the stability and positioning accuracy of the camera. In order to further improve the accuracy of photogrammetry, satellite 02 of ZY-3 is equipped with satellite to ground laser ranging load, which breaks through the core technology of domestic laser ranging calibration and data processing, and its ranging accuracy is better than 1 m, significantly improving the elevation accuracy of regional image.

"Gaofen" project has greatly promoted the rapid development of space optical remote sensor technology, and achieved the improvement of camera's spatial resolution, spectral resolution, and imaging stability.

GF-1 satellite [4] is the first star of China's "Gaofen" project. It is equipped with two high-resolution cameras and four wide-band cameras, which respectively realize the combination of medium high resolution (2 m/8 m) and wide-band (830 km) imaging capabilities on small satellites. High resolution-2 camera is the first civil optical remote sensing camera with spatial resolution better than 1 m in China, and the observation width reaches 45 km. GF-1 and GF-2 work together and are widely used in various fields of economy, science and technology, and national defense construction.

On April 17, 2017, SuperView-1/2 satellite [5] was officially put into commercial service, providing remote sensing data service and application system solution service for global users. The launch of SuperView-1/2 satellite marks that China's commercial remote sensing has entered the era of 0.5 m spatial resolution; the focal length of the camera reaches 10 m, and the resolution is better than 0.5 m/multispectral 2 m, which has the characteristics of high agility and realizes the capabilities of high coverage, multi-objective, multi-mode, and fast response (Typical images taken by superview-1/2 satellite are shown in Fig. 1).

The application of high-performance optical remote sensing camera in high orbit effectively improves the ability of large-scale monitoring and rapid revisit in China. In 2015, the launch of GF-4 geostationary orbit camera, which is the only high-resolution optical remote sensing camera in orbit in the world at present, has realized continuous and uninterrupted ground observation; it can obtain 400 m ground pixel resolution image of the visible near-infrared spectrum segment (panchromatic and multispectral), 50 m middle wave infrared spectrum segment of the point under the stars. Camera data has been applied in many fields such as disaster reduction, meteorology, earthquake, forestry, and so on. The key technologies such as high-precision thermal control technology and imaging stability control related to high orbit high-resolution camera have been broken through and verified.

With the development of high geometric resolution imaging technology, the spectral segments of remote sensing detection are continuously subdivided, and the level of quantitative detection is gradually improved. The camera can not only obtain the two-dimensional spatial information of the target, but also obtain the spectral radiation information with the wavelength distribution, forming a "data cube." Over the years, there have been breakthroughs in precision spectroscopic technology, large dynamic range and high sensitivity detection technology, large aperture and high efficiency Fourier transform interferometer technology, high-precision radiation and spectral calibration technology. Filter spectrum, instrument, dispersive spectrometer, and interference spectrometer are all successfully devel-

**Fig. 1** Image of exhibition center of Tibet Autonomous Region (shot by SuperView-1) [6]

oped, which are mainly used in agriculture, forestry, water color, atmospheric detection, and other fields. In 2017, China launched the polar orbiting meteorological satellite FY-3, which is equipped with a remote sensing load hyperspectral greenhouse gas monitor specially for greenhouse gases. It mainly has the capacity of total amount of greenhouse gases such as $CO^2$, CH4, and Co, with a spectral resolution of 0.27 $cm^{-1}$. It is used to study a series of scientific issues such as global greenhouse gas emissions, greenhouse gas source and sink analysis, and the relationship between greenhouse gases and climate change.

With people's attention to environment, natural disasters, and other ecological issues, we are also expanding the application of optical remote sensors in this field. In 2016, the FY-4 lightning imager successfully operated in orbit. The lightning imager is a high-speed real-time near-infrared camera used for lightning detection. The combination of ultra-narrow band imaging (center wavelength 777.4 nm, bandwidth 1.0 nm), spatial filtering, and frame background removal (500 frames/s high-speed imaging) is used to realize the enhancement and detection of transient multi-point source lightning signal.

Over the years, the development trend of space optical remote sensor is to achieve high spatial resolution, high temporal resolution, high spectral resolution, high radiation resolution and high positioning accuracy. In pursuit of measurement, key technologies such as system design, optical design and manufacturing, electronic information acquisition, precision thermal control, mechanism and control of space optical remote sensor have been broken through, supporting the research and development of a large number of advanced optical loads such as high-resolution observation, civil mapping, meteorological observation, etc., laying a solid foundation for the construction of major national projects.

## 3 Development of Optical Remote Sensor Technology

(a) Continuous improvement of the collaborative design ability of the integration of satellite and earth, and better system performance
With the in-depth study of target background characteristics, remote sensing camera design pays more attention to practical application. Camera system design no longer simply meets the technical indicators, but starts from the application, carries out the task analysis and design of the integration of heaven and earth, carries out the whole link simulation and imaging evaluation, and comprehensively optimizes the working mode and load indicators in combination with the application scenarios.
In recent years, through technical research, the whole link simulation has been continuously improved, and a relatively complete imaging link simulation model of optical satellite system has been established. The development of full link simulation is divided into several stages:
The first stage is the development from stand-alone product to camera system design. The optical system model, detector model, and circuit model of remote sensor are defined, and the influencing factors of signal and noise are modeled at system level. This stage is mainly applied to the optimal matching of all components of camera system, which promotes the development of large f-number system design and other technologies.
The second stage is the development from system design to intelligent manufacturing. From the perspective of system engineering, the optical mechanical thermal integration analysis platform of remote sensor is developed, and the simulation data source is expanded, including engineering nodes such as materials, manufacturing process error, measurement and inspection, assembly and integration, calibration and test. The quality information in manufacturing is effectively combined with design and simulation, and the model has the ability of quality evaluation.
In the third stage, the camera system is expanded to the on orbit imaging prediction of the whole system. A simulation platform for the design and verification of the integration of heaven and earth is established. The image quality parameter representation and quantitative evaluation methods are improved, covering the target, ground scene, optical remote sensor, satellite platform, data transmission; and in ground processing and other links, the system MTF model accuracy reaches 80%, and the signal model accuracy reaches more than 82%. Through simulation and prediction, the system has the ability to predict the on orbit imag-

ing quality and provide reference for the adjustment of on orbit imaging parameters.

In the fourth stage, the system optimization function is being established, and the optimization algorithm is integrated into the system, so as to analyze and evaluate the confidence degree of parameters and the sensitivity of influencing elements in the optical imaging link, and further study the multi-objective and multi-variable imaging quality enhancement model and the identification method of parameters, which can be applied to the efficiency simulation of multi-source data acquisition system.

The whole link simulation technology of space optical remote sensing is the basis of the conceptual design, scheme demonstration, system optimization, and efficiency evaluation of the optical remote sensing satellite system (The whole link simulation system of space optical remote sensing are shown in Fig. 2), which has been successfully applied to SuperView-1 and other projects. The continuous improvement of the collaborative design ability of the integration of heaven and earth is of great significance to optimize the performance of the imaging system, predict the imaging quality of the system, reduce the difficulty of the system development, and promote the continuous leap of the development ability of China's optical remote sensing satellite.

(b) Imaging stability is constantly improved, and target positioning is more accurate

In terms of spatial resolution, geometric positioning accuracy, and image radiation quality, China's high-resolution satellites and payload products have achieved leapfrog development. Atmospheric correction technology and long-life laser ranging technology continue to break through. It includes high stability integrated support technology, high precision time scale technology, high precision geometric calibration technology and so on. The positioning accuracy is effectively improved.

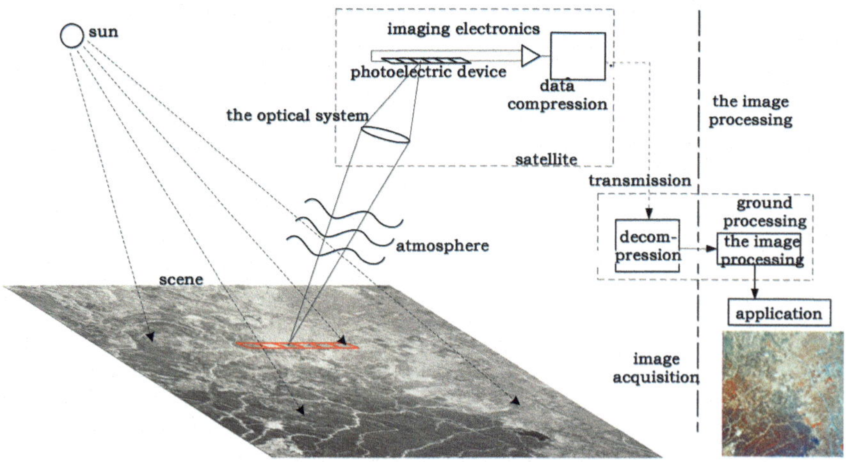


**Fig. 2** The whole link simulation system of space optical remote sensing

The improvement of remote sensor imaging stability has made breakthroughs in several aspects.

First of all, the stability of the camera's internal orientation elements is improved, the design of zero distortion optical system, and the adoption of high humidity and heat stability structure ensure the camera's high image quality, zero distortion, and high stability of the internal orientation elements, which can achieve the stability of the cameras in orbit orientation elements better than 0.3 pixels within half a year.

Secondly, in order to improve the accuracy of high-precision stereo mapping, laser technology can help improve the imaging accuracy. At the same time, integrated design and on orbit optical quality measurement and monitoring technology are applied. The application of high-precision laser pointing positioning technology makes the pointing positioning accuracy second stage. The height measurement data of laser rangefinder significantly improves the elevation accuracy of regional image.

In addition, with the development of large-diameter optical technology and the development of detection integration products, the directional correlation measurement system is applied to high-resolution imaging system. The directional correlation accuracy is improved from 5″~10″ to 0.5″ by using high-precision load directional angle measurement technology, which significantly improves the positioning accuracy of uncontrolled geometry.

(c) With the improvement of spectral resolution, the level of quantitative detection is gradually improved

At present, the detection spectrum segment is continuously subdivided, the spectral resolution is continuously improved, and the quantitative detection level is gradually improved. The infrared very high spectral resolution detector of atmospheric environment to be installed on "GF-5" adopts the technology of Fourier transform time modulation interferometer, with a spectral resolution of 0.03 $cm^{-1}$. It has reached the international advanced level; in terms of key technologies, it has broken through the large-scale immersion grating technology and reached the international leading level; the application of high-precision ground, satellite radiation, and spectral calibration technology has made the long-term in orbit calibration accuracy visible in near-infrared 2%, ultraviolet 3% (absolute calibration accuracy), and reached the international advanced level.

China has formed two series of products, grating spectrometer and Fourier transform spectrometer, which are applied to the measurement of global greenhouse gases and contribute to the national green and sustainable development.

(d) Continuous improvement of optical system development capability

In recent years, the design and development ability of space optical system have been developed by leaps and bounds. The demand of high spatial resolution and large width also promote the innovation and development of advanced optical design theory and method. The space optical system has been developing from the transmission type, the catadioptric type to the aspheric total reflection system, and then to the free-form surface reflection system. The volume of the remote sensor is 1/2 of the original. The compression ratio and field of view of the system are increasing, the distortion control ability is increasing, and the zero distortion system is realized.

In the aspect of optical processing, optical processing technology has developed from "traditional optical processing technology" which initially used manual polishing to "deterministic optical processing technology" which used intelligent polishing equipment. On the basis of continuous process exploration and product application, we have mastered high-precision ultrasonic lightweight processing technology, and the lightweight rate of aspheric mirror is better than 80%. The intelligent CNC polishing technology makes the process parameters highly integrated, achieves quantitative removal, and reduces the polishing time by more than 30%; the ion beam polishing technology realizes the non-contact, deterministic, and nano-level controllable removal of optical elements, and the machining accuracy can reach RMS $\leq \lambda/100$.

In the aspect of mounting and testing, optical mounting technology is closely combined with the development demand of remote sensor. In the early optical remote sensor, due to the limitation of processing ability and detection technology, in order to meet the requirements of optical imaging, the "reserved tolerance" mounting technology was used in optical mounting. With the development of computer technology, assembly technology has developed from "reserved tolerance" to "error compensation." With the improvement of the simulation accuracy, the whole process image quality prediction technology is widely used in the process of reflective and refractive optical system adjustment. And gradually become an effective method to diagnose and solve the problems in the process of optical system installation and adjustment. With the improvement of the intelligent level of assembly and detection, the adjustment factor of 1 m level optical system reaches 0.9 (Optical adjustment and test chart of 1 m aperture camera are shown in Fig. 3).

(e) The integration degree of imaging electronics is constantly improving and tends to be intelligent

The optical remote sensor with high performance and high reliability cannot do without the support of powerful electronic information equipment. Electronic information products have experienced three stages: from scratch, steady

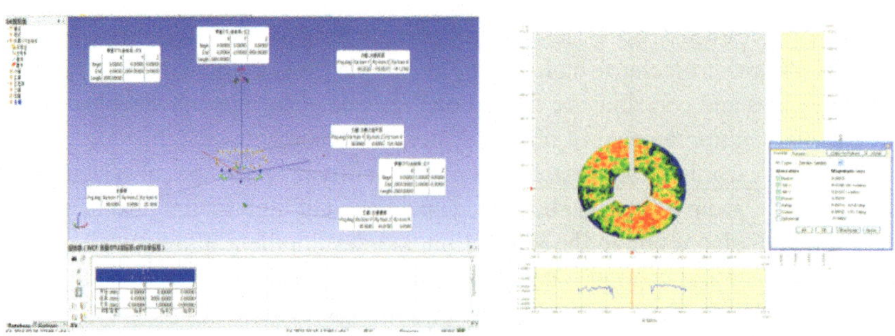

**Fig. 3** Optical adjustment and simulation

development, and quantity and quality increase. In recent years, China has developed rapidly in CCD and CMOS detectors. Large area array CMOS devices and long line time delay integrated complementary metal oxide semiconductor (TDICMOS) devices are becoming more and more mature. In view of the time sequence driving of the detector, the low noise processing of high-speed video signal, the detection of small signal, the relative radiation correction and compression of real-time image on the satellite, the control of intelligent management, and the low power consumption and lightweight of electronic information equipment, a breakthrough solution has been put into application. The data dynamic storage capacity of the single board is increased from 2 Gbit to 4 Gbit, and the output code rate is increased from 10–400 Mbit/s to 10.3–15.8 Gbit/s. The signal processing capacity of the remote sensor circuit has made a qualitative leap.

In general, great changes have taken place in electronic information technology of optical remote sensor in the past 10 years, from single data management technology to high-speed and high integrated information processing technology, from huge circuit structure to micro-packaging technology, MCM and ASIC design concepts, from simple function realization to excellent performance index, Electronic information products are gradually upgraded from "big, slow, and few" to "fast, small, and many." In addition, based on the integrated application of high-performance CPU, embedded operating system, high-speed data memory, and high-performance field programmable gate array (FPGA), as well as the system architecture design, processor design, hardware component design, software architecture, and parallel mapping, the integrated optimization design of on orbit processing architecture can meet the application requirements of multiple loads and on orbit processing.

# 4 Development Trend of Optical Remote Sensing System and Technology

(a) Integrated detection technology

With the development of camera multi-channel technology, full spectrum modular spectral imaging technology, and other technologies, the integrated multi-functional detection technology on board has been improved. The development direction of multi-functional integrated detection technology is mainly reflected in two aspects. First, it is "one type and multi-purpose," that is, cross spectral detection for high-precision target detection, multi-angle hyperspectral polarization multi-dimensional information integrated detection, etc. The second is "one type of multi-function," that is, integration of imaging and communication, integration of mapping and imaging communication, etc.

Satellites need to obtain multi-dimensional information of targets through satellite cluster to meet the needs of various user information services and information support tasks.

(b) Meet the national strategic needs and develop the assembly technology of super large caliber remote sensor
In order to meet the needs of high-resolution observation, China's demand for the construction of large-scale space optical facilities in the future is mainly for large-scale optical remote sensors, whose main mirror aperture of the optical system reaches the order of 10 m. At the same time, the construction of super large aperture remote sensor can meet the needs of large-scale scientific research fields such as remote dim target detection and astronomical observation.

(c) Micro-nano and intelligent load technology
The intelligent integrated micro-nano system with sensing, processing, control, communication, and other functions has been developing continuously. It is a trend to build an intelligent design system for remote sensor and develop an intelligent remote sensor with accurate sensing state information ability and independent adjustable imaging parameters. "Cloud" in orbit service mode is being actively demonstrated, and China's micro-nano payload and satellite innovation mode are also actively explored.

(d) Remote sensing big data mining and its application
With the continuous progress of remote sensing technology, the amount of remote sensing data is becoming larger and larger, the types are more and more, the distribution is more and more scattered, the complexity and personalization of remote sensing applications are also increasing, and remote sensing is moving towards the era of big data. Artificial intelligence technology is gradually applied in the field of remote sensing, which provides a feasible technical means to break the bottleneck of data and information, and is an important direction of remote sensing big data application in the future.
To sum up, remote sensing has great application potential in business services and other emerging applications. At the same time, emerging markets also pose new challenges to remote sensing. The optical remote sensing industry urgently needs innovation, including innovation platform, innovation load and innovative application; it needs to explore new development mode, build an open platform, cooperate widely in the frontier technology field, and provide better remote sensing technology services for the market.

## 5 Conclusion

China's space optical remote sensing technology focuses on major projects and major projects, and has broken through a large number of core technologies. As the country vigorously promotes the Internet, big data, and artificial intelligence, it will continue to innovate the application mode of remote sensing, realize the deep mining of remote sensing data, and provide users with high-quality information services.

## References

1. Chen, H., Li, Y.: 60 years technical achievements and prospects of BISME. Spacecr. Recovery Remote Sens. **39**(04), 1–9 (2018)
2. Qingjun, Z., Ma, S.: Achievements and progress of China-Brazil earth resource satellite. Spacecr. Eng. **18**(04), 1–8 (2009)
3. Deren, L.: China's first civilian three-line-array stereo mapping satellite: ZY-3. Acta Geod. Cartogr. Sin. **41**(03), 317–322 (2012)
4. Chunling, L., Rui, W.: GF-1 satellite sensing characters. Spacecr. Recovery Remote Sens. **35**(04), 67–73 (2014)
5. Han, G., Lingye, M.: SuperView-1 is offically commercial, and China's commercial remote sensing has entered the 0.5m era. Satell. Appl. **65**(05), 62–63 (2017)
6. SuperView-1 sent back the first batch of high-definition beautiful pictures. Are you there? https://www.sohu.com/a/220501146473428

# Research on Space-Based Optical Imaging Simulation Soft Credibility Evaluation Method

**Han Yi and Chen Ming**

**Abstract** This paper introduces the space-based optical imaging simulation system members and their capabilities, as well as the simulation principles. Then according to the involved uncertain and fuzzy factors, we introduce the evaluation method based on fuzzy analytic hierarchy process (FAHP) method and the fuzzy comprehensive evaluation method, and discuss the detailed process, mathematic model, and executive approach. Combing the idea and methods of verification and validation, we establish the hierarchical structure and framework of credibility evaluation. At last we accomplish the evaluation work and the result shows that method is available and the simulation system is credible. This research can provide references for future studies on evaluation methods of simulation system.

**Keywords** ISSOIA 2019 · Optical imaging system · System simulation · Credibility evaluation

## 1 Introduction

Space target space-based optical imaging system has huge advantages in the space coverage and observation timeliness; it can capture clear target images in closer distance under suitable conditions, obtain the target geometry features, and recognize target type. Therefore it has become the development direction of space target surveillance network. Due to the limitation of actual ground laboratory conditions, we need to utilize the Modeling and Simulation (M&S) technique to study the performance of space-based optical imaging system [1]. The main goals include two aspects: one is providing target optical signal source for optical observation system prototype in the ground experiments; the other is providing space-based optical imaging simulation results for the experiments of a variety of image processing

H. Yi (✉) · C. Ming
Beijing Aerospace Control Center, Beijing, China



H. P. Urbach, Q. Yu (eds.), *6th International Symposium of Space Optical Instruments and Applications*, Space Technology Proceedings 7,
https://doi.org/10.1007/978-3-030-56488-9_16

algorithm, such as image pre-processing and target recognition. Therefore we build and model the key elements of the end-to-end imaging chains for optical imaging system mathematically, and use VC++ and OpenGL to develop the software of space-based optical imaging simulation system (SOISS), which has the capabilities of the space scene visualization, target optical scattering properties analysis, imaging simulation, and system performance analysis. But whether the simulation model can replace the actual model and whether the simulation accuracy can meet corresponding requirements we cannot ignore, and the simulation credibility must be researched and evaluated. In fact, with the rapid development of advanced simulation technology, the continuous expansion of simulation application, the entire industry has urgent demand to evaluate the credibility of simulation system [2].

The simulation credibility, which is the degree that users believe in simulation results under certain conditions, is one of the most important parameters of simulation system [3]. The simulation application which is lack of enough credibility is meaningless. Research on the methods, tools, and application of simulation system credibility evaluation, can be meaningful to improve the correctness of the simulation results, reduce the risk of application, enhance the reusability of the system, and guarantee the quality of the simulation system. Developing the verification and validation (V&V) work for simulation system is effective and necessary to reduce the risk and improve the credibility of simulation [4]. Several commonly used evaluation methods include the analytic hierarchy process (AHP) method, the fuzzy evaluation method, the multiple similarity method, the CLIMB method, and the evaluation tree method, etc. Each method has its own background and application area and has its own advantages and disadvantages; there is not an absolute judgment criterion [5]. The correctness of the comprehensive evaluation of complicated system is not only related to evaluation experts and the detail index structure framework, but also related to the evaluation method.

In this paper we first introduce the members and simulation principle of SOISS, then analyze the theoretical model and implementing steps of our evaluation method, which is based on fuzzy analytic hierarchy process (FAHP) and the fuzzy comprehensive evaluation method [6]. During the M&S developing process, we also pay attention to the V&V work. On this basis, we construct the evaluation hierarchical index framework. According to the involved uncertain and fuzzy factors, we choose triangular fuzzy number (TFN) to represent experts' judgment information. This method can not only reflect the fuzziness in credibility evaluation, but also consider and integrate comprehensive evaluation opinions effectively; therefore the evaluation results can be more reasonable. At last we accomplish the evaluation experiment and the result shows that the SOISS is credible.

## 2 Space-Based Optical Imaging Simulation System

Under the Windows system, we design and develop the system simulation software tool, SOISS, based on the Socket programming by utilizing Visual C++, STK, and OpenGL. SOISS has the integrated capabilities of target optical scattering characteristics analysis, visible imaging system simulation and performance analysis, etc. By the way of software development, the mathematic simulation models of target scattering characteristics, space-based visible imaging chain can be translated into executable computer program. We utilize SOISS to research the process of space-based optical detection and imaging in the simulation experiment.

SOISS contains five members: space situational scene member, target optical characteristics analysis member, wide FOV detection member, narrow FOV imaging member, and comprehensive information processing and display member. They are running on high performance graphics workstations respectively, and form a local network through Ethernet cable. The relationship of five members and the transmission data flow is shown in Fig. 1. The peripheral hardware, including stereoscopic display equipment, high resolution LCD display, optical collimator, can provide service for SOISS to accomplish the semi-physical simulation experiment.

The transmission data above each arrow line in Fig. 1 is illustrated as follows: (1) high resolution 3D space scene; (2) real-time orbit and attitude parameters of target satellite and observation satellite, the real-time sun illumination parameters, etc.; (3) optical cross section (OCS) and equivalent visual magnitude of target, as well as all parameters of (2); (4) pristine radiance images with celestial background, and the target is just imaging as a point because of wide FOV; (5) simulated point target images with noise and blur effects, and the performance index parameters of wide FOV detection subsystem; (6) radiance images of target satellite, and the target may cover several pixels because of narrow FOV; (7) simulated target images with noise and blur effects, and the performance index parameters of narrow FOV detection subsystem; (8) parallel light signal of target satellite.

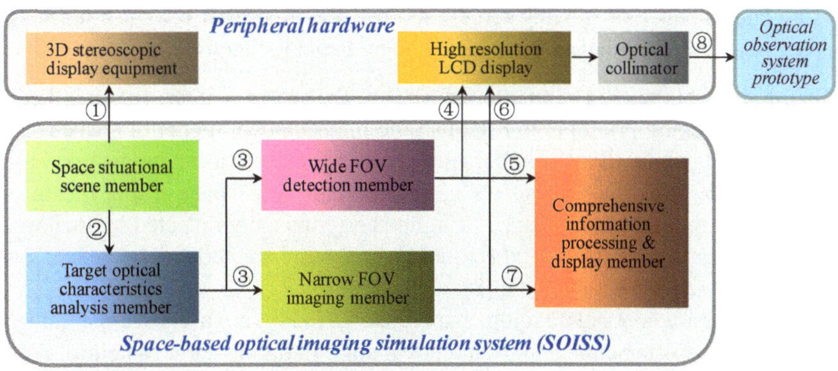


**Fig. 1** The relationship diagram of simulation members

Each member has high specialty, reliability, and upgrading. The main functions of each member are as follows:

1. Space situational scene member is to calculate and display space 3D and 2D scene of the moving imaging system and target in real time, according to initial configuration such as orbital elements, simulation time, and step time, etc. It also dispatches related parameters and output 3D space scene into stereoscopic display equipment.
2. Target optical characteristics analysis member is to calculate the illumination condition or geometric visibility of target, including complex target and simple shape target in 3DS format and their combination, and calculate the parameters of target OCS, visual magnitude, sun phase angle, etc. It can also analyze the influence of different factors such as flying attitude, working state condition, surface materials, and target geometry on OCS, by utilizing a fast visualization calculation method based on an OpenGL [7].
3. Wide FOV visible detection system member and narrow FOV visible imaging system member are to simulate the detection and imaging results captured by wide FOV and narrow FOV visible imaging subsystem, respectively. The former is mainly used to scan, search, detect, and track large-scale and long-distance space targets, acquire point target images, and guide the latter. The latter is mainly used for short-range imaging and precise tracking of space targets by utilizing the large aperture and long focal distance receiving optical system. The point target imaging simulation needs to calculate the irradiance of the target at the pupil of the detection system, and then calculate the gray value of the target pixel point, while the area target imaging simulation needs to analyze the distribution of the irradiance in the observation direction. They confirm the positions and brightness of stellar based on the Hipparcos catalogue and the FOV direction of optical system. They also output the pristine radiance images of target into the high resolution LCD display, and the display results can be translated into parallel light signal by optical collimator. The method can provide simulated signal source for the ground experiment of optical observation system prototype.
4. The comprehensive information processing member to process the sequence simulated images of moving target, and test the algorithm of real-time image processing, determination of target orbit and target recognition.

In the study of imaging simulation, we utilize the theory of linear filtering theory and optical imaging principle to establish the mathematical model of space-based optical imaging chain. Because photoelectric imaging system can be regarded as a mix consists of series subsystems with a certain frequency (space or time) features, so we mainly research the variance of light energy intensity reflected by the target in each link, as well as the spatial frequency transfer characteristics of optical system, detector, and relative motion. [8].

The space-based optical imaging chain is shown in Fig. 2. This process is categorized into six key elements, mainly including solar radiation, target satellite, space environment, motions of target and imaging platform, optical receiver system, and digital sensor. The influences of motion and space environment on imaging quality

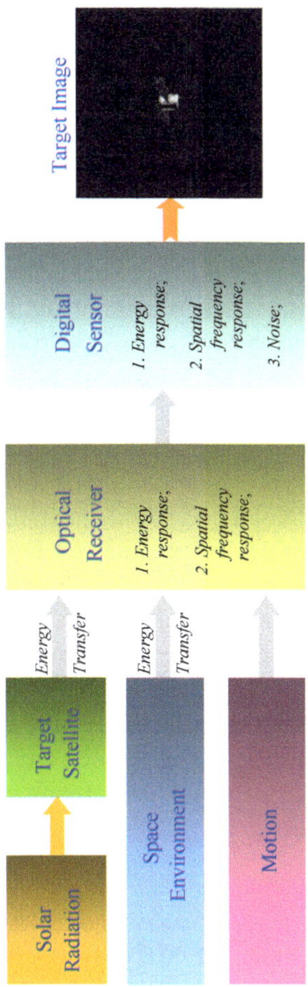


**Fig. 2** Diagram of space-based optical imaging chain

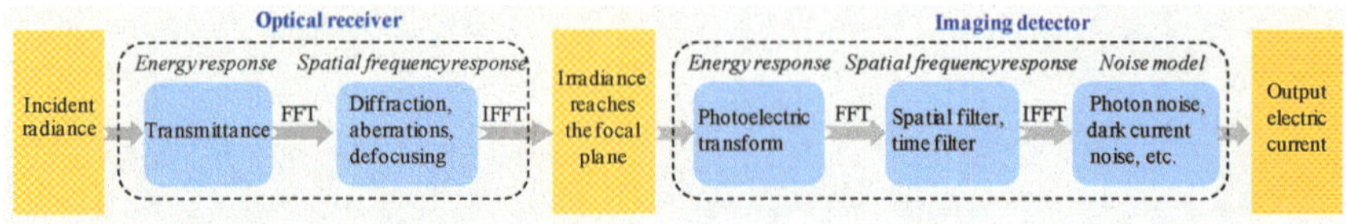


**Fig. 3** Diagram of simulation model of optical imaging system

are mainly shown by the image degradation and the stars background. The main input parameters of imaging simulation include: space target parameters (3D geometry model, surface materials and their optical scattering characteristic parameters, orbit parameters), imaging system parameters (optical system focal length, aperture, transmittance, and image detector pixel size, pixel number, pixel spacing, quantum efficiency, exposure time, dynamic range and noise parameters, etc.)

The compendious simulation model diagram of the visible light imaging system link is shown in Fig. 3. We mainly consider the energy response characteristics, the spatial frequency response characteristics of optical system and digital sensor, as well as the noise model and the star background in FOV. By utilizing the method of Fast Fourier Transform and Inverse Fast Fourier Transform, we precede the pristine radiance image, multiply it with each MTF expressions, and add noise step by step. We can obtain the target's irradiance image at the focal plane of receiver optical system, as well as the final simulated image output by the CCD sensor.

The simulation goal is to generate high fidelity target gray images, so we adopt the imaging simulation method based on OpenGL and 3D target model. By setting appropriate parameters in OpenGL, the pristine radiance image of target can be generated. Through the theoretical calculation we can obtain the number of pixels covered by different shape and size of simple targets. We also count the number of pixels covered by the same targets in the OpenGL simulated images. By comparing the theoretical values with the simulation values, the results show that this simulation method has high precision. In fact, the principle of geometry imaging of optical imaging system determines the image size, target's attitude display, and its parts pixels number; the imaging mathematic models determine the pixel gray values, the noise, and the blurring effects [8].

## 3 Method and Procedure of Credibility Evaluation

SOISS has five simulation members and the simulation models are complicated, and the credibility indexes are qualitative mostly and cannot be quantified exactly. It is not easy to judge the opposite and weight of the same level indexes. According to the actual conditions of SOISS, we decide to adopt the evaluation method based on the FAHP method and the fuzzy comprehensive evaluation method. The FAHP method can make decisions more reasonable and overcome the shortcomings of traditional AHP method, for example, people's subjective judgment, choices, and preferences have a great influence on the evaluation result. FAHP is the extension of AHP under the fuzzy conditions, and can make the calculated weight more scientific, reasonable, and easy to implement. The difference between FAHP and AHP is the judgment matrix. The core of FAHP is to construct the judgment matrix by utilizing the pair-wise comparison judgment method, which is expressed by TFN. It is convenient and flexible to make quantitative analysis on the non-quantitative events [9]. Aiming at actual complicated problems, the fuzzy comprehensive evaluation method can consider multiple effect factors comprehensively, and evaluate the subjection rank conditions of evaluation object appropriately. Theoretically speaking this method is feasible and the evaluation effect of multilayer complex problem is better.

The fundamental procedure of the evaluation method used in this paper is summarized as follows:

## 3.1 Establish the Hierarchical Structure Framework

Establishing a multi-criteria systematic framework is the premise for solving multi-attribute decision-making problem. The decision is based on the satisfaction degree of each alternative in every criterion. The hierarchical framework should cover all the important criteria in the whole life of the systems.

Use the AHP method to divide the various influence factors of the system hierarchically. First divide the actual problem into several factors, take these factors into different groups based on their properties, and then establish the hierarchical structure. The structure generally can be divided into the highest, middle, and bottom layers. The highest layer, that is, the target layer, namely is the goal of the problem. The middle layer is the criterion layer, and the elements are the rules to judge the scheme is good or bad. This layer can be divided into criterion layer and sub-criterion layer according to the size and complexity of the problem. The lowest layer is the detail index layer or plan layer [10].

## 3.2 Construct the Evaluation Matrix of Single Factor

Based on the fuzzy transformation principle, experts compare the evaluation index and objects respectively and structure the judgment matrix by using TFN. For each fuzzy judgment matrix, we calculate the fuzzy comprehensive degree of each element. In comparison, the experts can use equally important, somewhat important, obviously important, very important, extremely important to describe qualitatively the relative importance degree of one factor than another. In practice, the comparison matrix usually employs 1–9 scales defined by Salty or the exponential scale method. The TFN theory will not be described in detail in this article.

### 3.3 Calculate the Weight Vector

The degree of possibility for a convex fuzzy number to be greater than $k$ convex fuzzy numbers $M_i$ ($i = 1, 2, \ldots, k$) can be defined as:

$$\vee\left(M \geq M_1, M_2, \cdots, M_k\right) == \min \vee\left(M \geq M_i\right), \quad i = 1, 2, \cdots k \tag{1}$$

Let $d'(A_i) = \min \vee (S_i \geq S_k)$, for $k = 1, 2, \ldots, n$, $k \neq i$, $S_i$ denotes the value of fuzzy synthetic extent of each element compared with other elements, and $A_i$ denotes the $i$th element. Then the weight vector is given by:

$$\boldsymbol{W}' = \left(d'\left(A_1\right), d'\left(A_2\right), \cdots, d'\left(A_n\right)\right)^T \tag{2}$$

After normalized calculation, the final weight vector is given by:

$$\boldsymbol{W} = \left(d\left(A_1\right), d\left(A_2\right), \cdots, d\left(A_n\right)\right)^T \tag{3}$$

While comparing two TFNs if the condition appears that $l_1$ is larger than $u_2$, the elements of judgment matrix should be normalized first.

### 3.4 Fuzzy Comprehensive Evaluation

Fuzzy comprehensive evaluation method is based on the theory of fuzzy mathematics and on the basis of fuzzy relations with their synthetic operations, with the help of function in fuzzy theory to express the state of the factors. Let $\boldsymbol{U} = \{u_1, u_2, \ldots.., u_n\}$ be an object factor set, and $\boldsymbol{V} = \{v_1, v_2, \ldots.., v_m\}$ be a remark set. While evaluating single factor, we should structure fuzzy evaluation matrix $\boldsymbol{R} = (r_{ij})_{m \times n}$, where $r_{ij}$ is the membership degree subject to remark $v_j$ from the viewpoint of $u_i$. The comprehensive evaluation result $\boldsymbol{B}$ is:

$$\boldsymbol{B} = \boldsymbol{W} \circ \boldsymbol{R} = \left\{b_1, b_2, \cdots, b_m\right\} \tag{4}$$

Where $\circ$ represents the fuzzy arithmetic operator, and $\boldsymbol{B}$ should be normalized. While processing the multistage fuzzy comprehensive evaluation, factors can be divided into several categories according to certain properties of the elements in the set $U$. First carry out the comprehensive evaluation of the factors of low layer, and use the evaluation results vector of the lower layer as a fuzzy vector, so that multiple lower-level fuzzy vectors may constitute a fuzzy evaluation matrix of a higher layer. Then perform the fuzzy synthesis operation of this relationship matrix and can obtain the comprehensive evaluation value in the higher layer [11].

# 4 Credibility Evaluation of Soiss

## *4.1 Hierarchical Structure of Evaluation*

To make sure SOISS has certain credibility and accuracy, we have done some verification and validation work in developing process. The relationship of V&V and M&S in the development process is shown in Fig. 4.

The main work of verification is to check and determine the correct degree of system concept and technical requirements. This work mainly focuses on the integrity, consistency, and traceability of simulation system. The major contents include the verification of M&S requirements, the verification of simulation system overall design, the verification of subsystem detailed design, the verification of system implementation, and the verification of system integration. The objective of validation is to estimate the adaptability of simulation system on behalf of the real world. This work mainly focuses on the validity, fidelity, and veracity of simulation system. The major contents include the validation of conceptual model and the validation of simulation results.

In the evaluation of credibility, we should divide the index system and construct the hierarchical structure and framework from the whole point of view, as well as according to the correlation and effect of various factors. The highest level of evaluation framework is the credibility of the simulation system. The intermediate level is index layers. They are the major ingredient of index system and the main factors determining the simulation system's credibility. The lowest level is sub-indexes, which are the embodiment of the main evaluation index.

According to above principles, we establish the credibility evaluation hierarchical structure based on the basic framework of M&S and V&V process model, as shown in Fig. 5. We divide the simulation system credibility evaluation contents into the conceptual model evaluation, subsystem simulation model evaluation, the

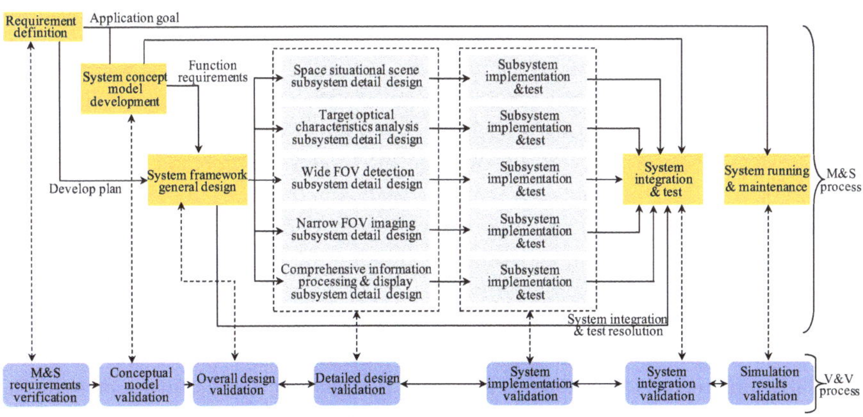


**Fig. 4** The relationship of V&V and M&S of simulation system

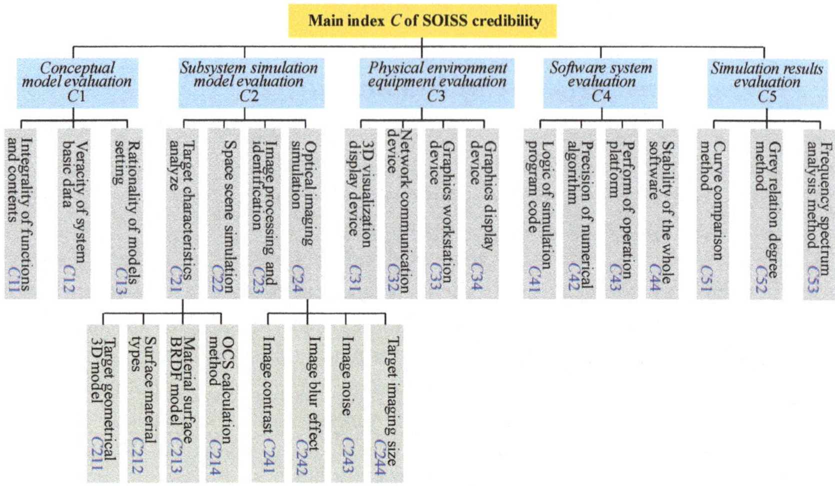


**Fig. 5** The hierarchical structure of credibility evaluation

physical environment equipment evaluation, software system evaluation, and simulation results evaluation, respectively.

The conceptual model evaluation mainly includes the description verification of the simulation research purposes and the research objects. The research objects include the optical imaging system itself and the simulated space environment and the received data, and so on. This step is very important, because if the conceptual models are not correct, we may solve wrong problems next and the simulation is meaningless.

The subsystem simulation model evaluation includes the verification and evaluation of all subsystem model type and modeling methods. The mathematical models of each subsystem are the abstraction of the real object, and many assumptions are needed in the process of mathematical modeling. Therefore it is necessary to check the rationality of these assumptions, as well as to validate the types of conventional model and methods [12].

The physical device evaluation includes analyzing the precision of physical devices, and evaluating the influence of various components' accuracy on the precision of the physical simulation model. By comparing the actual results with theoretical results, or other experiment results, the effectiveness of model can be verified.

The software system evaluation includes analyzing the performance of computer operation platform, the logic of simulation program and the correctness and precision of numerical algorithm, as well as estimating the running performance, reliability, and stability of the whole software.

The simulation results evaluation refers to the comparison of the simulation results and actual test data, as well as the error evaluation, by utilizing different qualitative and quantitative analysis methods, such as the curve comparison method, the gray relation degree method, and the frequency spectrum analysis method.

### 4.1.1 Evaluation Result of Simulation System

After determining the evaluation indexes, we construct the judgment matrix of corresponding elements next according to the working process. For the main index $C$ of simulation system, we request subject domain experts (SME) to construct the fuzzy judgment matrix $D$ about its five sub-indexes $C_1$, $C_2$, $C_3$, $C_4$, $C_5$ by using triangular fuzzy number, as shown in Table 1. We adopt the 1–9 scaling method while evaluating.

For the sub-index $C_1$, the fuzzy judgment matrix $D_1$ about its three sub-indexes $C_{11}$, $C_{12}$, $C_{13}$ can be calculated. Similarly, for the sub-index $C_2$, $C_3$, $C_4$, $C_5$, their fuzzy judgment matrix $C_2$, $C_3$, $C_4$, $C_5$ can be calculated. First we calculate each value of fuzzy synthetic extent of five indexes $C_1$, $C_2$, $C_3$, $C_4$, $C_5$, respectively, then we can obtain that $\vee(S_1 \geq S_j)$ is 0.402, 0.438, 1, 0.827, respectively; $\vee(S_2 \geq S_j)$ is 1, 1, 1, 1, respectively; $\vee(S_3 \geq S_j)$ is 1, 0.856, 1, 1, respectively; $\vee(S_4 \geq S_j)$ is 0.853, 0.296, 0.297, 0.695, respectively; $\vee(S_5 \geq S_j)$ is 1, 0.628, 0.715, 1, respectively. At last the weight vector can be calculated: $\boldsymbol{W}' = (0.402\ 1\ 0.856\ 0.296\ 0.628)$. After normalized, the weight vector is $\boldsymbol{W} = (0.126\ 0.314\ 0.269\ 0.093\ 0.197)$. Similarly, the weight vectors of each level index can be calculated. The results are listed as follows: the normalized weight vector of three sub-indexes of both index $C_1$ and index $C_5$ is $\boldsymbol{W}_1 = \boldsymbol{W}_5 = (0.168\ 0.411\ 0.431)$; the normalized weight vector of four sub-indexes of $C_3$ and $C_4$ is $\boldsymbol{W}_3 = \boldsymbol{W}_4 = (0.511\ 0.261\ 0.099\ 0.129)$. After the weight vectors of each level index are confirmed, we evaluate their credibility according to five-level remark set. To emphasize the main influence factors chiefly and also give attention to the other ordinary factors, we choose the fuzzy arithmetic operator $M(\cdot, \oplus)$.

The primal work is fuzzy evaluation of single factor. For example, for the three sub-indexes of $C_{21}$ and the three sub-indexes of $C_{24}$, the experts construct corresponding judgment matrixes and do normalized transformation. Then the fuzzy evaluation result of sub-index $C_{22}$ estimated by experts is $\boldsymbol{B}_{22} = (0.36\ 0.32\ 0.24\ 0.06\ 0.02)$, and the evaluation result of sub-index $C_{23}$ is $\boldsymbol{B}_{23} = (0.41\ 0.29\ 0.17\ 0.09\ 0.04)$. So that the fuzzy evaluation result of $C_2$ can be calculated as: (0.43 0.27 0.18 0.08 0.04). Similarly, we precede the fuzzy evaluation of the other sub-indexes $C_1$, $C_3$, $C_4$, $C_5$ and obtain the evaluation results $\boldsymbol{B}_1$, $\boldsymbol{B}_3$, $\boldsymbol{B}_4$, $\boldsymbol{B}_5$. At last the fuzzy comprehensive evaluation result of the simulation system is $\boldsymbol{B} = (0.34\ 0.39\ 0.21\ 0.04\ 0.02)$. Because the largest number of element in $\boldsymbol{B}$ is 0.39, and its corresponding credibility level is more credible, so that based on the maximum membership degree law, the credibility evaluation level of simulation system is very credible.

**Table 1** The judgment matrix $D$ of main index $C$

| Sub-index | $C_1$ | $C_2$ | $C_3$ | $C_4$ | $C_5$ |
|---|---|---|---|---|---|
| $C_1$ | (1, 1, 1) | (1/4, 1/3, 1/2) | (1, 2, 3) | (1/3, 1/2, 2/3) | (2/3, 1, 3/2) |
| $C_2$ | (2, 3, 4) | (1, 1, 1) | (3/2, 3, 4) | (2, 4, 5) | (1/3, 1, 3/2) |
| $C_3$ | (1/3,1/2,1) | (1/4, 1/3, 2/3) | (1,1,1) | (4, 5, 6) | (5/2, 3, 7/2) |
| $C_4$ | (3/2, 2, 3) | (1/5, 1/4, 1/2) | (1/6, 1/5, 1/4) | (1, 1, 1) | (1/4, 1/3, 1) |
| $C_5$ | (2/3, 1, 3/2) | (2/3, 1, 3) | (2/7, 1/3, 2/5) | (1, 3, 4) | (1, 1, 1) |

## 5 Conclusions

In this paper we introduce the composition of SOISS, the corresponding functions of each subsystem, and the simulation principles and method. We combine the M&S developing process and V&V, build the evaluation system framework, and adopt the evaluation method based on FAHP and the fuzzy comprehensive evaluation method. The evaluation experiment result shows that SOISS is credible. Our method and the hierarchical structure of credibility evaluation is reasonable and feasible, it can overcome the adverse effects of human subjective judgment and preference to decision in some degree. Therefore the method has certain application value. The research on simulation credibility can increase the correctness of the simulation results and reduce the risk of applying optical imaging simulation system; it can promote the in-depth research on software engineering, system testing and evaluation, simulation accuracy, and other issues.

We have already completed a number of simulation tests by using SOISS, and the simulation results are reasonable and reliable. The design and realization of SOISS can provide reference for the development of space target photoelectric imaging software which is homemade, practical, and multifunctional. As a credible and flexible simulation tool, SOISS can be helpful to the spread and application of corresponding research.

In fact, space-based optical imaging system is a complex, sophisticated satellite payload. By now it is not easy to build the correct and complete mathematical model of the random factors, which may lead to the degradation of image quality in actual imaging process. Therefore the simulation credibility remains to be further improved. In addition, how to find out the important influencing factors and weak parts of SOISS credibility, by utilizing this evaluation method, and how to improve the instructional and reference values of the evaluation work on the maintenance and upgrade of SOISS, are also main contents of the next research work.

## References

1. Han, Y., Sun, H.Y.: Advances in space target space-based optical imaging simulation. Infrared aLaser Eng. **41**, 3372–3378 (December 2012)
2. Wang, J.F., Chen, Z.T., Hu, X.Y., et al.: Research on method of high-precision 3D scene optical remote sensing imaging simulation. J. Mod. Opt. **66**, 1859–1870 (November 2019)
3. Ahmed, F., Kilic, K.: Fuzzy analytic hierarchy process: a performance analysis of various algorithms. Fuzzy Sets Syst. **3621**, 110–128 (May 2019)
4. Wu, J., Wu, X., Gao, Z.: Fuzzy synthesis evaluation of modeling & simulation credibility for complex simulation system. Comput. Integr. Manuf. Syst. **16**, 287–292 (February 2010)
5. Liu, Y., Ni, W., Ge, Z.Q.: Fuzzy decision fusion system for fault classification with analytic hierarchy process approach. Chemom. Intell. Lab. Syst. **16615**, 61–68 (July 2017)
6. Alaqeel, T.A., Siddharth, S.: "A fuzzy analytic hierarchy process algorithm to prioritize smart grid technologies for the Saudi electricity infrastructure", *Sustainable Energy*. Grids and Networks. **13**, 122–133 (March 2018)

7. Han, Y., Sun, H.Y., Li, Y.C., Guo, H.C.: Fast calculation method of complex space targets' optical cross section. Appl. Opt. **52**, 4013–4019 (May 2013)
8. Han, Y., Lin, L., Sun, H.Y.: Modeling space-based optical imaging of complex target based on the pixel method. Optik. **126**, 1474–1478 (April 2015)
9. Bologa, O., Breaz, R., Racz, S.: Using the analytic hierarchy process (AHP) and fuzzy logic to evaluate the possibility of introducing single point incremental forming on industrial scale. Proc Comput Sci. **139**, 408–416 (2018)
10. Ponciano, L., Brasileiro, F.: Agreement-based credibility assessment and task replication in human computation systems. Futur. Gener. Comput. Syst. **87**, 159–170 (October 2018)
11. Junior, F.R.L., Osiro, L., Carpinetti, L.C.R.: A comparison between fuzzy AHP and fuzzy TOPSIS methods to supplier selection. Appl. Soft Comput. **21**, 194–209 (August 2014)
12. Petkovic, J., Sevarac, Z., Jaksic, M.L., Marinkovic, S.: Application of fuzzy AHP method for choosing a technology within service company. Tech Technol Educ Manag. **7**, 332–341 (January 2012)

# Technical Characteristics and Application of Visible and Infrared Multispectral Imager

**Yanhua Zhao and Yan Li**

**Abstract** The Visible and Infrared Multispectral Imager (VIMI), an important earth observation payload onboard the GF-5 satellite, has the characteristics of more bands, wide spectral range, higher spatial resolution, and higher radiometric calibration precision. The imager has an 8 years design life on orbit and operates in a push-room imaging fashion to acquire earth data from 0.45 μm to12.5 μm over 60 km swath. The nadir spatial resolution is 20 m and the absolute radiometric calibration precision is less than 2.6% for reflected solar bands (B1–B6). For thermal emissive bands(B7-B12), the nadir spatial resolution is 40 m and the absolute radiometric calibration precision is less than 0.9 K@300 K. The high integration and advanced technical indicators have greatly improved China's multispectral optical remote sensor in respects of spectral number, spatial resolution, radiation resolution, and radiation calibration accuracy. This paper summarizes the technical characteristics of the VIMI and the on-orbit test of the applied products.

**Keywords** Hyperspectral · Visible and infrared multispectral imager · Remote sensing · GF-5 satellite · Application

## 1 Preface

The main purpose of the "GF-5" satellite is to obtain high-resolution remote sensing data of spectrum from ultraviolet to long-wave infrared with six remote sensing instruments loaded on the satellite. As applications in national pollution reduction, environmental quality supervision, air composition and climate change monitoring, land and resources investigation, and remote sensing monitoring are carried out as demonstration. Objects monitored include pollution gases, greenhouse gases,

Y. Zhao (✉) · Y. Li
Beijing Key Laboratory of Advanced Optical Remote Sensing Technology, Beijing Institute of Space Mechanics and Electricity, Beijing, China



H. P. Urbach, Q. Yu (eds.), *6th International Symposium of Space Optical Instruments and Applications*, Space Technology Proceedings 7,
https://doi.org/10.1007/978-3-030-56488-9_17

regional ambient air quality, water environment and ecological environment, air composition, climate change, geological and mineral resources surveys, etc.

The Visible and Infrared Multispectral Imager (VIMI) is one of the main payloads of the "GF-5" satellite. The payload is acquired with spatial resolution of 20/40 m and swath of 60 km. With spectral radiation information from visible light to long-wave infrared, comprehensive monitor of water, ecological environment, and land resources can be achieved to meet the needs of environmental protection, monitoring, supervision, emergency response, evaluation, planning, and resource survey.

## 2 Technical Characteristics and Advancement

The technical characteristics and advancement of VIMI are as follows:

(a) The system is complex with high indexes for the first time in China. The complexity is shown in the following aspects. To realize spectral subdivision, several methods are adopted including splitting optical field of view, setting dichroic films in the converging light path, and using micro-combination filter on the focal plane.

The system has three focal planes for in total 12 spectrum bands, and several calibration methods are adopted including full optical path calibration, full aperture diffuse reflector calibration, and full aperture blackbody calibration. Multiple focal planes and calibration call for high reliability and very complex system implementation. The spatial resolution is high, especially in long-wavelength infrared spectral band acquired with 40 meters resolution, which is the highest resolution for payloads of the same field in the world. The VIMI is highly competitive compared with similar foreign loads. The VIMI has high radiation resolution up to better than 0.2 K owing to detection in four split windows in the long-wave channel. High temperature inversion accuracy can be achieved by the method. The payload has wide spectral coverage from visible to thermal infrared (0.45um ~ 12.5um) and this is the widest spectral range in China's high-resolution multispectral remote sensing.

(b) For the first time, an off-axis three-mirror system of visible and infrared waveband is realized with characteristics of shared front-end optics, re-imaging, and three-mirror astigmatism technology. Utilizing methods like splitting field of view, dichroic film, and combined filter, a highly integrated system is realized which has three channels and 12 spectral wave bands covering visible light, short wave infrared, medium wave infrared, and long-wave infrared. By mainly using mirrors instead of lens, system transmittance is high, with 0.90 for visible channel and 0.80 for infrared channel (Fig. 1).

(c) Breakthrough has been made in key technical bottlenecks of infrared detectors and pulse tube refrigerators. For the first time, multispectral (four bands in 8–12.5 μm) and long linear (1536 pixel) HgCdTe IR detector and large cooling

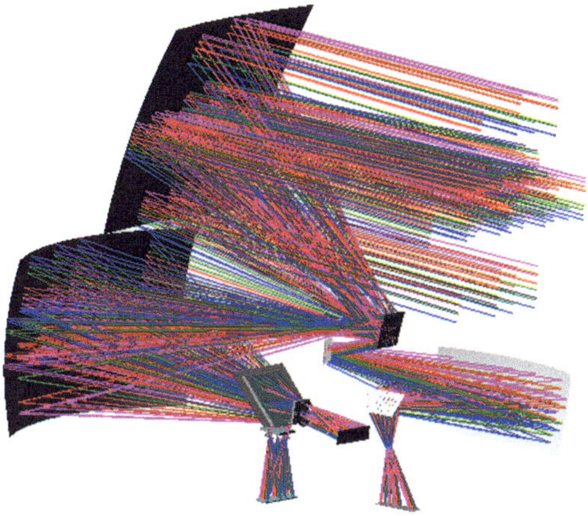

**Fig. 1** The optical system of VIMI

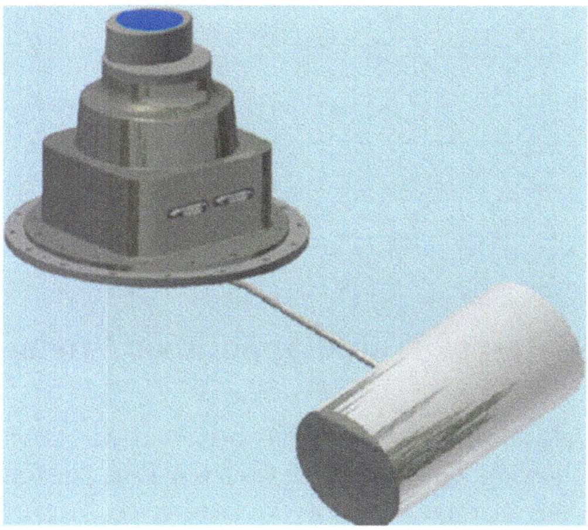

**Fig. 2** The three-dimensional model diagram of refrigerator for the infrared detectors

capacity (2.75 W@60 K) pulse tube refrigerator with eight-year life made in China start work on orbit, which lays a solid foundation for the development of infrared optical remote sensing technology in China (Fig. 2).

(d) Adopting the diffuse reflector and the blackbody onboard to achieve high-precision radiation calibration on orbit of different spectral bands. Full path full diameter on-orbit radiation calibration by using diffuse reflector is realized for the first time in China. The calibration accuracy is 2.6%, which is currently internationally advanced. Radiometer has been set to monitor the change of the diffuse reflector and emergency exit mechanism is set to solve the problem that the diffuse reflector may block the optical path (Fig. 3).

**Fig. 3** The model of diffuser mechanism

## 3 Test of Typical Application Products on Orbit

On orbit test evaluation has been carried out on application products focusing on following items: key lake reservoir water quality, inland large-scale water quality, water thermal pollution, vegetation coverage, lithologic structure interpretation, mineral and anomaly information extraction, mine surface feature classification, typical glacier groups and background snow monitoring of the Qinghai-Tibet Plateau, drought information extraction, local high temperature monitoring. The evaluation is based on comparison with similar mainstream loads (Landsat/TM, OLI, TIRS, MODIS, ASTER, GF-1 16 m camera) at home and abroad, and the method of authenticity testing in the field of typical application demonstration area.

### *3.1 Monitoring of Ambient Temperature and Drainage Around Nuclear Power Plants (Fig. 4)*

The images above are the remote sensing monitoring results of the temperature and intensity of the warm drainage water near the Dayawan nuclear power plant. Since the long-wave four-segment split window detection with a spatial resolution of

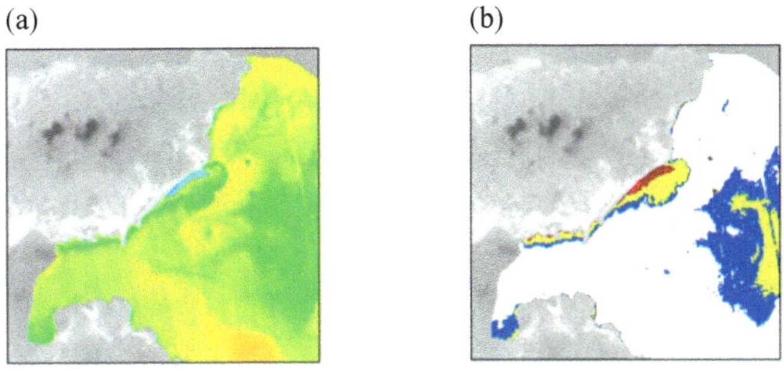


**Fig. 4** Nuclear power plant temperature drainage. (**a**) Distribution of sea surface water temperature near Dayawan Nuclear Power Plant. (**b**) Distribution of temperature and drainage intensity of Dayawan Nuclear Power Plant

40 m is realized for the first time in the world, water temperature inversion accuracy is improved and diffusion area of each level can be clearly identified. The temperature rise level recognition ability is better than that of Landsat-8/TIRS.

## 3.2 *Vegetation Coverage Information Extraction (Fig. 5)*

The above picture shows the vegetation coverage of the Poyang Lake National Nature Reserve in Jiangxi obtained by VIMI in September 2018. Vegetation coverage in the reserve is accurately reflected, and the coverage information extraction capability is equivalent to the GF-1/16 m imager.

## 3.3 *Lithology-Structural Interpretation (Fig. 6)*

Lithology is distinguished according to different image tones, and structural interpretation is carried out according to information such as homochromatic geological dislocation. From the lithology-structural interpretation results, it can be seen that the interpretation of the geological body boundaries, veins, and fault structure information is consistent with geological map of 1:200,000 scale, indicating that the data is great for lithologic structure interpretation.

## 3.4 *Extraction of Minerals and Anomalies (Fig. 7)*

Based on the thermal infrared band data, the relative abundance map of silica was inverted according to the absorption characteristics of silica near 8.5 μm. The principal component analysis method was used to extract the rock and mineral information

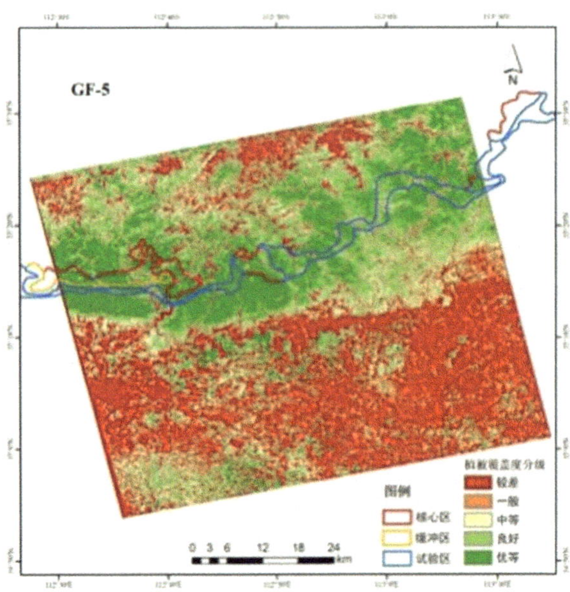


**Fig. 5** Remote sensing information extraction of vegetation coverage

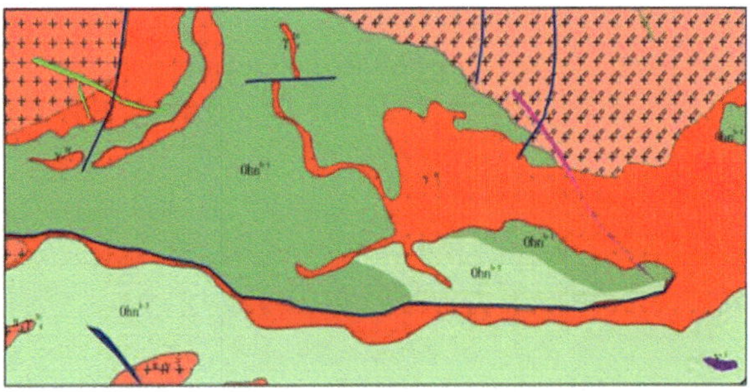

**Fig. 6** Lithology-structural interpretation

from the visible-near red data of the full spectrum, and iron ore information such as iron dyeing, hydroxylation, and solicitation is obtained. Compared with ASTER, the information extraction ability for hydroxyl minerals is almost the same, and information extraction ability for iron dyeing is better. The relative abundance of silica is consistent with result obtained by ASTER, with more refined and richer details.

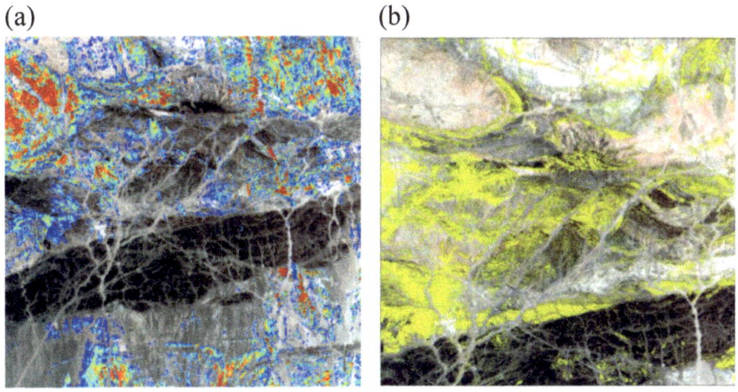


**Fig. 7** Extraction of mineral information from the Yingmao area in Gansu Province. (**a**) Inversion of silica content. (**b**) Alteration information (hydroxyl) extraction map

**Fig. 8** Dexing copper mine classification

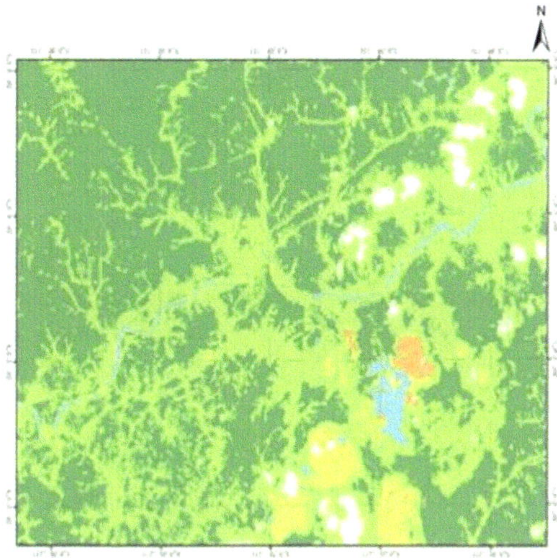


## 3.5 *Classification of Mine Features (Fig. 8)*

The VIMI data has clear details. Based on the data, the classification effect is good, and the accuracy is slightly better than that of landsat8/OLI. It means that the VIMI has the same ability as the international similar load, and can be used for classification of mine features.

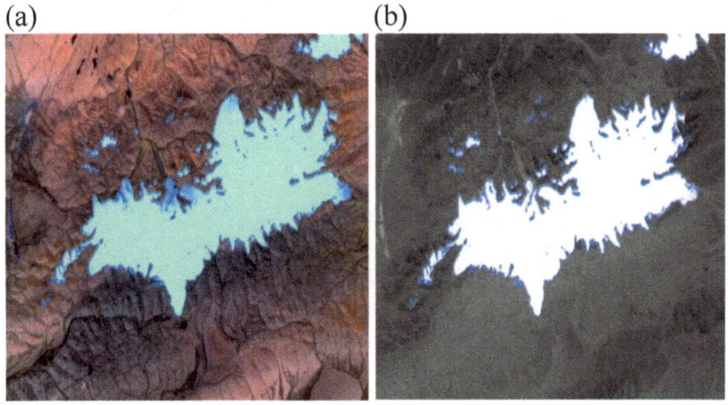

**Fig. 9** Typical Glacier Groups on the Tibetan Plateau and Background Snow Monitoring. (**a**) B5-B4-B3 band composite map. (**b**) Snow coverage and border vector

## *3.6 Typical Glacial Groups and Background Snow Monitoring in the Qinghai-Tibet Plateau (Fig. 9)*

Using VIMI's solar reflection band, the information on the Qinghai-Tibet Plateau glacial group and the background snow can be extracted, and the misjudgment of the water body is avoided, which has good applicability.

The information of glacier group and background snow cover is extracted from data of solar reflection band in the whole spectrum, which avoids misjudgment caused by water body and has good applicability.

## *3.7 Greenland Glacier Monitoring (Fig. 10)*

The above two figures are monitoring results of the Greenland glaciers obtained by VIMI. The amount of broken ice and the area of the glacial lake are accurately monitored in the left picture. Picture on the right shows the brightness temperature map of the broken ice, which can realize the full-day crushed ice and glacial lake monitoring. Data supports local weather forecasts, production, and living. VIMI can realize sea ice observations in the Arctic region for four times a day, with a single observation capability throughout the Arctic region, and can obtain remote sensing data of long-wavelength infrared spectrum in full-time, high spatial resolution, and the data may support Arctic channel monitoring and navigation decisions.

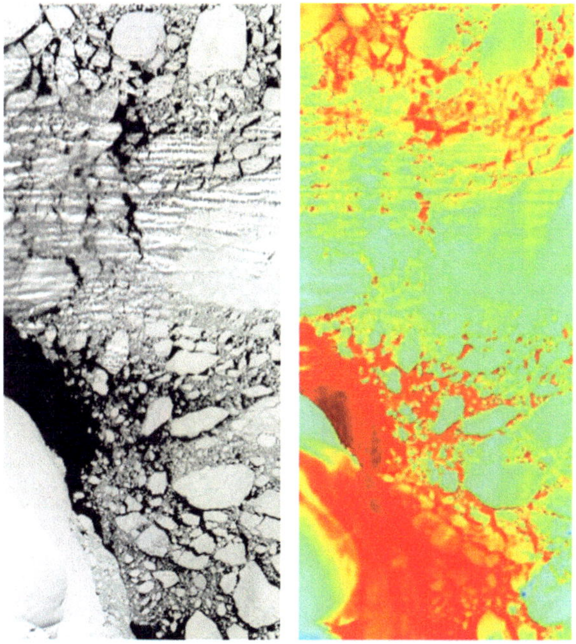

**Fig. 10** Greenland crushed ice true-color map

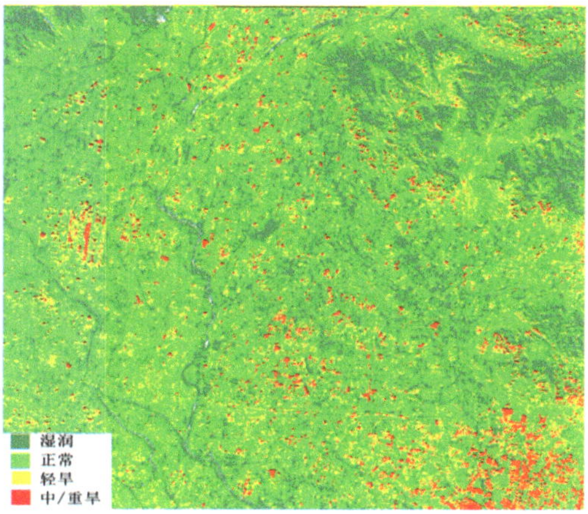


**Fig. 11** Drought remote sensing information extraction

## *3.8 Drought Remote Sensing Information Extraction (Fig. 11)*

The above picture shows the results of VIMI's drought monitoring in southeastern Beijing on October 2, 2018. The load can simultaneously obtain visible-thermal infrared images. The drought information extracted is in good agreement with Landsat/TM. The visible and thermal infrared images can be used to monitor the drought degree of land surface effectively.

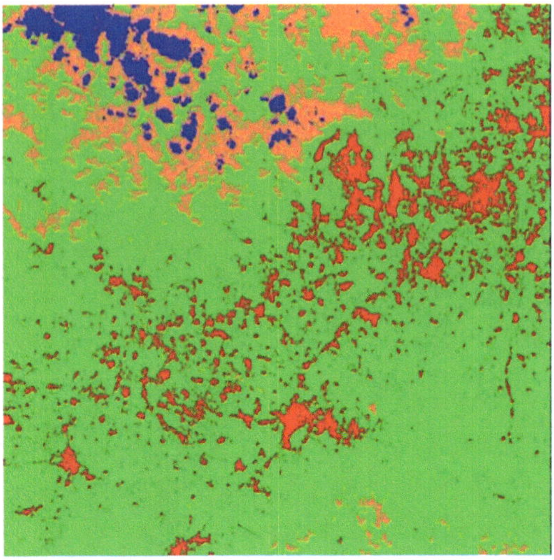

**Fig. 12** Local high temperature monitoring

### *3.9 Local High Temperature Monitoring (Fig. 12)*

The above picture shows the southwestern Beijing brightness temperature distribution map of 40 m spatial resolution obtained by VIMI. The grade distribution is clearly identified. Different color stands for different temperature: blue for areas under 292 K, orange for 292 to 294 K, green for 294–297 K, and red for over 297 K. Red represents high temperature in the image, which can effectively monitor local high temperature area.

### *3.10 Urban Heat Island Monitoring (Fig. 13)*

The spatial resolution of VIMI in thermal infrared spectrum is 40 m, which is the highest index in international civil satellites. It has the ability to observe surface temperature throughout the day. The images obtained during the day and night can provide high-precision quantitative remote sensing data for urban heat island effect monitoring, and serve the urban ecological environment monitoring and other fields.

## 4 Conclusion

VIMI captures remote sensing data in 12 spectral bands simultaneously, from visible to very long-wave infrared (0.45 ~ 12.5 μm). It is the first multispectral imager in China that covers bands including VIS/NIR/SWIR/MWIR/LWIR, and VLWIR. The

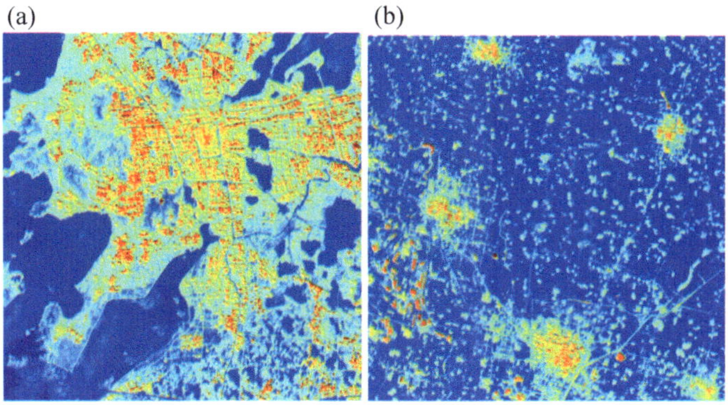


**Fig. 13** Urban heat island monitoring. (**a**) Suzhou (daytime). (**b**) A town in Tangshan (at night)

resolution at thermal infrared is 40 m, an international leading level and it is capable of all-day earth observation.

The spectral radiation information from visible light to long-wave infrared can comprehensively monitor water, ecological environment, and land resources to meet the needs of environmental protection, monitoring, supervision, emergency, evaluation, planning, and resource survey.

**About the Author** Zhao Yanhua, female, born in 1977, holds a master's degree in aircraft design from China Academy of Space Technology. Currently working at the Beijing Institute of Space Electromechanical Research as the chief designer of the high-resolution five-spectrum spectral imager. The research direction is the overall design of the aerospace optical remote sensor and radiation calibration technology research. Email: zhaoyh304@sina.com.

## References

1. Ke, Y., Im, J., Lee, J.: Characteristics of Landsat 8 OLI-derived NDVI by comparison with multiple satellite sensors and in-situ observations. Remote Sens. Environ. **164**, 298–313 (2015)
2. Barsi, J., Schott, J., Hook, S.: Landsat-8 thermal infrared sensor (TIRS) vicarious radiometric calibration. Remote Sens. **6**(11), 11607–11626 (2014)
3. Reuter, D.C., Richardson, C.M., Pellerano, F.A.: The thermal infrared sensor (TIRS) on Landsat 8: design overview and pre-launch characterization. Remote Sens. **7**(1), 1135–1153 (2015)
4. Estoque, R.C., Murayama, Y.: Classification and change detection of built-up lands from Landsat-7 ETM+ and Landsat-8 OLI/TIRS imageries: a comparative assessment of various spectral indices. Ecol. Indic. **56**, 205–217 (2015)
5. Yang, L., Cao, Y., Zhu, X., Zeng, S., Yang, G., He, J., Yang, X.: Land surface temperature retrieval for arid regions based on Landsat-8 TIRS data: a case study in Shihezi, Northwest China. J. Arid. Land. **6**(6), 704–716 (2014)

6. Zhu, L., Xingfa, G., Liangfu, C.: Comparison of LST retrieval precision between sigil-channel and split-window for high-resolution infrared camera. J Infrared Millimeter Waves. **27**(5), 346–353 (2008)
7. Gu, X.F., Tian, G.L., Tao, Y.: Radiation calibration principle and method of space optical remote sensors, pp. 172–175. Science Press, Beijing (2013)
8. Zhao, Y., Dai, L., Bai, S.: Design and implementation of full-spectrum spectral imager system. Aerospace Return Remote Sens. (3) (2018)

# Research on the Influence of Connecting Component on Dynamic Performance of Space Focusing Platform

**Penghui Cheng and Mengyuan Wu**

**Abstract** Dynamic characteristics of focusing mechanism affects greatly the reliability and accuracy of the space camera. It ensures the image quality of the camera and compensates for the defocus caused by complicated space environments. The goal of the research is to achieve dynamic simulation analysis of the focusing platform from a model space camera study the effect of connection component on the dynamics performance of the focusing platform. In this research, the modeling of the guide joint and the focusing platform is completed by finite element method. The dynamic parameters are obtained by the analytical formula based on elasticity theory. The accuracy of dynamic model is verified by simulation and sinusoidal vibration test. Finally, this research accomplished the dynamic analysis of space focusing platform with different coupling, researched the basis influence regular pattern of connection component on the dynamic characteristics of the focusing platform.

**Keywords** Space camera focusing platform · Dynamic analysis · Finite element method

P. Cheng
Xian Institute of Optics and Precision Mechanic, Chinese Academy of Sciences, Xi'an, Shaanxi, China

University of Chinese Academy of Sciences, Beijing, China
e-mail: chengpenghui2016@opt.cn

M. Wu (✉)
Xian Institute of Optics and Precision Mechanic, Chinese Academy of Sciences, Xi'an, Shaanxi, China
e-mail: wmy@opt.ac.cn



H. P. Urbach, Q. Yu (eds.), *6th International Symposium of Space Optical Instruments and Applications*, Space Technology Proceedings 7,
https://doi.org/10.1007/978-3-030-56488-9_18

# 1 Introduction

Focusing mechanism is a significant component to ensure the image quality of the camera, and compensate for the defocus caused by space environments. Its dynamic performance affects greatly the reliability and accuracy of the focusing mechanism. In addition, connection components have a very important influence on the dynamic characteristics of the focusing mechanism due to variety of motion parts and complex mechanical environment [1]. Understanding the influence regular pattern is of great significance to ensure the safety of space loads.

Coupling is the significant connection component in the focusing platform. In this research, the influence pattern of the coupling on dynamic characteristics of focusing platform is obtained by dynamic comparative analysis. In order to achieve dynamic analysis of platform, the dynamic parameters of different motion parts are integrated into the guide joint, and the dynamic model of guide joint is built by spring damping element. The dynamic parameters are obtained with the analytical formula based on elasticity theory. The reliability of the model is verified by experimentation. Finally, the research accomplished the dynamic simulation analysis of the focusing platform with different couplings, and obtained the influence regular pattern of the coupling and the dynamic characteristics of the focusing mechanism.

# 2 Create the Finite Element Model

## *2.1 Modeling of Focusing Platform*

The research's object is the space focusing platform of the focusing mechanism in a model space camera. The platform transmits the input signal decelerated to the ball screw through the coupling, converts the rotation into translation by the ball screw, the rolling guide as the guiding mechanism to carry the focal components for the specified displacement to accomplish focusing precisely. This paper builds the finite element model of the focusing platform with Hypermesh and MSC. Patran software. The finite element model of each motion parts keeps the entity exterior. The material properties of the model are shown in Table 1. The modeling results of the focusing platform are shown in Fig. 1.

**Table 1** Material property

| Material | Density (kg/m$^3$) | Elastic meodulus (GPa) | Poisson ratio |
|---|---|---|---|
| Ti-6Al-4V | 4429 | 104 | 0.31 |
| LY12-CZ | 2800 | 73 | 0.33 |
| 9Gr18-SUS304 (Stainless steel) | 8049 | 194 | 0.3 |

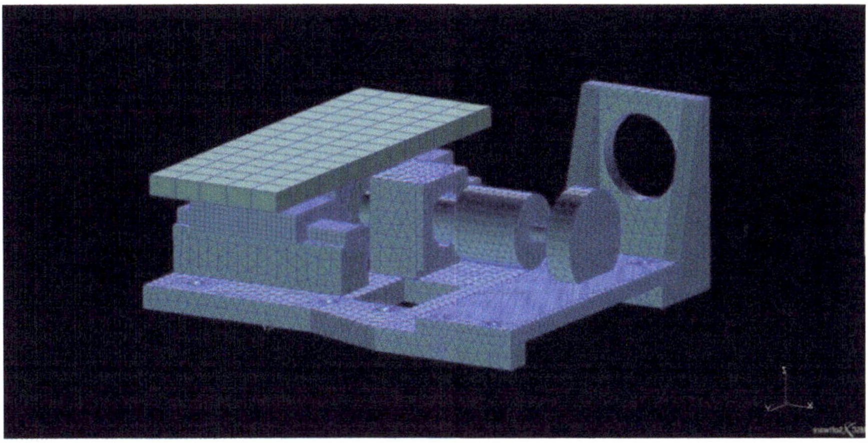

**Fig. 1** Simulation model of focusing platform

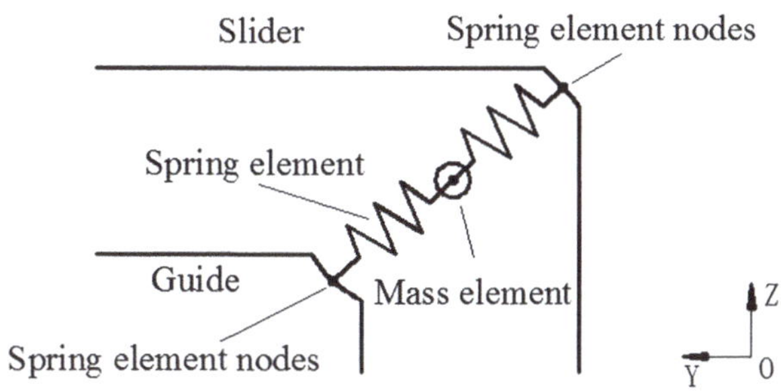


**Fig. 2** Framework of guide joint dynamic model

## 2.2 *Modeling of Guide Joint*

The guide is an entry point of research in this article. The method of modeling on guide joint affects the accuracy and reliability of the overall dynamic model greatly. Common modeling methods include analytical expression method, spring damper element method, contact element simulation method, joint element established method, and hypothetical material simulation method [2]. The spring damping element method is the most commonly used in motion joint model, the elastic damping characteristics of the joint can be simulated better, but it cannot simulate the mutual coupling characteristics between the balls, and it is difficult to simulate the nonlinear characteristics of the joint. However, this method still has high precision and reliability for the micro-guides for space applications.

The research simulates a single ball with 1D spring element and OD mass element. Building spring damping elements for each set of connection points on three translational degree of freedom. The X direction of the global coordinate is the slider movement direction, the Y direction is perpendicular horizontally to the slider movement direction, and the Z direction is perpendicular to the platform. The modeling of each connection points in the guide joint is shown in Fig. 2. Considering the similarity of contact forms, this research selects 12 sets of connection points for each guide [3].

# 3 Acquisition of Dynamic Parameters

## 3.1 Ball Stiffness in the X Direction of Freedom

The ball stiffness in the X direction of freedom depends on the system stiffness of the ball screw and the coupling in this platform. When the screw and the coupling are regarded as two elastic systems, the dynamic model is shown in Fig. 3.

The system stiffness of screw and coupling are equivalent to the joint of the guide. The stiffness $K_1$ of the single ball in the X direction of freedom can be obtained via (1) due to relationship of elastic systems. Where $K_1$ is the axial stiffness of the ball screw system, $K_2$ is equivalent translational stiffness of the coupling system.

$$\frac{1}{24K_X} = \frac{1}{K_1} + \frac{1}{K_2} \tag{1}$$

Axial stiffness of the ball screw system is based on the following assumptions: The axial dynamic deformation error and stiffness caused by the bending deformation of the screw weight is not considered in the axial stiffness of the ball screw system [4]. The effect of surface roughness on the stiffness of the joint is ignored in the ball screw system. The elastic deformation of the contact area is much larger than the surface roughness of the raceway or ball under the preload or working load. In addition, the ball and the raceway only produce elastic deformation, and the Hook theorem is obeyed. The deformation of the joint is within the material elastic range, and the contact point elastic deformation does not exceed one ten thousandth of the rolling element diameter, which is much smaller than the elastic deformation of the contact area [5]. Therefore, the axial stiffness of the ball screw system can be expressed via (2):

$$\frac{1}{K_1} = \frac{1}{K_\alpha} + \frac{1}{K_\beta} + \frac{1}{K_\gamma} \tag{2}$$

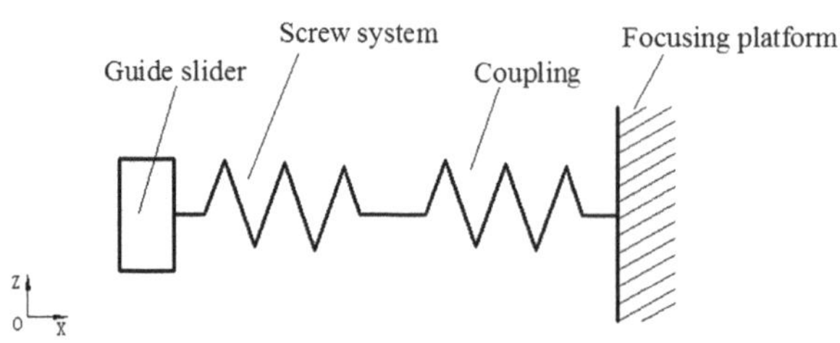


**Fig. 3** Framework of focusing platform dynamic model

where $K_1$ is the axial stiffness of the screw system; $K_1$ is the axial stiffness of the screw shaft; $K_1$ is the contact stiffness of the ball and the raceway; $K_1$ is the axial stiffness of the support bearing. This can be calculated via (3).

$$K_\alpha = \frac{A \times E}{1000 \times L}, \quad K_\beta = K\left(\frac{F_a}{0.3C_a}\right)^{\frac{1}{3}} \times 0.8, \quad K_\gamma \approx \frac{20}{3} Z \sin^2 \alpha D_\alpha^{\frac{1}{3}} Q^{\frac{2}{3}}, \quad \text{where}: Q = \frac{F_{a0}}{Z \sin \alpha} \tag{3}$$

where, $A$ is the cross-sectional area of the screw shaft; $E$ is the Young's modulus of the material; $L$ is the installation spacing; $F_a$ is the axial load of the screw; $C_a$ is the basic dynamic load rating of the screw; $Q$ is the axial load of the bearing; $D_\alpha$ is the diameter of the bearing ball; $Z$ is the number of balls of the bearing; $F_{a0}$ is the preload of the bearing; $\alpha$ is the initial contact angle of the bearing.

It can be known from (2) and (3) that $K_1$ is a constant value consequently. This focusing platform has small load and low preload, and the bearings are angular contact ball bearing. $K_1$ is calculated by checking the technical manual [6]. $K_1 = 2531460.8$ N/m.

The presentation of the stiffness of the coupling system is the torsional stiffness. The unit of torsional stiffness is N.m/rad. The physical meaning of this dimension is the torque value required when the torsion angle is 1 radians. It can also be understood that the force applied to a point on the coupling edge produce the torque and the elastic deformation of the arc length corresponding to the unit radians. In this way, the torque calculation formula and the arc length relationship are used to convert the torsional stiffness $K_\mu$ into a translational stiffness $K_2$, that is the (4). The equivalent translational stiffness dimension is N/m.

$$K_2 = \frac{K_\mu}{R^2} \tag{4}$$

## 3.2 Ball Stiffness in the Y Z Direction of Freedom

The ball stiffness of the guide in the Y Z direction of freedom mainly depends on the structural parameters of the guide and the elastic deformation of the ball. Therefore, dynamic parameters of the ball in the Y–Z direction of freedom are obtained by analytical formula based on elasticity theory. Assuming the contact area between the ball and the raceway of the guide is only deformed elastically. The steel ball normal stiffness can be obtained via (5).

$$K = \frac{dp}{d\delta} = \frac{3}{2}\left(\frac{\pi\left(0.6E^*\right)}{\sqrt{2}}\left(R'R''\right)^{\frac{1}{4}}\right)\delta^{\frac{1}{2}} \tag{5}$$

$$\text{where } E^* = \frac{1}{\frac{1-V_1^2}{E_1} + \frac{1-V_2^2}{E_2}};\ R' = R_2;\ R'' = \frac{1}{\frac{1}{R_2} - \frac{1}{R_1}} + \frac{R_1 R_2}{R_1 - R_2}$$

where $R_1$ and $R_2$ are the groove radius and ball radius, respectively, $E_1$, $E_2$, $V_1$, $V_2$ are the elastic modulus and Poisson's ratio of the ball and the guide, respectively, and $\delta$ is the ball deformation [7]. Normal contact stiffness $K$ can be calculated by checking the guide technical manual. The initial contact angle of double-row ball rolling guide is 45°; therefore, the stiffness of the guide ball in the Y–Z translational freedom is calculated: $K_Y = K_Z = 1.089 \times 10^7$ N/m.

## 4 Verification of Finite Element Model

Dynamic characteristics are inherent characteristics of the structure, including natural frequency, damping, and mode shapes. They are only related to the mass, stiffness, and material of the structure. Therefore, the dynamic characteristics of the structure can fully reflect the performance of the structure under difference dynamics condition. Verifying the finite element model by the means of comparing dynamic characteristics analysis results of simulation and sinusoidal vibration test. The stiffness of the guide ball in the Y–Z direction of freedom affects greatly the modal frequency of the focusing platform. And the stiffness in the X direction of freedom affects greatly the modal fundamental frequency. Therefore, the target of verification is the fundamental frequency of the focusing platform in this research.

Because the stiffness of the screw system can be controlled well. We have customized an aluminum alloy rigid coupling to verify the accuracy of finite element model better. The stiffness of rigid coupling can be regarded as extremely large. Therefore, the $K_X$ mainly depends on the stiffness of the screw system $K_1$ due to the relationship between $K_1$ and $K_2$ in the (1). $K_X$ is calculated by (1): $K_X = 1/24\ \ K_1 = 210955$ N/m.

Finally, the fundamental frequency of the sinusoidal vibration test of focusing platform when connecting the rigid coupling was 371 Hz, and the fundamental frequency of the model simulation was 385 Hz with an error of 3.7%. It shows that the model simulation has higher accuracy. The sinusoidal vibration test of the focusing platform is shown in Fig. 4.

**Fig. 4** Sinusoidal vibration test of the focusing platform

**Table 2** Database of couplings

| Coupling model | Type | Outer diameter (mm) | Length (mm) | Static torsional stiffness (N.m/rad) | Equivalent translational stiffness (N/m) | Mass(g) |
|---|---|---|---|---|---|---|
| LR-D-20*25 | Helical coupling | 20 | 25 | 52 | 520000 | 11.5 |
| SRB-16C | Parallel coupling | 16 | 21.5 | 75 | 1171875 | 8.2 |
| SRB-19C | Parallel coupling | 19.1 | 23 | 150 | 1662049 | 12 |
| WQ-C26L26 | Single membrane | 26 | 26 | 650 | 4082840 | 24 |

# 5 Dynamic Analysis of Space Focusing Platform

The dynamic parameters of motion parts are equivalent to the guide joint. The dynamic parameters of the finite element model can be obtained by the formula (1)–(5). Accomplishing dynamic simulation analysis of the focusing platform by MSC. Patran and Nastran software.

In the space focusing platform, the coupling is a key component for transmitting signals, and it is also a frail part for anti-vibration, which has great influence on the reliability of the focusing platform. There have been cases of failure in the past experiments. Therefore, the relationship between the coupling and the dynamic characteristics of the focusing platform is discussed and studied in this section.

The dynamic analysis result of focusing platform connected with different types and different stiffness couplings were compared, The dynamic analysis result of focusing platform connected with different types and different stiffness couplings were compared, including model LR-D-20*25 helical coupling, model SRB-16C, SRB-19C parallel couplings and model WQ-C26L26 single membrane coupling. The data of couplings is shown in Table 2.

**Table 3** Result of simulation analysis

| Coupling model | $K_1$(N/m) | $K_2$(N/m) | $K_X$(N/m) | $K_Y$, $K_Z$ (N/m) | Fundamental frequency(Hz) |
|---|---|---|---|---|---|
| LR-D-20*25 | 2531460.8 | 520000 | 64999 | $1.089 \times 10^7$ | 218 |
| SRB-16C | | 1171875 | 95500 | | 261 |
| SRB-19C | | 1662049 | 109390 | | 279 |
| WQ-C26L26 | | 4082840 | 130217 | | 305 |

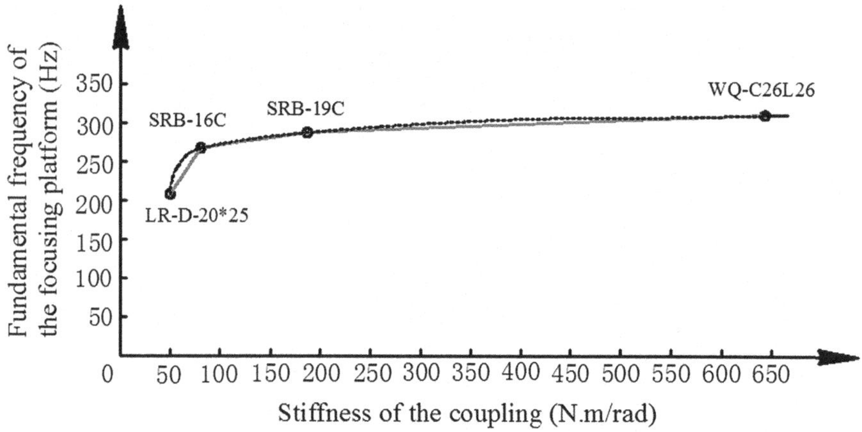


**Fig. 5** Relationship of couplings and fundamental frequency of the focusing platform

Table 3 shows the results of simulation analysis of the focusing platform when installed different couplings. The system stiffness of the coupling and the screw affect greatly first-order frequency of the focusing platform. The coupling stiffness is major factor that affects first-order frequency of the focusing platform, which is much greater than the effect of type, size, and mass of the coupling.

The relationship between the stiffness of couplings and the fundamental frequency of the focusing platform can be drawn according to the simulation results. It can be seen easily from Fig. 5 that the relationship between the couplings stiffness and the fundamental frequency of the focusing platform is positively correlated, and expressing the apparently nonlinearity. The impact on the platform's fundamental frequency tends to be flat with coupling stiffness increase. It has a drastic effect when the stiffness of the coupling is small. This regular pattern has important significance for the practical application of the space focusing platform. It provides a theoretical basis for the control of fundamental frequency of focusing platform.

## 6 Summary

The influence regular pattern that coupling on the dynamic characteristics of the space focusing platform is obtained by the dynamic comparative analysis of the focusing platform in this research. A dynamic analysis method for the linear focusing platform is proposed and verified including simulation method of guide ball, and equivalent method of different motion parts. The dynamic parameters of the guide and its relational expression are obtained comprehensively with the analytical formula based on elasticity theory and the relationship of elastic system. Verifying the simulation model by the experimentation. The research accomplished dynamic analysis of the focusing platform under different condition, and obtained the basic relationship between different couplings and the fundamental frequency of the focusing platform.

The stiffness of the coupling has an obvious non-linear positive correlation with fundamental frequency of the focusing platform is positively correlated and apparently nonlinearity. It has a drastic effect when the stiffness of the coupling is small, but the impact tends to be flat with coupling stiffness increase. The research provides a new basis and methodologies for further control of the dynamic stability of the focusing platform. The methodologies have a great guiding significance for controlling the fundamental frequency of the product, protecting the payload of the focusing platform effectively, and providing an important basis for the reliability and stability of the space camera focusing mechanism.

## References

1. Chen, S., Yang, B., Wang, H.: Missile and Aerospace Series - Satellite Engineering - Space camera design and experiment, vol. 1, 3rd edn, Aerospace Press, Reston, VA (2003)
2. Sun, W., Lu, M., Wang, B.: Research on analysis method of dynamics characteristics for linear rolling guide. Manufact. Technol. Mach. Tool. **3**, 48–53 (2011)
3. Zhao, Z.: Research on static stiffness modeling and dynamic characteristics analysis of linear rolling guideNortheastern University, Shenyang (2011)
4. Yang, F., Fan, Y.-x.: Calculation and analysis of ball screw drive rigidity based on ABAQUS. Mach. Building Automat. **40**, 27–30 (2011)
5. Jiang, S., Zhu, S.: Dynamic characteristic parameters of linear guideway joint with ball screw. J. Mech. Eng. **1**, 92–99 (2010)
6. Zhu, X.: Practical Rolling Bearing Manual1st edn, Shanghai Science and Technology Press, Shanghai (2010)
7. Yong-Sub, Y., Yong Young, K., Jae Seok, C.: Dynamic analysis of a linear motion guide having rolling elements for precision devices. Mech. Sci. Technol. **22**, 50–60 (2008)

# Design of Binary Optical Element Applied in Laser Shaping

**Xun Liu and Wei Li**

**Abstract** In this paper, the binary optical element (BOE) for excimer laser shaping is studied. The Gerchberg–Saxton (GS) algorithm is compiled to confirm the function of the pure-phase components. The basic unit of BOE is the Fresnel zone plate, and the structure is the single-layer compound eyes. According to the application, the parameters of BOE are determined as follows: the effective size of BOE is $33 \times 33$ mm$^2$, the number of arrays is $80 \times 80$, the step number is 8, and the cell aperture is 0.4125 mm. The related experiments have been made to measure its transmissivity and the ability of excimer laser shaping. Its energy transmissivity is about 85%, and it plays an obvious role in laser shaping, which makes the uniformity of laser energy better.

**Keywords** Uniformity · Laser shaping · Binary optical element · Plasma etching

## 1 Introduction

In Lidar technology and MEMOS technology, the excimer laser micro-processing has become a good means of micro-processing because of its many advantages [1]. In lithography mask experiment used by excimer laser, lithography resolution is the key factor to determine system performance. An effective line width is a function of object intensity, so the laser beam uniformity is one of the most basic factors. On the other hand, the uniform distribution of laser beam will directly impact on energy distribution of sensitive-light surface, which determines the quality of the production, so uniform distribution of laser beam is very important in the exposure system [2, 3].

However, because of the inherent characteristics of excimer laser beam, the uniformity of laser beam is very poor. The distribution of energy in X and Y directions is different, which is similar to a flattened distribution in X direction and a Gaussian distribution in Y direction. Therefore, the research on energy uniformity of excimer laser has been a hot issue concerning people's attentions [4], and a large number of

X. Liu · W. Li (✉)
Beijing Key Laboratory of Advanced Optical Remote Sensing Technology, Beijing Institute of Space Mechanics and Electricity, Beijing, China



H. P. Urbach, Q. Yu (eds.), *6th International Symposium of Space Optical Instruments and Applications*, Space Technology Proceedings 7,
https://doi.org/10.1007/978-3-030-56488-9_19

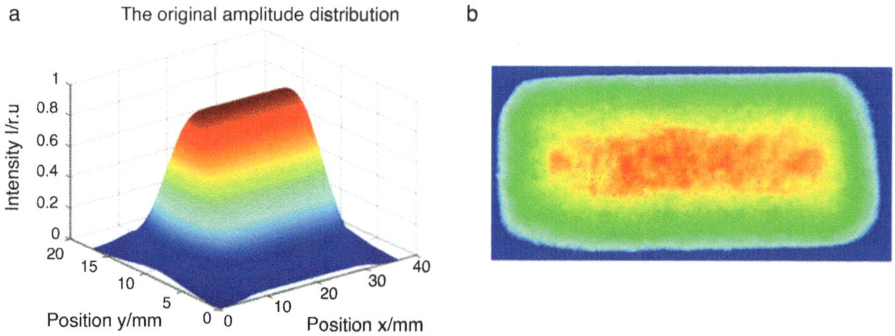


**Fig. 1** Wave surface and energy distribution of original spot. ((**a**) Wave surface, (**b**) Energy distribution)

theoretical and experimental works has been done based on several methods, for example, prism, waveguide, cylindrical lens, kaleidoscope, and Fly's-eye lens arrays.

In this paper, we put up the new method to design and make the Fresnel band wave lens based on the binary optical theory. The lens can be used in wave front shaping of excimer laser, which plays a vital role in microelectronic processing techniques (Fig. 1).

## 2 Theoretic Study

In the first part of this paper, a binary optic element is designed theoretically. The target of design is to obtain an optical field of Butterworth distribution. This problem can be summed up as in Fig. 2. The phase distribution function of the binary optical element is $\phi(x, y)$ (the transmission function is $\exp(i\varphi(x, y))$), so the incident laser $E(x, y)$ can be modulated to the output laser $E(x', y')$ by using the BOE. There are several algorithms based on this thought, such as Y-G algorithm, simulating annealing algorithm, Gerchberg–Saxton (GS) algorithm, and so on [5].

The GS algorithm is employed for realizing the optimization and MATLAB® was utilized to simulate the change of optical field. And we simulated the shaping result when the number of iteration was 106, also obtained the phase distribution of binary optic element. In Fig. 3, the results present the distribution of emergent optic field is Butterworth [6].

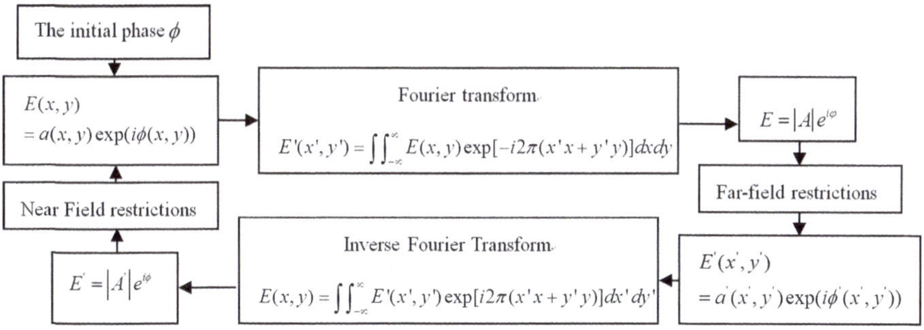


**Fig. 2** Flow graph of GS algorithm

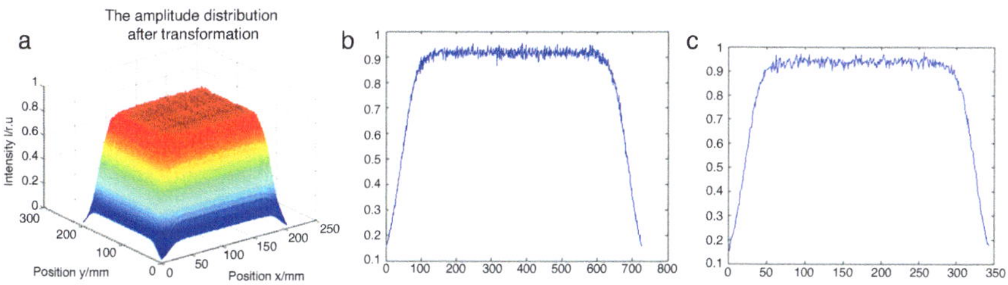


**Fig. 3** Analog results after 1,000,000 times iterative operation. ((**a**) Simulation result, (**b**) X direction's energy distribution, (**c**) Y direction's energy distribution)

# 3 Experiments

## *3.1 Construction of BOE*

Fresnel Zone Plates was selected for the basic unit. The structure was single-layer of compound eye. In order to be compatible with the excimer laser lithography mask system, that is, in the laboratory excimer laser beam will be expanded to the spot size of 30 mm × 30 mm, so the effective size of components was identified 33 mm × 33 mm, the number of arrays was 80 × 80, the number of main's band is 4.5, the number of steps is 8, the main focal length is 16.0 mm, the diameter of unit aperture is 0.4125 mm. To match up the KrF excimer laser UV wavelength, the quartz (JGS1) was selected as lens' material.

In the process of component's production, a procedure was compiled for calculation of radius value of various wave bands in unit array firstly. And then based on calculations, the professional software LEDIT® was used to draw three mask plates and finally produced as in Fig. 4. In this process, the number of main's band is important for the performance of binary optical element. When the number increases from 4.5 to 5, the more radiuses are added in the element, which makes the whole element have more functional zone and less blank zone. To obtain the 8-step con-

**Fig. 4** Scheme of three masks by LEDIT®

struction, all the three masks were used in the process, and the patterns of 2-step, 4-step, and 8-step were realized on the plates after employing the masks like left, middle, and right images, respectively, as shown in Fig. 4.

## *3.2 Fabrication of BOE*

In order to ensure the accuracy of mask alignment, decreasing the impacts brought by alignment's error to the maximum extent, alignment marks were added in both sides of all the mask plates. The alignment mark is composed of seven basic patterns on both sides of mask as shown in Fig. 5. Then, after repeated trials, the best exposure time is 13 s for quartz material. In this process, the photoresist cannot be adhered to the JGS1's surface, because the quartz is one kind of hydrophilic material. For resolving this problem, the Hexamethyl disilylamine is employed to adhere to the JGS1's surface advanced and then the photoresist is adhered to the Hexamethyl disilylamine. The exposure results are showed in Fig. 6.

Plasma etching technology was used to transfer the mask graphics to the quartz substrate eventually. The parameters of plasma etching were researched, such as the etching rate and etching time. According to the requirements of component's structure, the depth of etching was determined, and further the etching time was defined. After repeated experiments, etching rate of quartz by plasma is about 50 nm/min, the resultant etching time of three versions are 9 min, 4.5 min, and 2.25 min, respectively [1].

The BOE's pattern was observed by three-dimensional profiler instrument (WYKO NT1100, Veeco) in Fig. 7 and microscope in Fig. 8. We can see in Fig. 7 that the different colors stand for the various heights of surface. It is clear that each unit has been divided into five ring-belts, and there are eight steps of surface in each ring-belt. In Fig. 8, the pictures show that the Microscope photos of surface with different magnifications. The surface also was observed by SEM after steaming with 6 nm gold. From the observation's results. In Fig. 9, the structure of binary optical element is integrity, and the lines are clear, which shows that the fabrication of binary optical element is good.

**Fig. 5** The alignment mark in this design

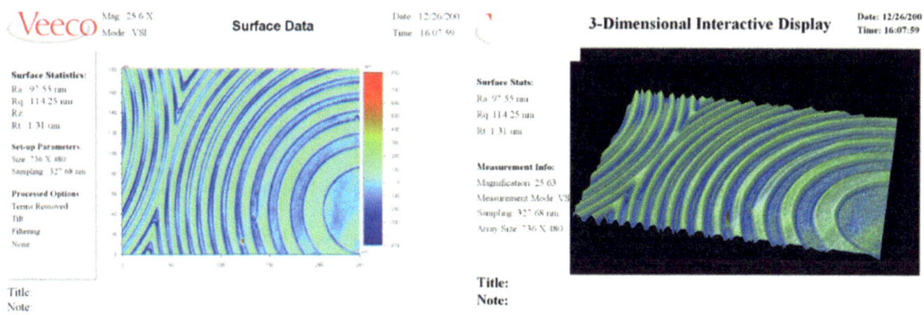


**Fig. 6** Results of parameters' experiment at 13 s of time

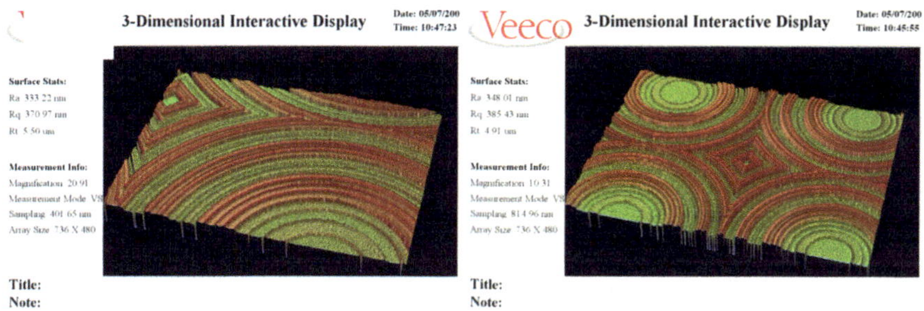


**Fig. 7** Results of third etching

## 3.3 *Measurement of BOE*

In order to understand the performance of BOE, its energy transmission was measured and compared with quartz rod arrays. It is found that the original energy transmission used by quartz rod arrays is about 77.0%, and the energy transmission of BOE is about 85.0%, with a significant increase [7]. The energy distribution in direction X and Y of original excimer laser beam is showed in Fig. 10. According to the characteristics of beam, for example multi-mode distribution, the X direction is a flattened distribution approximately, and the Y direction is similar to a Gaussian distribution.

Through experiment, it is found that the BOE has a good working performance. When the binary optical element is used in the lithography masking system, it plays an important role in the energy uniformity. Figure 11 shows the energy distribution of X and Y directions.

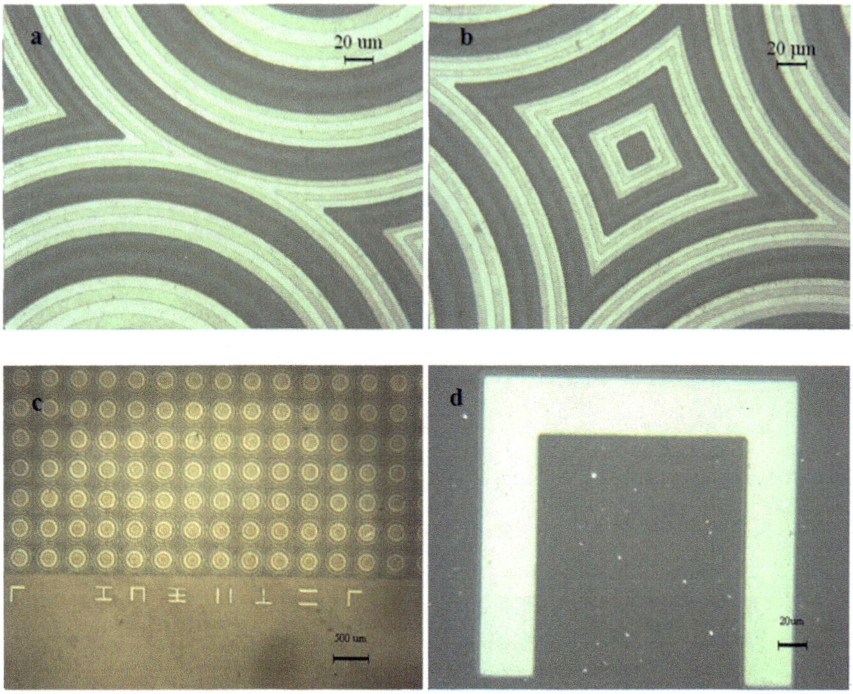


**Fig. 8** Pictures of BOE under microscope. ((**a**). Tangent area (**b**). Center area (**c**) Alignment mark series (**d**) Alignment mark)

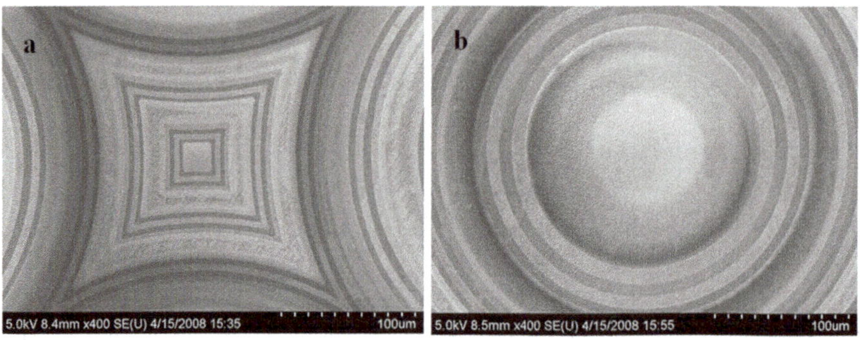


**Fig. 9** Pictures of BOE used SEM with 6 nm gold film

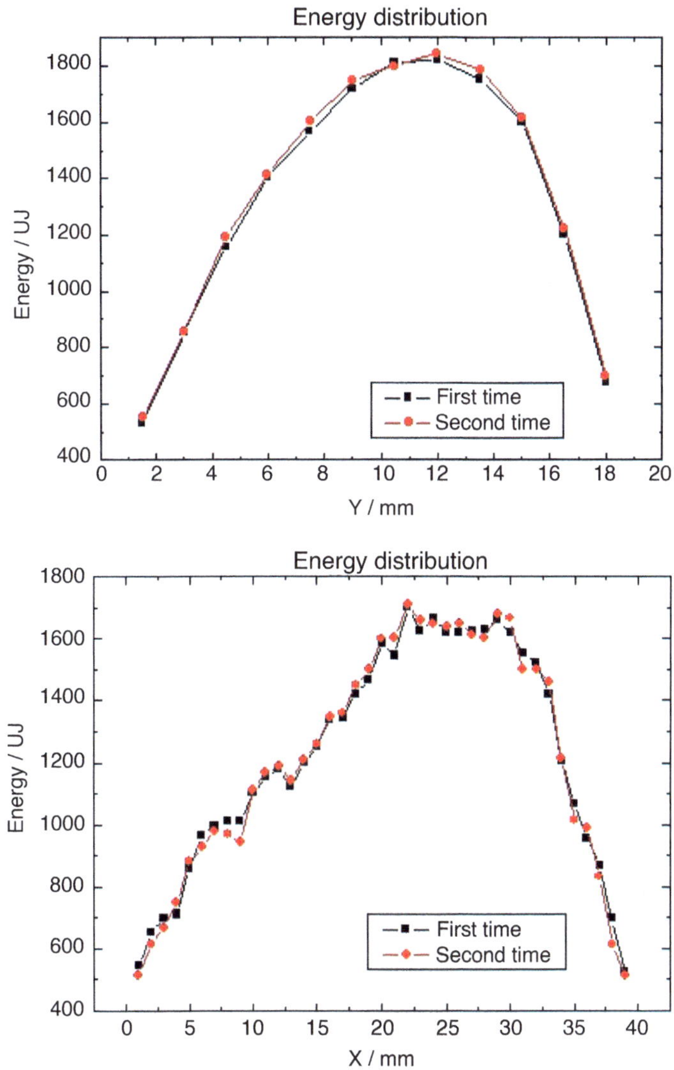


**Fig. 10** The intensity distribution of excimer laser in center line

We effect on the component's performance and make. The error analysis are made for knowing the influence of error involved in the process. There are four kinds of errors including system's etching error, random etching depth error, line-width error, and alignment error. The system's etching error and the alignment error are important factors to influence the component's performance. The system's etching error is caused by the inhomogeneous of line width and limitation of etching technology. The alignment error is caused by the deviation between masks in the

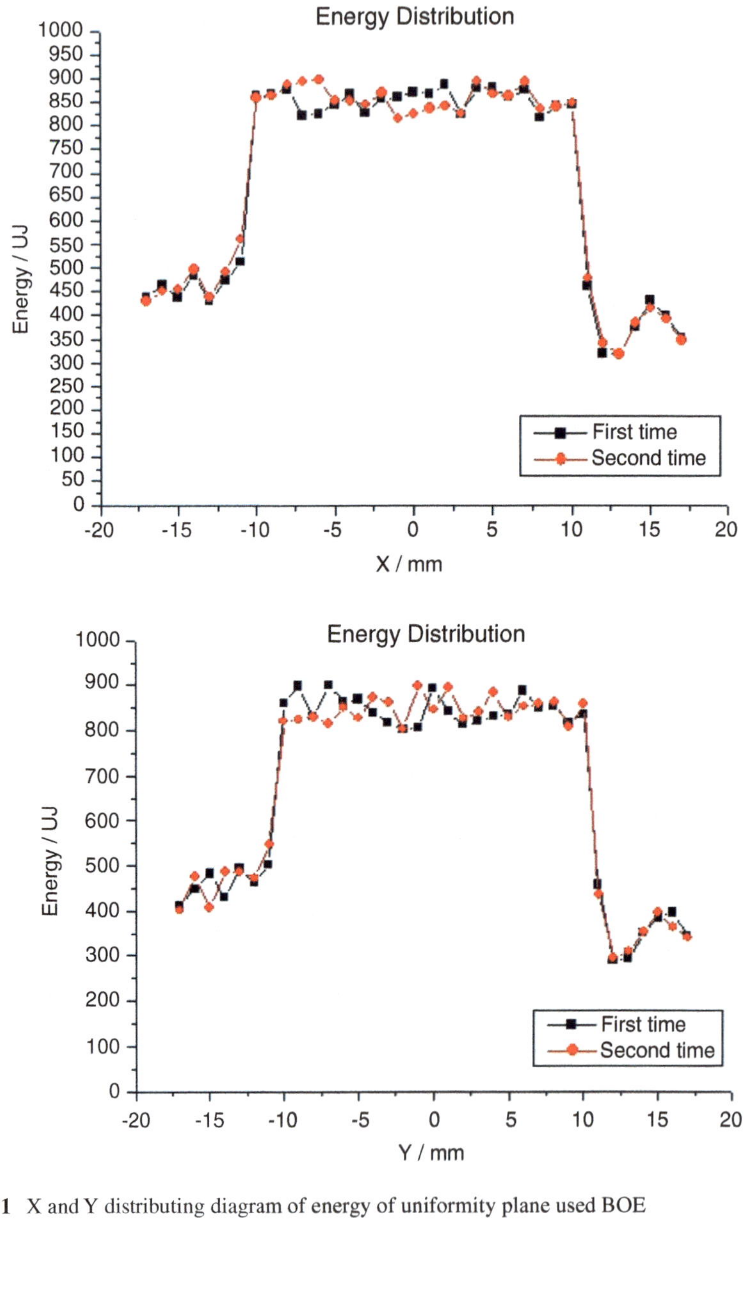


**Fig. 11** X and Y distributing diagram of energy of uniformity plane used BOE

process of alignment. In the actual production, the practical ways should be found to minimize error's impact, and realize the fabrication of high-precision components [8, 9].

# 4 Conclusion

In this paper, the BOE used in laser shaping has been researched. The BOE has higher energy efficiency and plays an obvious role in laser shaping. It also has advantages in the mediation, assembly convenience, and cost, which can be employed in Lidar and MEMOS. As the principle of binary optics is not the same as the traditional optics, the stability of BOE requires further study.

**Acknowledgement** This work is financially supported by The National Key Research and Development Program (2018YFB0504400).

# References

1. Cui, Z.: Overview of micro/nanofabrication technologies and applications. Physics. **35**(1), 34–39 (2006)
2. Cui, Z.: Micro-Nanofabrication Technologies and ApplicationsHigher Education Press, Beijing (2005) (in Chinese)
3. Chengde, L., Tao, C., Tiechuan, Z.: Design of fly's eye homogenizer for excimer laser micromachining. Chinese J. Lasers. **A26**(6), 560–564 (1999)
4. Guojun, D., Tao, C., Tiechuan, Z.: Lens array homogenizer for excimer laser. J. Otoelectronics·Laser. **16**(3), 279–281 (2005)
5. Guofan, J., Tan, Q., Yingbai, Y., et al.: Binary optics used in high power laser shaping. Eng. Sci. **2**(6), 27–32,39 (2000)
6. MATLAB. The Language of Technical Computing, Version 6.1.0.450 Release 12.1. May 18, 2001
7. Allen Cox, J., Werner, T.R., Lee, J.C., Nelson, S.A., et al.: Diffraction efficiency of binary optical elements. Proc. SPIE. **1211**, 116–124 (1990)
8. Farn, M.W., Goodman, J.W.: Effect of VLSI fabrication errors on kinoform efficiency. Proc. SPIE. **1211**, 125–136 (1990)
9. Li, W., Liu, X., Ruan, N., Su, Y., Yang, B.: Wireless power transmission of space solar power by using plasma channel. **35**(2), 87–93 (2014)

# Analysis of Straylight for the Green House Gas Instrument

**Long Ma**

**Abstract** The Green House Gas Instrument (GHGI) is an imaging spectrometer, which has four spectrum bands within 0.75 ~ 2.38 μm. This paper discussed the stray light performance of GHGI. Because the straylight requirement of GHGI is high, below 1%, the analysis considered not only the ghost and scattering of structure, but also the roughness and particle contamination of optic surface, overlapping of orders, and thermal radiation. According to the analysis results, we determined the straylight requirements of optical elements and structure parts.

**Keywords** Straylight · Ghost · Scattering of structure · The roughness · Particle contamination · Thermal radiation

## 1 Introduction

The Green House Gas Instrument (GHGI) is used to detect concentration distribution of global greenhouse gases, including $CO_2$, $CH_4$ and CO. It has four bands, B1:0.754 ~ 0.77 μm, B2:1.591 ~ 1.621 μm, B3:2.041 ~ 2.081 μm, B4:2.33 ~ 2.38 μm. The instrument prototype is developed by BISME and will be completed by the end of 2019. The schematic diagram and prototype are shown in Fig. 1.

The straylight requirement is below 1%, it is high. Because the requirement of general remote sensing cameras is below 3%. And most straylight came from leak light, ghost and scattering of structure. We should consider the roughness and particle contamination of optic surface, overlapping of orders, and thermal radiation.

L. Ma (✉)
Beijing Key Laboratory for Advanced Optical Remote Sensing Technology, Beijing Institute of Space Mechanics and Electricity, Beijing, P. R. China



H. P. Urbach, Q. Yu (eds.), *6th International Symposium of Space Optical Instruments and Applications*, Space Technology Proceedings 7,
https://doi.org/10.1007/978-3-030-56488-9_20

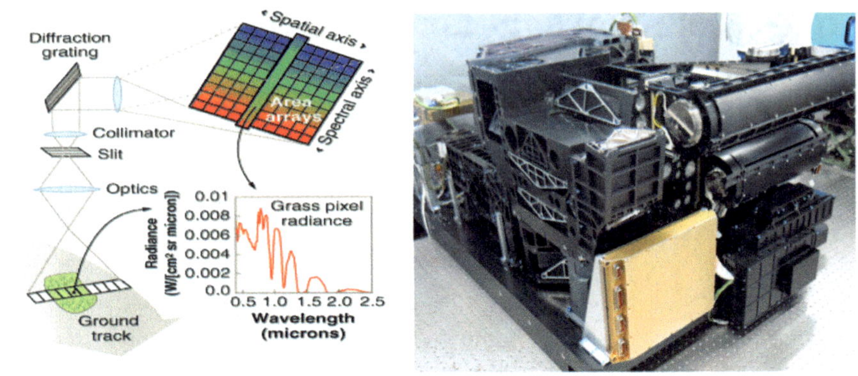


**Fig. 1** Schematic diagram and prototype of GHGI

**Table 1** B1 unwanted overlap wavelength and orders

| | Overlap order | | Wanted order | Overlap order | | | | |
|---|---|---|---|---|---|---|---|---|
| Grating order | −7 | −8 | −9 | −10 | −11 | … | −33 | −34 |
| Wavelength (nm) | 969.429 | 848.25 | 754 | 678.6 | 616.909 | … | 205.636 | – |
| | 990 | 866.25 | 770 | 693 | 630 | … | 210 | 203.824 |

## 2 Suppression of Overlapping of Orders

Each band in the optical system has a grating. Optical grating has high diffraction efficiency of multi-orders for multi-wavelength. In the range of optical material and the response of detector, more than one unwanted spectrum overlaps with operating wavelength, and unwanted spectrum has high efficiency at overlap orders. We can calculate the overlap orders at each band. For example, the spectral ranges of B1 detector is 200 nm–1100 nm, so overlap orders at B1 are shown in Table 1.

Before calculating the straylight of operating order, we need to suppress the straylight from overlap order. We coat the band-pass coating on every collimator lens of each band, as shown in Fig. 2. In order to reduce processing difficulty, high pass coating and low-pass coating are coated, respectively, on two surfaces of collimator lens. The results of design for B1 band-pass coating are given, as shown in Fig. 3. The band-pass coating will be imported to straylight simulation software.

## 3 Ghost

We simulate the ghost of optical system. The wavelength of light source inside the instrument is 300 nm ~ 3000 nm, and FOV is 0° or 3.6°. The ghost in B2 detector at FOV=3.6° is shown in Fig. 4. Other detectors have similar results. When the lens reflectivity of each band is 2%, 1%, 0.5%, the ratio between max illuminance of

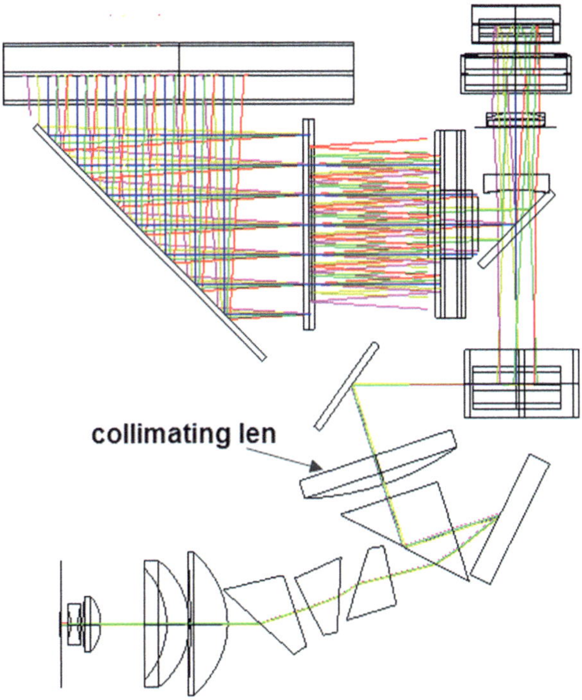


**Fig. 2** Band-pass coating on B1 collimator lens

**Fig. 3** The results of design for band-pass coating

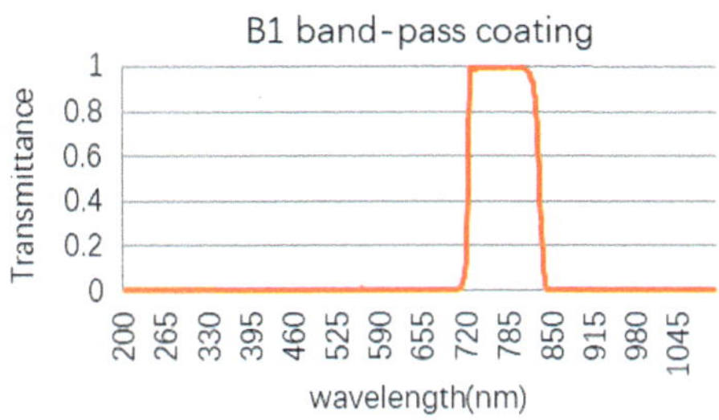


ghost and normal image is calculated as shown in Table 2. As it turns out, the lens reflectivity below 0.5% is necessary for suppressing ghost.

## 4 Scattering of Structure

We simulate stray light of scattering from structure. Light inside instrument is one FOV, such as 0*FOV, ±0.3*FOV, ±0.7*FOV, ±1*FOV. The wavelength is 300 nm ~ 3000 nm. Optical properties of structure surface are BSDF, which is black anodized aluminum. The range of reflectance is 4% ~ 18%. The simulation model

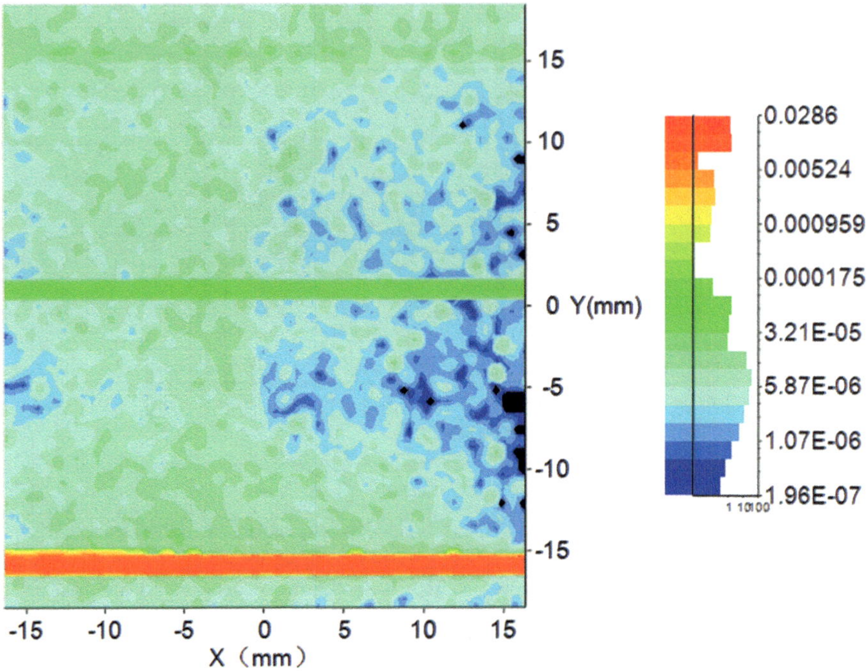


**Fig. 4** Simulated images on B2 detector at FOV = 3.6°

**Table 2** Ratio between max illuminance of ghost and normal image

| Band | 2% | 1% | 0.5% |
|---|---|---|---|
| B1 | 1.39% | 0.583% | 0.295% |
| B2 | 1.19% | 0.486% | 0.271% |
| B3 | 1.13% | 0.46% | 0.262% |
| B4 | 1.14% | 0.5% | 0.265% |

is shown in Fig. 5. And the result of simulation is shown in Table 3. As it turns out, B1 and B2 have a few FOV beyond 0.5%. Except for adding baffle to prevent scattering light, as shown in Fig. 6, we also can use new black paint, which the range of reflectance is below 2%.

## 5 Scattering of Surface Roughness

Scattering of surface roughness can be described by Harvey model or ABg model [1]. 2 nm and 5 nm roughness ABg parameters are shown in Table 4. Because TIS become smaller with increase of wavelength, so B1 is the maximum scattering band

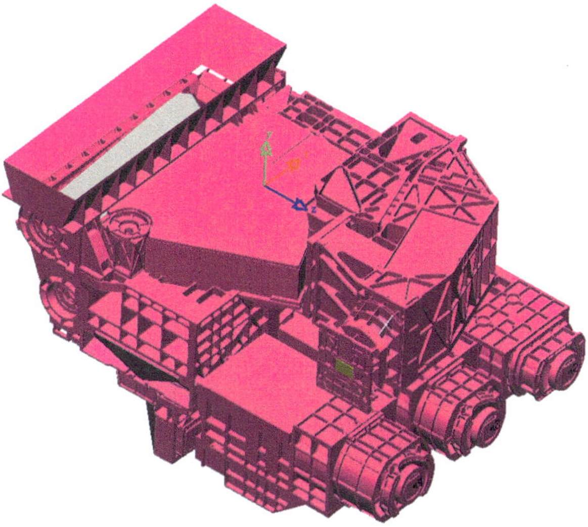

**Fig. 5** Simulation model of scattering from structure

**Table 3** Ratio between stay light and normal light on each band

| Band | 1*FOV | 0.7*FOV | 0.3*FOV | 0*FOV | +0.3*FOV | +0.7*FOV | +1*FOV |
|---|---|---|---|---|---|---|---|
| B1 | 0.37% | 0.52% | 0.49% | 0.56% | 0.48% | 0.46% | 0.37% |
| B2 | 0.20% | 0.34% | 0.40% | 0.70% | 0.42% | 0.46% | 0.28% |
| B3 | 0.19% | 0.27% | 0.30% | 0.49% | 0.42% | 0.26% | 0.18% |
| B4 | 0.06% | 0.12% | 0.08% | 0.33% | 0.08% | 0.11% | 0.06% |

of four. And we calculate the illuminance ratio of scattering to imaging of B1, as shown in Table 5. As the result turns out, the roughness of optical elements should be below 2 nm.

## 6 Scattering of Particle Contamination

Particle contamination can be described by two corrected Harvey model or ABg model [2–5]. CL230 and CL500 ABg parameters are shown in Table 6. Because TIS become smaller with increase of wavelength, so B1 is the maximum scattering band of four. And we calculate the illuminance ratio of scattering to imaging of B1, as shown in Table 7. As the result turns out, CL of optical elements should be below CL230.

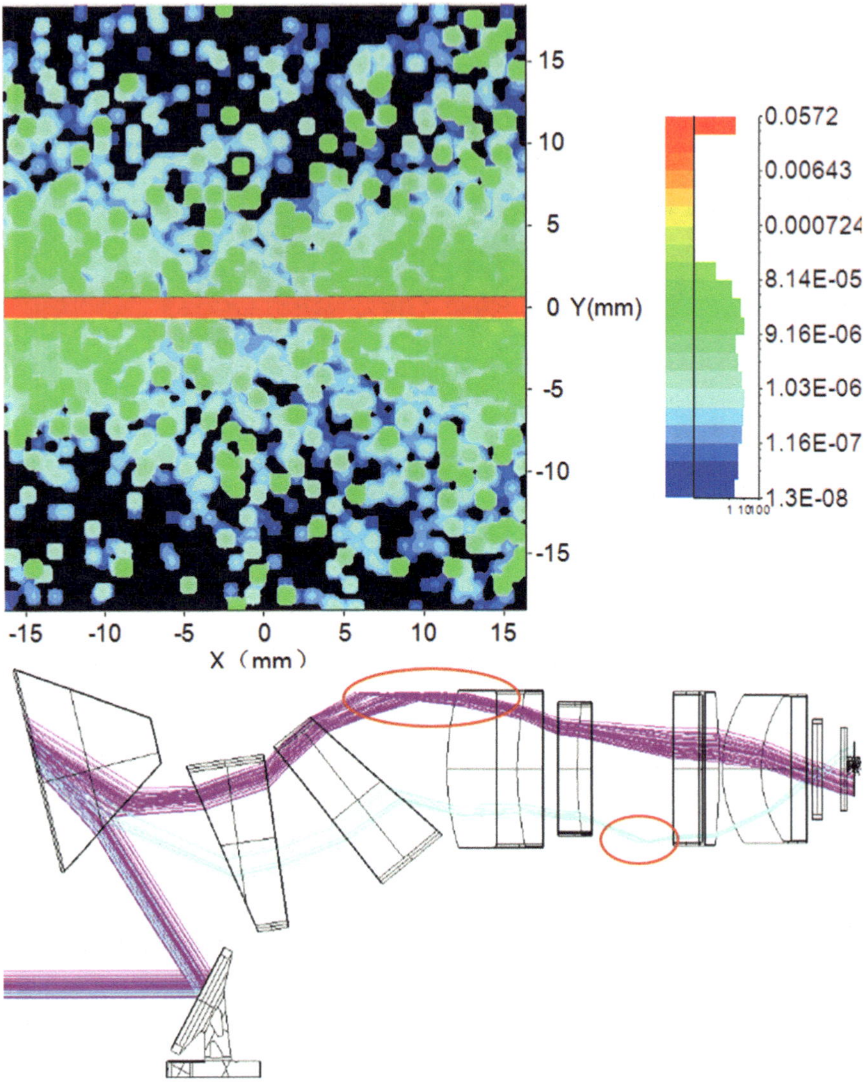


**Fig. 6** Simulation energy distribution of detector and the straylight path

**Table 4** ABg parameters of different roughness

| Roughness(nm) | A | B | g | TIS |
|---|---|---|---|---|
| 2 | 6.355e-5 | 1.188e-5 | 1.689 | 0.00011156 |
| 5 | 0.0003972 | 1.188e-5 | 1.689 | 0.00697 |

**Table 5** Illuminance ratio of scattering to imaging of B1

| Roughness(nm) | scattering | imaging | ratio |
|---|---|---|---|
| 2 | 1.6e-5 | 0.002638 | 0.6% |
| 5 | 9.74e-5 | 0.002646 | 3.8% |

**Table 6** ABg parameters of different Cleanliness Levels(CL)

| CL | A | B | g | TIS |
|---|---|---|---|---|
| 230 | 4.40e-6 | 2.549e-5 | 2.097 | 0.000182 |
| 500 | 3.848e-5 | 1.188e-5 | 1.689 | 0.00159 |

**Table 7** Illuminance ratio of scattering to imaging of B1

| CL | scattering | imaging | ratio |
|---|---|---|---|
| 230 | 3.56e-6 | 0.002741 | 0.13% |
| 500 | 3.09e-5 | 0.002693 | 1.1% |

## 7 Thermal Radiation

B2 ~ B4 are infrared spectrum, so we must take thermal radiation into account. Because the spectral response range of detector is 1 μm ~ 3 μm, we calculate the thermal radiation of each structure part near detector in 1 μm ~ 3 μm. And we add a low-pass filter to reduce thermal radiation on detector. The coating is high transmittance of incident light at 0° ~ 45°, as shown in Fig. 7. Then we simulate the thermal radiation on detector. According to the results of simulation, we change some key surfaces of mechanical structure to shine, as shown in Fig. 8. In this way, we decrease the surfaces emissivity, and reduce the radiation from the surfaces to detector. When the thermal radiation on detector is below $10^{-12}$, the temperature met the SNR requirements. According to the results of simulation, as shown in Table 8, when temperature is below 293 K, the thermal radiation on B2 detector will below $10^{-12}$; when temperature is below 243 K, the thermal radiation on B3 detector will below $10^{-12}$; when temperature is below 223 K, the thermal radiation on B4 detector will below $10^{-12}$.

## 8 Conclusions

Straylight analyses have been performed, and we determined the straylight requirements of optical elements and structure parts. Obviously, these requirements increase the difficulty of manufacture and test. If the straylight requirement becomes more stringent, below 0.1% for example, it is very difficult to achieve in the hardware.

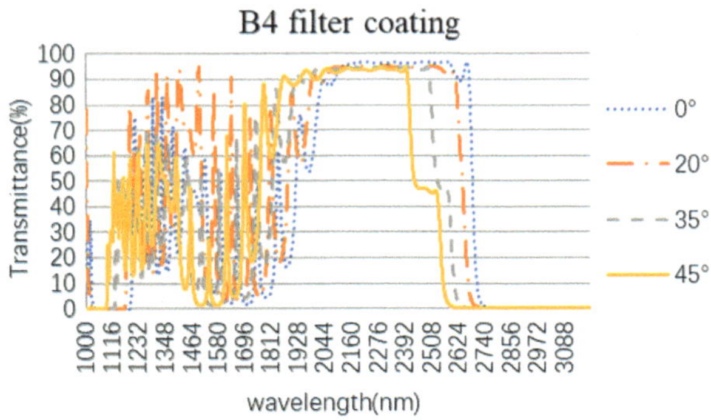


**Fig. 7** Low-pass filter to reduce thermal radiation

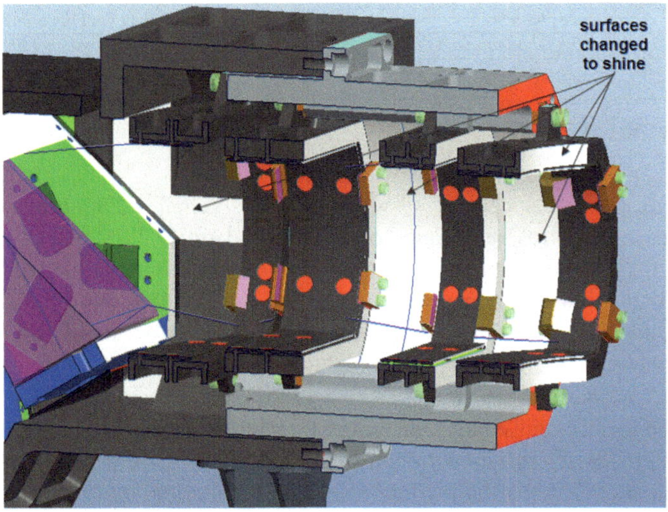


**Fig. 8** Change parts surfaces to shine

**Table 8** Simulation results of thermal radiation

| Band | Temperature (K) | Max illuminance (W/mm$^2$) |
|---|---|---|
| B2 | 293 | 2.05E-13 |
| B3 | 273 | 7.66E-12 |
| | 263 | 3.18E-12 |
| | 253 | 1.24E-12 |
| | 243 | 4.44E-13 |
| B4 | 253 | 1.43E-11 |
| | 243 | 5.77E-12 |
| | 233 | 2.16E-12 |
| | 223 | 7.39E-13 |

## References

1. Bennett, H.E.: Scattering Characteristics of Optical Materials. Opt. Eng. **17**(5), 480–488 (1978)
2. Young, R.P.: Mirror scatter degradation by particulate contamination. Proc. SPIE. **1329**, 246–254 (1990)
3. Kunjithapatham Balasubramanian, Stuart Shaklan, Amir Give'on. Stellar Coronagraph Performance Impact due to Particulate Contamination and Scatter. SPIE, 2009, 7440, 74400T
4. S. Ellis, R. N. Pfisterer. *Advanced Technology Solar Telescope (ATST)*: Stray and Scattered Light Analysis[DB/OL]. (2003)[2003-05-01]
5. R. Hubbard. *M1 Microroughness and Dust Contamination*[DB/OL], (2013)

# Design of Compact Long-Slit Wynne-Offner Spectrometer for Hyperspectral Remote Sensing

**Jiacheng Zhu and Weimin Shen**

**Abstract** This paper focuses on the compact long-slit Wynne-Offner imaging spectrometer suitable for wide-swath high-resolution remote sensing applications. Through tracing the chief ray, astigmatism of Wynne-Offner spectrometer was theoretically analyzed, and the expression of astigmatism was deduced for any object point. It was pointed that the anastigmatic region of Wynne-Offner spectrometer is much larger than that of the classic Offner spectrometer. Steps for designing a Wynne-Offner spectrometer are shown and two examples of long-slit Wynne-Offner spectrometers are given. One works in the wavelength range of 0.5–1.0 μm and another works in 1.0–2.5 μm. The image quality of both spectrometers was close to the diffraction limit, and distortions were negligible. Such Wynne-Offner spectrometers have the advantages of long slit, compact structure, and high imaging quality.

**Keywords** Spectrometer · Long slit · Remote sensing

## 1 Introduction

Imaging spectrometers are widely used in mineral exploration, environmental monitoring, agriculture, forestry, etc. For a large observation range, the imaging spectrometer is required to have long slit to increase swath width. In this paper, we are concerned with a particular spectrometer with long slit, compact structure, high imaging quality, and negligible distortions: the Wynne-Offner spectrometer.

Wynne-Offner spectrometer was developed from the classic Offner spectrometer. In 1973, the unit magnification Offner scanner was proposed by Abe Offner for

J. Zhu (✉)
Key Lab of Modern Optical Technologies of Education Ministry of China, Soochow University, Suzhou, Jiangsu, China

W. Shen
Key Lab of Advanced Optical Manufacturing Technologies of Jiangsu Province, Soochow University, Suzhou, Jiangsu, China



H. P. Urbach, Q. Yu (eds.), *6th International Symposium of Space Optical Instruments and Applications*, Space Technology Proceedings 7,
https://doi.org/10.1007/978-3-030-56488-9_21

lithography, it consists of two concentric concave mirrors and has optimal imaging performance in an annual field [1–3]. However, aberrations will increase if its slit is located outside of this annual field, so only short slit is allowed. In 1989, Wynne [2] widened the optimal imaging region by adding a concentric meniscus lens in front of the convex mirror to correct the spherical aberration of chief rays. Lobb [4] used this feature to design a compact spectrometer with wide spectral dispersion [5]. In 2014, X. Prieto-Blanco et al. proved that Wynne-Offner spectrometer can meet the telecentric condition at two different off-axis locations, and they pointed out that this spectrometer has the advantages of long slit, compact structure, and low distortions [6]. Recently, freeform spectrometers [7–12] with strong aberration correcting ability have become a new focus. However, processing and testing of the freeform gratings are quite difficult and hard to be widely applied in the short term. Wynne-Offner spectrometers contain only spherical optics. They can be developed more easily with lower cost and have been applied in hyperspectral remoting sensing [13, 14]. Current reports on Wynne-Offner spectrometers are mostly about optimized results of optical design software, but no analytical design based on the aberration theory has been reported. In Sect. 1 of this paper, we discussed the astigmatism of Offner and Wynne-Offner spectrometers through tracing the chief ray. It was pointed that the latter is suitable for long slit and its design procedure is given. In Sect. 2, two examples of long-slit Wynne-Offner spectrometers are given, which were proved superior to the classic Offner spectrometers in correcting aberrations and reducing volume. Summary and conclusions are given in Sect. 3.

## 2 Theoretical Analysis

To demonstrate the main aberration anastigmatism of the Wynne-Offner spectrometer, we derived the astigmatism expressions of Offner and Wynne-Offner spectrometers through tracing the chief ray. Astigmatism of these two spectrometer forms was compared and the anastigmatism advantage of the Wynne-Offner spectrometer was proved.

### *2.1 Astigmatism Analysis of Offner Spectrometer*

The Offner spectrometer comprises a primary mirror M1, a diffraction grating G, and a third mirror M3 with a common center C. Figure 1 shows the chief ray light path tracing from any object point on the slit, and the Cartesian coordinate system is established with point C as the origin. If the point O is not the midpoint of slit, the incident plane $y_{O'}Cz$ in the object side in Fig. 1a is non-coplanar with the emergent plane $y_{I'}Cz$ in the image side in Fig. 1b. In Fig. 1a, $\varphi$ is defined as the object squint angle. $I_{M1}$ is the meridional image and $I_{S1}$ is the sagittal image of M1. $\varphi'_{S1}$ is defined as the sagittal image squint angle. In Fig. 1b, $I_{S2}$ is the sagittal object of *M*3. $\varphi'_{S2}$ is

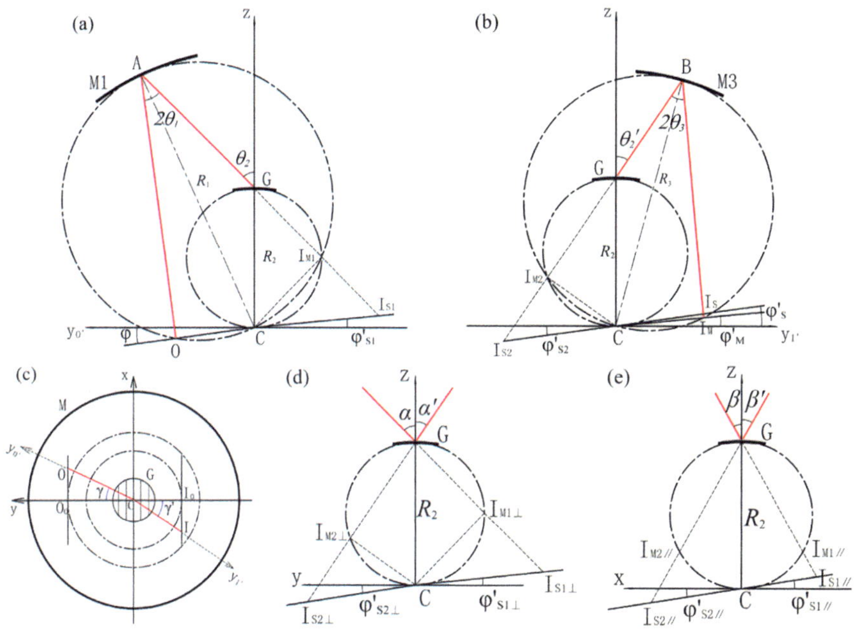

**Fig. 1** Lightpath for the chief ray of Offner spectrometer. (**a**) Path of incident light from any object point O on slit. (**b**) Path of emergent light showing the meridional and sagittal images. (**c**) Top view. (**d**) Projection in the grating's principal section. (**e**) Projection perpendicular to the grating's principal section

defined as the sagittal object squint angle of $M3$. $I_M$ is the meridional image and $I_S$ is the sagittal image of M3. $\varphi'_M$ is the meridional squint angle and $\varphi'_S$ is the sagittal image squint angle. Incident angles of the chief ray on each surface are indicated in Fig. 1a and b. According to Young's formula [15]:

$$\begin{aligned} \varphi'_{S1} &= \varphi \\ \varphi'_{S2} &= \varphi'_S. \end{aligned} \tag{1}$$

Radii of $M1$, $G$, and M3 are $R_1$, $R_2$, and $R_3$, respectively. When the meridional images meet Roland's condition.

$$OC = R_1 \sin\theta_1 = R_2 \sin\theta_2, \tag{2}$$

$$CI_M = R_2 \sin\theta_2' = R_3 \sin\theta_3, \tag{3}$$

$$\varphi = \theta_2 - 2\theta_1, \tag{4}$$

$$\varphi'_M = 2\theta_3 - \theta_2'. \tag{5}$$

Figure 1c shows the top view of the spectrometer. The incident and the diffractive azimuths on the grating are $\gamma$ and $\gamma'$, respectively. $O_0$ is the center of the slit. The grating equations are.

$$\begin{aligned} \sin\theta_2' \cos\gamma' - \sin\theta_2 \cos\gamma = mg\lambda \\ \sin\theta_2' \sin\gamma' - \sin\theta_2 \sin\gamma = 0, \end{aligned} \tag{6}$$

where $m$ denotes the diffraction order, $g$ denotes grating density, and $\lambda$ is the wavelength. As shown in Fig. 1d, e, diffraction on the grating can be, respectively, projected onto the principal section and the plane perpendicular to the principal section. The incident angle in Fig. 1d is $\alpha$, diffraction angle is $\alpha'$, meridional object is $I_{M1\perp}$, meridional image is $I_{M2\perp}$, sagittal object is $I_{S1\perp}$, sagittal image is $I_{S2\perp}$, sagittal object and image squint angles are $\varphi'_{S1\perp}$ and $\varphi'_{S2\perp}$, respectively. The incident angle in Fig. 1e is $\beta$, reflection angle is $\beta'$, meridional object is $I_{M1\|}$, meridional image is $I_{M2\|}$, sagittal object is $I_{S1\|}$, sagittal image is $I_{S2\|}$, sagittal object and image squint angles are $\varphi'_{S1\|}$ and $\varphi_{S2\|}$, respectively. Angles in Fig. 1e have the relationships:

$$\beta = \beta', \tag{7}$$

$$\begin{aligned} \tan\varphi'_{S1\|} = \tan\varphi'_{S1} \sin\gamma \\ \tan\varphi'_{S2\|} = \tan\varphi'_{S2} \sin\gamma', \end{aligned} \tag{8}$$

$$\sin\beta \tan\varphi'_{S1\|} = \sin\beta' \tan\varphi'_{S2\|}. \tag{9}$$

With Eqs. (1) and (7–9), the relationship between $\varphi'_S$ and $\varphi$ is

$$\tan\varphi'_S = \tan\varphi \cdot \frac{\sin\gamma}{\sin\gamma'}. \tag{10}$$

Astigmatism of Offner spectrometer can be expressed as:

$$\text{astig}_{\text{Offner}} = CI_M \tan\left(\varphi'_S - \varphi'_M\right). \tag{11}$$

From Eq. (11), if both $\varphi'_S$ and $\varphi'_M$ are 0, the Offner spectrometer is anastigmatic and means telecentric in the object space and in the image space. The coordinate of any object point in the object plane is defined as $O$ ($x$, $y$), and $k = \sqrt{x^2 + y^2}\ /R_2$, $k_1 = R_1/R_2$. According to Eqs. (2–6) and (10), with $k$ and $k_1$ as variates, we can deduce $\text{CI}_M$, $\varphi'_S$, and $\varphi'_M$ in Eq. (11) as the following expressions:

$$\text{CI}_M = kR_2 \frac{\sin\gamma}{\sin\gamma'}, \tag{12}$$

$$\varphi'_S = \arctan\left[\frac{\sin\gamma'}{\sin\gamma}\cdot\tan\left(\arcsin k - 2\arcsin\frac{k}{k_1}\right)\right], \tag{13}$$

$$\varphi'_M = 2\arcsin\left(\frac{k}{k_1}\cdot\frac{\sin\gamma}{\sin\gamma'}\right) - \arcsin\left(k\cdot\frac{\sin\gamma}{\sin\gamma'}\right). \tag{14}$$

For a specific Offner spectrometer, we assumed that $R_1 = R_3$, $g = 200$ lp/mm, $m = -1$, $\lambda = 600$ nm. We preferred $k_1$ and the off-axis magnitude of slit. When $k_1 = R_1/R_2 = 1.91$, $y/R_2 = 0.605$, it is anastigmatic at the center of slit, but its astigmatism increases rapidly as the slit length increases, as shown in Fig. 2.

## 2.2 *Astigmatism Analysis of Wynne-Offner Spectrometer*

The Wynne-Offner spectrometer consists of a concave mirror *M*, a convex diffraction grating *G*, and a meniscus mirror *L* with a common center *C*. *G* is directly etched into the convex surface of *L*. The concave and convex radii of L are $R_1$ and $R_2$, respectively, and the radius of M is $R_3$. Chief ray light path tracing from any object point on the slit is shown in Fig. 3. In Fig. 3a, $O_1$, $O_2$, and $O_3$ are Rowland circles of *L*'s concave surface, convex surface, and *M*. *E* and *F* are virtual images

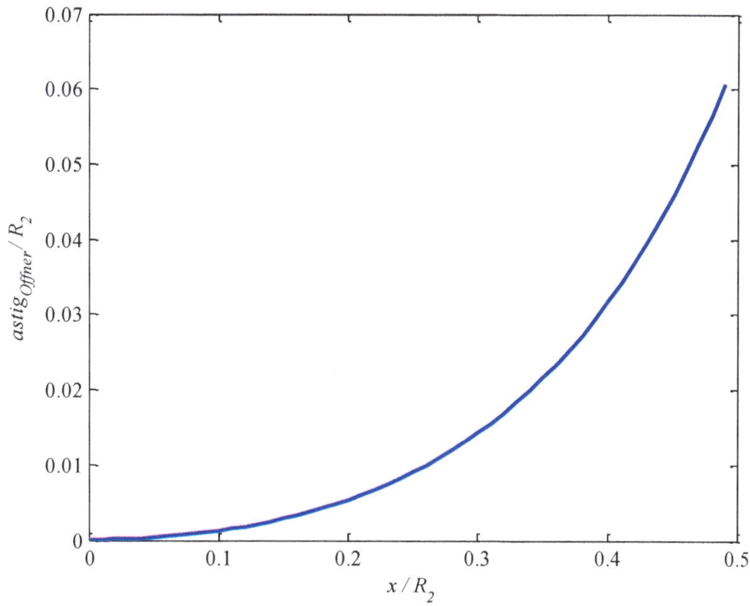


**Fig. 2** The dependence of astigmatism of the Offner spectrometer with its slit length

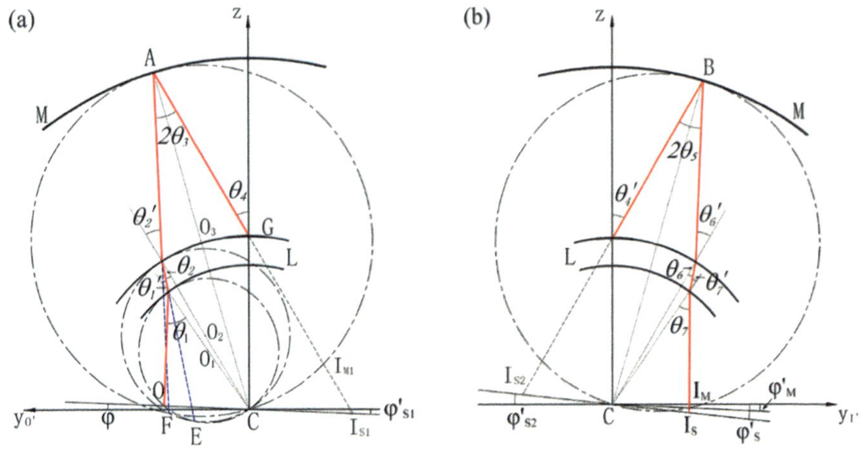


**Fig. 3** Lightpath for the chief ray of Wynne-Offner spectrometer. (**a**) Path of incident light from any point O on the slit. (**b**) Path of emergent light showing the meridional and sagittal images

formed by concave and convex surfaces of $L$. $I_{M1}$ is the meridional image of $M$ and $I_{S1}$ is the sagittal image. $\varphi$ and $\varphi'_{S1}$ are object squint angle and sagittal image squint angle, respectively. Through geometrical relationships between incident angles and refraction angles on each surface, the following equation can be deduced:

$$\varphi = 2\theta_3 + \theta_2 + \theta_1 - \theta_4 - \theta'_2 - \theta'_1, \tag{15}$$

According to Snell's law and Rowland's condition [15],

$$OC = R_1 \sin\theta_1 = R_1 n \sin\theta'_1 = R_2 n \sin\theta_2 = R_2 \sin\theta'_2 = R_3 \sin\theta_3 = R_2 \sin\theta_4, \tag{16}$$

where $n$ is the refractive index of $L$. In Fig. 3b, $I_M$ is the meridional image and $I_S$ is the sagittal image. $\varphi'_M$ is the meridional and $\varphi'_S$ is the sagittal image squint angle. $CI_M$, $\varphi'_M$, and $\varphi'_S$ can be expressed as:

$$\varphi_M' = \theta_4' + \theta_6' + \theta_7' - 2\theta_5 - \theta_6 - \theta_7, \tag{17}$$

$$CI_M = R_1 \sin\theta_7 = R_1 n \sin\theta'_7 = R_2 n \sin\theta_6 = R_2 \sin\theta'_6 = R_3 \sin\theta_5 = R_2 \sin\theta_4', \tag{18}$$

$$\tan\varphi'_S = \tan\varphi \cdot \frac{\sin\gamma}{\sin\gamma'}, \tag{19}$$

where $\gamma$ and $\gamma'$ meet the grating equations:

$$\begin{aligned} \sin\theta_4' \cos\gamma' - \sin\theta_4 \cos\gamma &= mg\lambda \\ \sin\theta_4' \sin\gamma' - \sin\theta_4 \sin\gamma &= 0, \end{aligned} \tag{20}$$

Astigmatism of Wynne-Offner spectrometer can be expressed as:

$$\text{astig}_{\text{Wynne-Offner}} = \text{CI}_M \tan\left(\varphi'_S - \varphi'_M\right). \tag{21}$$

We define the coordinate of any object point as O $(x, y)$ and $k' = \sqrt{x^2 + y^2}\ /R_2$, $k_1' = R_3/R_2$, $k_2' = R_1/R_2$. According to Eqs. (15–20), with $k'$, $k_1'$, $k_2'$ as variates, we can deduce $\text{CI}_\text{M}$, $\varphi'_\text{S}$, and $\varphi'_\text{M}$ in Eq. (21) as the following specific expressions:

$$\text{CI}_M = k' R_2 \frac{\sin\gamma}{\sin\gamma'}, \tag{22}$$

$$\varphi'_S = \arctan\left\{\frac{\sin\gamma'}{\sin\gamma}\tan\left[2\arcsin\left(\frac{k'}{k_1'}\right) + \arcsin\left(\frac{k'}{n}\right) + \arcsin\left(\frac{k'}{k_2'}\right) - 2\arcsin k - \arcsin\left(\frac{k'}{nk_2'}\right)\right]\right\}, \tag{23}$$

$$\begin{aligned}\varphi'_M = {} & 2\arcsin\left(k'\cdot\frac{\sin\gamma}{\sin\gamma'}\right) + \arcsin\left(\frac{k'}{nk_2'}\cdot\frac{\sin\gamma}{\sin\gamma'}\right) - 2\arcsin\left(\frac{k'}{k_1'}\cdot\frac{\sin\gamma}{\sin\gamma'}\right) \\ & - \arcsin\left(\frac{k'}{n}\cdot\frac{\sin\gamma}{\sin\gamma'}\right) - \arcsin\left(\frac{k'}{k_2'}\cdot\frac{\sin\gamma}{\sin\gamma'}\right).\end{aligned} \tag{24}$$

For a specific Wynne-Offner spectrometer with the same $\lambda$, $m$, and $g$ as the Offner spectrometer discussed in Sect. 2.1, we preferred its $k_1'$ and $k_2'$ to maximize the anastigmatic region with $k_1' = 2.15$ and $k_2' = 0.814$. Figure 4a shows the astigmatism distribution of the Wynne-Offner spectrometer in the quarter object plane. It is anastigmatic in a circle field with a threshold value of 0.002, which allows long slit with low astigmatism.

The astigmatism distribution of the Offner spectrometer discussed in Sect. 2.1 is shown in Fig. 4. It is anastigmatic in an annular field, and only short slit with low astigmatism is allowed in such situation. Figure 5 shows the variation curves of astigmatism with slit length for Offner and Wynne-Offner spectrometers. The

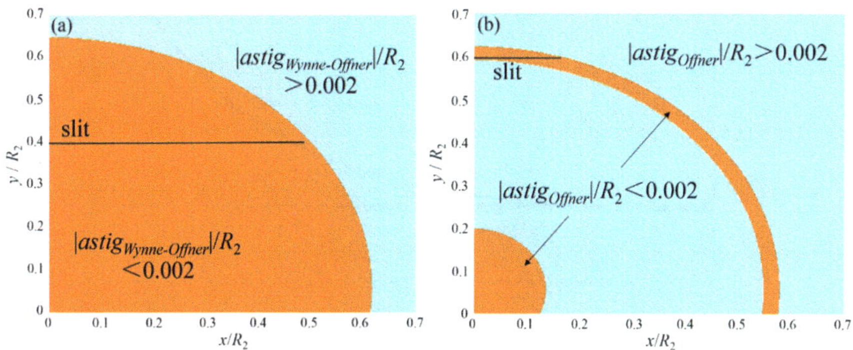


**Fig. 4** Astigmatism distribution of the object plane of (**a**) Wynne-Offner spectrometer (**b**) Offner spectrometer

horizontal axis shows the normalized slit length, and the vertical axis shows the normalized astigmatism. As shown in the figure, Wynne-Offner spectrometer is anastigmatic for a long slit, much longer than that of Offner spectrometer.

## 2.3 Design Procedure of Wynne-Offner Spectrometer

With Eqs. (21–24), when the Wynne-Offner spectrometer is anastigmatic, the relationship between the half-length of its slit and its structural factors $k_1'$ and $k_2'$ is shown in Fig. 6. Different curves represent different $k_2'$, the longer the curve is parallel to the horizontal axis, the longer the corresponding anastigmatic slit.

Inputting the slit length, $k_1'$ and $k_2'$ can be initially determined from Fig. 6. Relationship between the grating radius and system specifications of Offner spectrometer is given in [15], and it is also suitable for Wynne-Offner spectrometer:

$$R_2 = \frac{h_{\text{spec}}}{mg\Delta\lambda}, \tag{25}$$

where $h_{\text{spec}}$ is the dispersion width on the image plane and $\Delta\lambda$ is the width of the wavelength range. Steps for designing a Wynne-Offner spectrometer are shown in

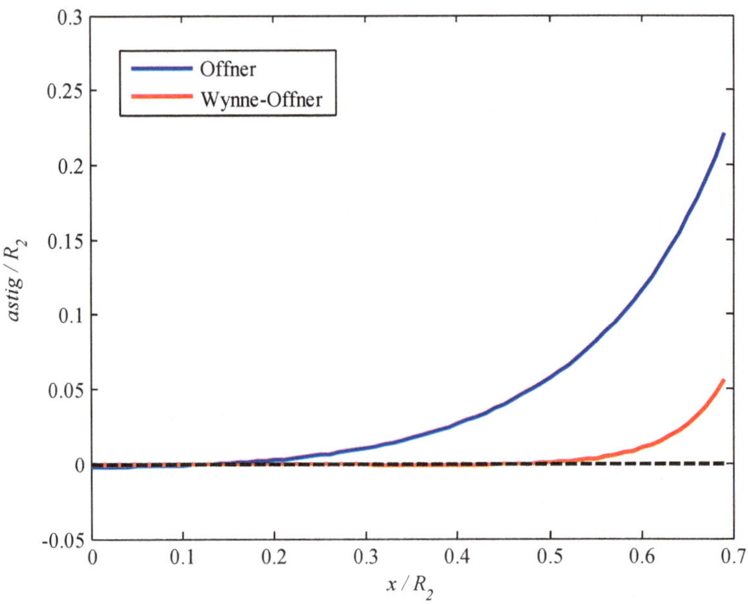


**Fig. 5** Variation curves of astigmatism with slit length for Offner and Wynne-Offner spectrometer. Off-axis magnitude is $0.6R_2$ for Offner spectrometer and $0.4R_2$ for Wynne-Offner spectrometer

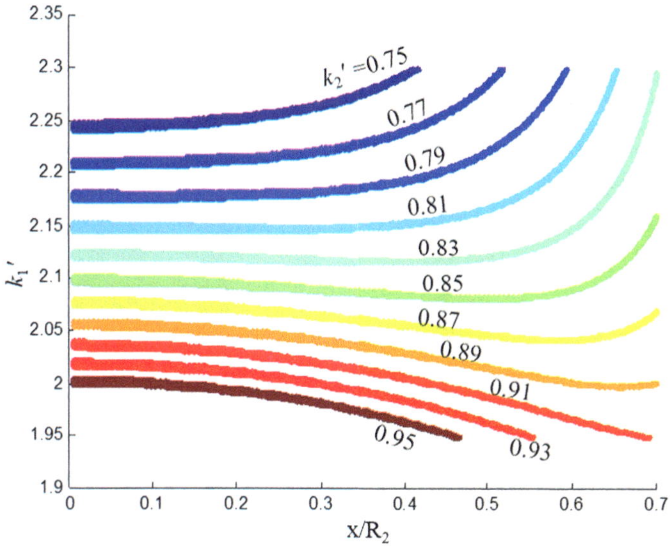


**Fig. 6** Relationships between $x$, $k_1'$ and $k_2'$ when the Wynne-Offner spectrometer is anastigmatic with $y = 0.4R_2$

Fig. 7. Structural parameters of the system, i.e. $y$, $R_1$, $R_2$, and $R_3$, can be determined by the input specifications, i.e. $m$, $h_{spec}$, $\Delta\lambda$, $g$, and slit length.

## 2.4 Design Examples

In this section, two design examples are presented to verify the anastigmatism performance of Wynne-Offner spectrometer. One works in the wavelength range of 0.5–1.0 μm (VNIR) and another works in 1.0–2.5 μm (SWIR). We obtained their initial structures with the procedure shown in Fig. 7 and optimized the systems by using ZEMAX optical design software. The optical systems of the two spectrometers are shown in Fig. 8, and their main specifications and performance are given in Table 1.

Figure 8a, b shows an optical layout and an axial view of the VNIR spectrometer, respectively. It is planar symmetrical and its slit length is 70 mm with a large relative aperture and no obstruction. According to the data in Table 1, off-axis magnitude of its slit is 0.534, and structural factors $k_1' = 2.105$, $k_2' = 0.839$. Its imaging quality is close to the diffraction limit with negligible smile and keystone distortions, and its maximal RMS diameter of spot diagram is 6 μm. Figure 8c, d shows an optical layout and an axial view of the SWIR spectrometer, respectively. It has two slits S1 and S2 with a total length of 70 mm and a distance of 7 mm between them. According to the data in Table 1, off-axis magnitude of S1 is 0.439 and of S2 is 0.505, and $k_1' = 2.121$, $k_2' = 0.828$. The designed spectrometers have compact

**Fig. 7** Steps for designing a Wynne-Offner spectrometer

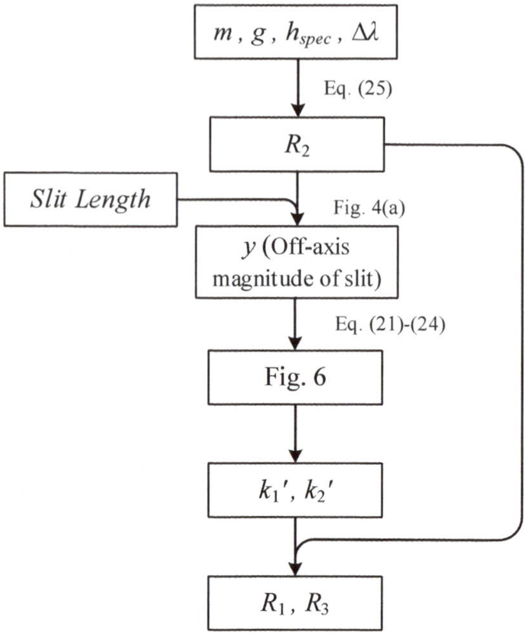


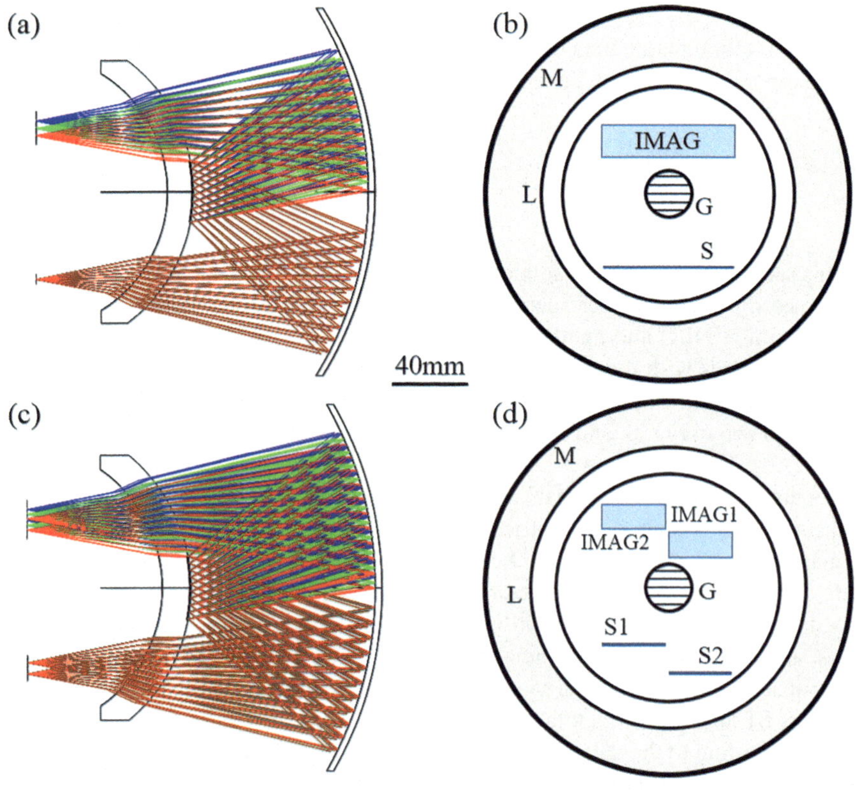


**Fig. 8** VNIR and SWIR spectrometers: (**a**) VNIR spectrometer; (**b**) axial view of the VNIR spectrometer; (**c**) SWIR spectrometer; (**d**) axial view of the SWIR spectrometer

**Table 1** Specifications and performance of the VNIR and SWIR spectrometers

| Specifications and performance | VNIR | SWIR |
|---|---|---|
| Wavelength range/μm | 0.5–1.0 | 1.0–2.5 |
| Slit length/mm | 70 | 35 × 2 |
| Interval of slits/mm | – | 7 |
| Dispersion width/mm | 9 | 6 × 2 |
| F/# | 2.7 | 2.7 |
| Groove density/lp·mm$^{-1}$ | 170 | 37.7 |
| Diffraction order | −1 | -1 |
| $R_1$/mm | 82.75 | 87.90 |
| $R_2$/mm | 98.62 | 106.11 |
| $R_3$/mm | 207.64 | 225.05 |
| Concentricity/mm | 0.06 | 0.14 |
| Off-axis magnitude/mm | 52.7 | 46.6, 53.6 |
| Size/mm | Φ111 × 207 | Φ114 × 225 |
| RMS diameter of spot diagrams/μm | <6 | <16 |
| Smile/μm | <0.053 | <0.12 |
| Keystone/μm | <0.053 | <0.10 |

structure and high concentricity, which allow easy alignment by using the Point Source Microscope [16].

In order to demonstrate the capability of correcting aberrations for Wynne-Offner spectrometers, we also designed a classic Offner spectrometer with the same specifications as the designed VNIR spectrometer for comparison. We used the spectral full-field display (SFFD) [7] to visually show the aberrations and performance (RMS wavefront error) of the spectrometers. When the volume is kept the same, the RMS wavefront error SFFD and the astigmatism SFFD of the Offner spectrometer are shown in Fig. 9a, b, respectively. Those of the Wynne-Offner spectrometer are shown in Fig. 9c, d, respectively. As shown in these figures, the wavefront error and the residual astigmatism of the Offner spectrometer are quite large when the slit lengthens, and they are extremely nonuniform in the full-field. Its maximum RMS wavefront error is 1.56λ, and its maximum residual astigmatism is 131 μm. The wavefront error and the residual astigmatism of the Wynne-Offner spectrometer are small and relatively uniform in the full-field. Its maximum RMS wavefront error is 0.11λ, only 7% of the Offner spectrometer, and its maximum residual astigmatism is 13 μm, only 10% of the Offner spectrometer. To prove that the Wynne-Offner spectrometer is more compact, we also designed a classic Offner spectrometer with equal imaging quality, which results in an increase of 3.5× in volume. In summary, the designed Wynne-Offner spectrometer can greatly reduce the volume and keeps high imaging quality, and it is quite important in hyperspectral remote sensing.

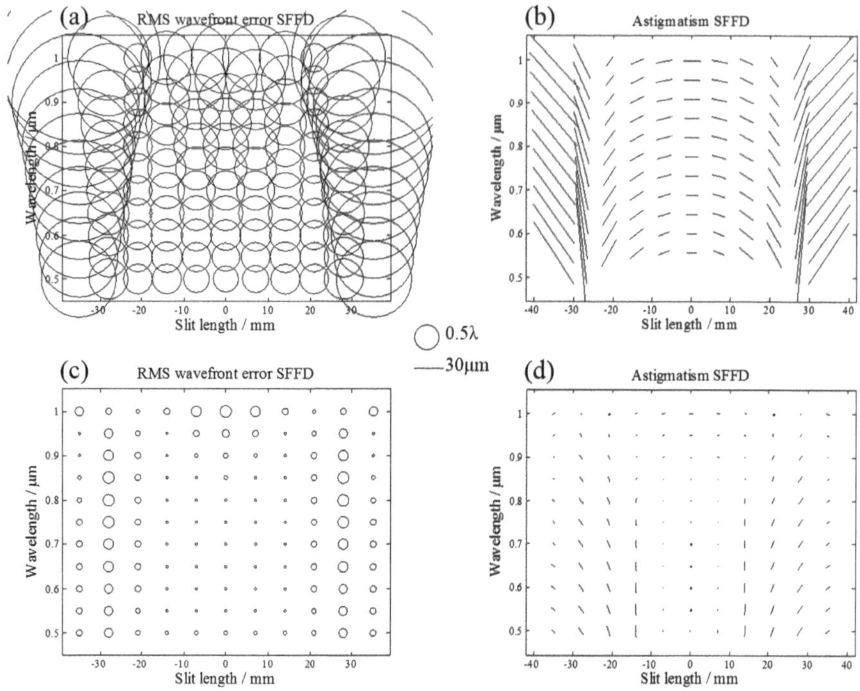


**Fig. 9** Spectral full-field display (SFFD) of spectrometers. (**a**) RMS wavefront error of Offner spectrometer. (**b**) Astigmatism of Offner spectrometer. (c) RMS wavefront error of Wynne-Offner spectrometer. (**d**) Astigmatism of Wynne-Offner spectrometer

# 3 Conclusion

In this paper, astigmatism of Offner and Wynne-Offner spectrometers was theoretically analyzed through tracing the chief ray. We compared their astigmatism characteristics and pointed out that the anastigmatic circle reign of Wynne-Offner spectrometer is larger than that of Offner spectrometer, and the former allows longer slit. By means of the anastigmatism condition, we gave the procedure of designing a Wynne-Offner spectrometer. Through optimization and comparison, it was verified that the Wynne-Offner spectrometer is significantly superior to the classic Offner spectrometer in correcting aberrations and reducing volume. The designed Wynne-Offner spectrometers have the advantages of long slit, compact structure, and high imaging quality, and they can be widely used in remote sensing missions requesting for wide swath and small size.

## References


1. Offner, A.: Unit power imaging catoptric anastigmat. US Patent 3748015, 1973
2. Wynne, C.G.: Optical imaging systems. US Patent 4796984, 1989
3. Chrisp, M.P.: Convex diffraction grating imaging spectrometer," US Patent 5880834, 1999
4. Lobb, D.R.: Theory of concentric designs for grating spectrometers. Appl. Opt. **33**, 2648–2658 (1994)
5. Dan, R.L.: Design of a spectrometer system for measurements on earth atmosphere from geostationary orbit. Proc. SPIE. **5249**, 191–202 (2004)
6. Prieto-Blanco, X., de la Fuente, R.: Compact Offner–Wynne imaging spectrometers. Opt. Commun. **328**, 143–150 (2014)
7. J. Reimers, E. M. Schiesser, K. P. Thompson, K. L. Whiteaker, D. Yates, and Rolland, J.P., : Comparison of Freeform Imaging Spectrometer Design Forms Using Spectral Full-Field Displays. In [C]//*OSA Technical Digest: Freeform Optics*, p. FM3B.3 (2015)
8. Wei, L., Feng, L., Zhou, J., Jing, J., Li, Y.: Optical design of Offner-Chrisp imaging spectrometer with freeform surfaces. Proc. SPIE. **10021**, 100211P (2016)
9. Risse, S., Krutz, D.: Design of an imaging spectrometer for earth observation using freeform mirrors. In: [C]//*ISCO: International Conference on Space Optics*, p. 161 (2016)
10. A. Z. Marchi and B. Borguet, "Freeform grating spectrometers for hyperspectral space applications: status of ESA programs," [C]//*OSA Technical Digest: Freeform Optics*, 2017: JTh2B.5
11. Reimers, J., Bauer, A., Thompson, K.P., Rolland, J.P.: Freeform spectrometer enabling increased compactness. Light Science & Applications. **6**, e17026 (2017)
12. Cheng, D., Yang, T., Wang, Y.: Freeform imaging spectrometer design using a point-by-point design method. Appl. Opt. **57**, 4718–4727 (2018)
13. Whyte, C.E., Leigh, R.J., Lobb, D., Williams, T., Remedios, J.J., Cuter, M., Monks, P.S.: Assessment of the performance of a compact concentric spectrometer system for atmospheric differential optical absorption spectroscopy. Atmos. Meas. Tech. **2**, 789–800 (2009)
14. Coppo, P., Taiti, A., Pettinato, L., Francois, M., Taccola, M., Drusch, M.: Fluorescence imaging spectrometer (FLORIS) for ESA FLEX Mission. Remote Sens. **9**, 649–666 (2017)
15. Prieto-Blanco, X., Montero-Orille, C., Couce, B., de la Fuente, R.: Analytical design of an Offner imaging spectrometer. Opt. Express. **14**, 9156–9168 (2006)
16. Parks, R.E., Kuhn, W.P.: Optical alignment using the point source microscope. Proc. SPIE. **5877**, 58770B (2005)

# Design of Compact Space Optical System with Large Relative Aperture and Large Field of View Based on Mangin Mirror

**Xuyang Li and Mengyuan Wu**

**Abstract** This paper designs an imaging optical system with a large relative aperture and large field of view based on the improved Mangin mirror. And it proposes a design method by analyzing the aberrations of Mangin mirror. The system uses the structure of a combination of catadioptric optical system and modified Mangin mirror. The system works in 450–850 nm band with 4° × 4° field of view, 1.8 F number, and 380 mm effective focal length. The imaging detector is a CMOS with detector with 2 μm × 2 μm pixels. The modulation transfer function value is greater than 0.50 at the Nyquist frequency 250 lp/mm and close to the diffraction limit. The secondary mirror of the system is designed, which is based on the theory of achromatic lens and Mangin mirror. The focal power of the achromatic Mangin mirror is solved by PW method which is based on the spherical aberration and off sine aberration obtained from the preliminary optimization analysis of the system initial structure. The focal power and radius of three surfaces of the achromatic Mangin mirror can be calculated by the achromatic conditions and residual chromatic aberration of the system. The system has good imaging quality with small monochromatic aberrations and chromatic aberrations.
**OCIS Codes** 220.3620 080.4035 120.0280 350.1260

**Keywords** Optical design · Large relative aperture · Large field of view · Mangin mirror · Catadioptric optical system

X. Li (✉) · M. Wu
Space Optics Lab, Xi'an Institute of Optics and Precision Mechanics, Chinese Academy of Sciences, Xi'an, Shaanxi, China


H. P. Urbach, Q. Yu (eds.), *6th International Symposium of Space Optical Instruments and Applications*, Space Technology Proceedings 7,
https://doi.org/10.1007/978-3-030-56488-9_22

# 1 Introduction

Space remote sensing optical system plays an important role in ground observation, meteorological monitoring, and other fields [1, 2]. As the relative aperture of the imaging optical system increases, it means higher system resolution and image surface illuminance, which in turn improves system sensitivity. Space remote sensors with large relative aperture, large field of view, and high sensitivity have become an important development direction [3–5].

When a small pixel size detector is used to receive an image in a large relative aperture optical imaging system, there is a difficulty that the modulation transfer function at the Nyquist frequency is difficult to approach the diffraction limit, resulting in high difficulties in coexisting with spatial frequency and high MTF [6]. At present, off-axis three mirrors or four mirrors are currently used to achieve large relative aperture optical system design [7–9], but it is often accompanied by problems such as difficulty in assemble and adjustment, increased launch difficulty caused by complex space structures, and high processing costs. Therefore, it is urgent to design a compact optical remote sensing system that can still get high MTF at Nyquist frequency.

According to the application characteristics of high sensitivity and large field of view, a space optical imaging system with a large relative aperture is designed, which solves the problem that large relative aperture is difficult to approach the diffraction limit at high spatial frequency of 250 lp/mm to obtain high MTF value. The initial structure of the system is analyzed and calculated; the residual aberration of the system is obtained by preliminary optimization; an achromatic Mangin reflector is proposed to correct the residual aberration of the system by improving the Mangin reflector; the initial structure of the achromatic Mangin reflector is calculated based on PW method and substituted for the preliminary optimized system structure. An imaging optical system with F/1.8, field of view angle $4^\circ \times 4^\circ$, 450–850 nm spectrum band, 380 mm effective focal length, and no less than 15 mm working distance was designed.

# 2 Optical System Structure and Principle

## 2.1 Initial Structure Selection

According to the effective focal length and F/# of the optical system, the optical aperture of the optical system is obtained. The reasonable choice of the relative aperture of the primary mirror is bound to take into account the spatial structure of the system and the difficulty of processing the lens. The overall length of the system is determined according to the system function, and the working distance is selected. According to the third-order aberration theory, the magnification and curvature radius of each surface are determined by the focal lengths of the system and the

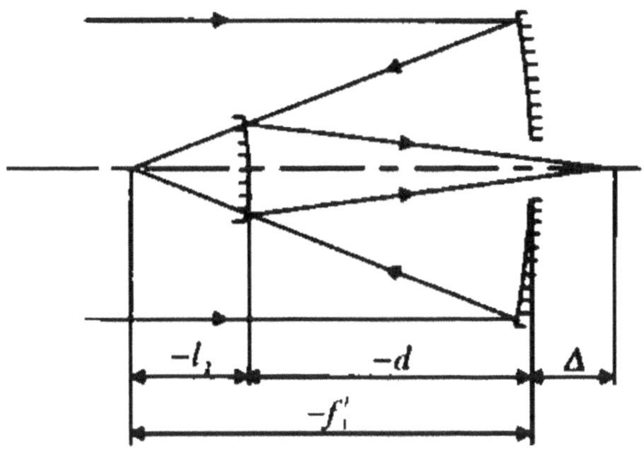


**Fig. 1** Initial configuration of catadioptric optical system

main mirror. Finally, the primary and secondary mirror coefficients are determined by the conditions of spherical aberration and coma, and we obtain the initial structure through choosing the appropriate structure and optimizing the parameters [10] according to the obstruct ratio, as shown in Fig. 1.

The distance between secondary mirrors and the system obstruct ratio is satisfied.

$$l_2 = \frac{-f_1' + \Delta}{\beta - 1} \tag{1}$$

$$\alpha = \frac{l_2}{f_1'} \tag{2}$$

The sum of curvature radius of the vertex of primary and secondary mirrors and the interval between the two mirrors satisfy

$$R_1 = 2 \times \frac{D}{\left(D / f_1'\right)} \tag{3}$$

$$R_2 = \frac{\alpha\beta}{\beta + 1} R_1 \tag{4}$$

$$d = f_1'\left(1 - \alpha\right) \tag{5}$$

The primary and secondary mirror shape parameters are obtained by the coma condition of spherical aberration elimination

$$e_1^2 = 1 + \frac{2\alpha}{\left(1 - \alpha\right)\beta^2} \tag{6}$$

$$e_2^2 = \frac{2\beta/(1-\alpha)+(1+\beta)(1-\beta)^2}{(1+\beta)^3} \tag{7}$$

## 2.2 Principle of Mangin Mirror

The FOV of the traditional refractive optical system is generally smaller than the FOV, and the corresponding distortion and magnification chromatic aberration are generally smaller, so it can be neglected. When FOV is F/1.8, the large aperture of optical system corresponds to large aperture aberration-spherical aberration and chromatic aberration. The aberration and magnification chromatic aberration are proportional to the cubic and the first power of the field of view, respectively, and the introduction of large field of view will increase the aberration and magnification chromatic aberration. It is related to aperture and field of view, so large relative aperture and large field of view will also introduce large coma. Large aberrations will cause the MTF of the system to be much lower than the diffraction limit. Therefore, aberrations such as chromatic aberration and distortion must be corrected. A Mangin reflector is introduced to correct the aberrations of the system.

The traditional Mangin mirror is a spherical aberration mirror, which has three surfaces. The form is shown in Fig. 2a, and the total focal power of the corresponding thin Mangin mirror is given.

$$\Phi = 2\Phi_1 + \Phi_2 \tag{8}$$

For the first side $\Phi_1$, the second side $\Phi_2$.

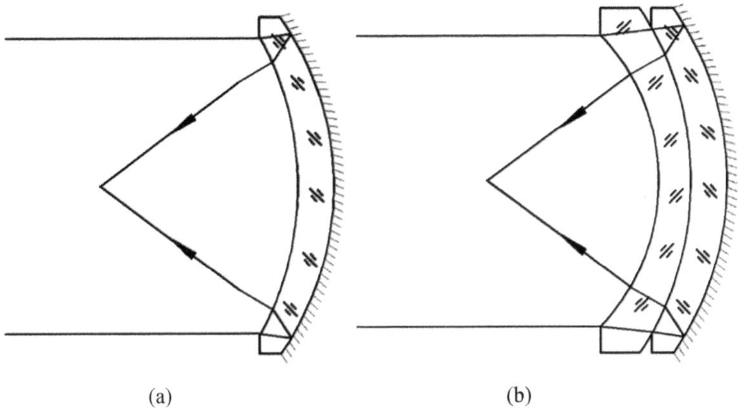


**Fig. 2** The schematic of Mangin mirror and achromatic Mangin mirror. (**a**) Traditional Mangin mirror (**b**) Achromatic Mangin mirror

When the diaphragm coincides with the Mangin mirror, the main aberrations corresponding to the mirror-spherical aberration and sinusoidal aberration satisfy [11].

$$LA_0' = \frac{1}{8}\frac{\left(8n^2 - 13n - 11\right)}{4n^2\left(n+3\right)}h^2\Phi \tag{9}$$

$$OSC_0' = \frac{1}{8}\frac{\left(-n^3 + 3n^2 - 3n - 3\right)}{2n^2}h^2\Phi \tag{10}$$

It can be seen from formula (9) that the positive and negative spherical aberration limits are only related to the molecule of the optical focus and factor of the Mangin mirror. When the relative aperture is fixed, the spherical aberration symbols can be changed by changing the refractive index to further correct the spherical aberration of the refractive system, and the coma aberration can also be corrected to a certain extent. Because no distortion and magnification chromatic aberration are introduced when the aperture coincides with the mirror, the residual aberration of the system can be corrected by adding a certain correction mirror behind the Mangin mirror.

## 2.3 *Optimization of Mangin Mirror*

In order to compress the axial space of optical system, lenses are often added before secondary mirrors. But at the same time, the lens introduces chromatic aberration while correcting spherical aberration and distortion of the system. Based on the aberration analysis of the Mangin reflector, the traditional Mangin reflector does not have the ability to correct the chromatic aberration. According to Smith [12], achromatic double-ply lens can be used to replace negative mirror to realize achromatic function. The structure is shown in Fig. 2b.

There are five areas for contribution correction of the new Mangin mirror, and the corresponding light focus of the thin achromatic double glued Mangin mirror is as follows:

$$\Phi' = 2\Phi_1' + 2\Phi_2' + \Phi_3' \tag{11}$$

It is the light focus of the left transmission surface, the middle bonded surface, and the right reflection surface. For double glued thin Mangin mirrors

$$2\mathrm{h}^2\left(\frac{\Phi_1' + \Phi_2'}{\upsilon_1} + \frac{\Phi_2' + \Phi_3'}{\upsilon_2}\right) + \mathrm{C} = 0 \tag{12}$$

It is the semi-pass aperture of the lens, the Abbe number of the glass with two lenses from left to right, and the residual chromatic aberration after optimizing the initial structure of the system.

Lenses are numbered from left to right from small to large. The refractive index of the medium is in turn, and the surface curvature radius is in turn be

$$\Phi_1^{'} = \frac{n_1 - n_0}{r_1}; \tag{13}$$

$$\Phi_2^{'} = \frac{n_2 - n_1}{r_2}; \tag{14}$$

$$\Phi_3^{'} = -2\frac{n_2}{r_3} \tag{15}$$

The range of aberration correction for optical systems with large relative aperture and large field of view increases with the increase of the range of optical focus. The achromatic Mangin reflector not only has spherical aberration and sinusoidal aberration correction ability of the traditional Mangin reflector, but also has chromatic aberration correction ability. Moreover, the spherical aberration and sinusoidal aberration correction ability of the system is greater than that of the traditional Mangin reflector.

## 3 Design of Optical System with Achromatic Mangin Mirror

### *3.1 Initial Structure Design of Imaging Optical System*

According to the main technical indicators of the system, as shown in Table 1.

According to the effective focal length EFL = 380 mm and F/# = 1.8, the optical aperture of the optical system is 210 mm. In order to compress the whole length of the optical system to achieve ultra-compact optical structure, the relative aperture of the primary mirror is selected as 1/1, and the working distance is_ = 50 mm. On the

**Table 1** Main specifications of imaging optical system

| System parameters | Value |
|---|---|
| Focal length/mm | 380 |
| Field of view /(°) | 4° × 4° |
| F/# | 1.8 |
| Working distance/mm | >15 |
| Spectral bands/nm | 450–850 |
| Pixel size | 2 μm × 2 μm |

**Table 2** Initial structural parameters of optical system

| Surface | Radius/mm | Quadratic coefficient | Thickness/mm |
|---|---|---|---|
| Primary mirror | −420.00 | 1.4893 | −117.14 |
| Secondary mirror | −417.86 | 24.8548 | 167.14 |

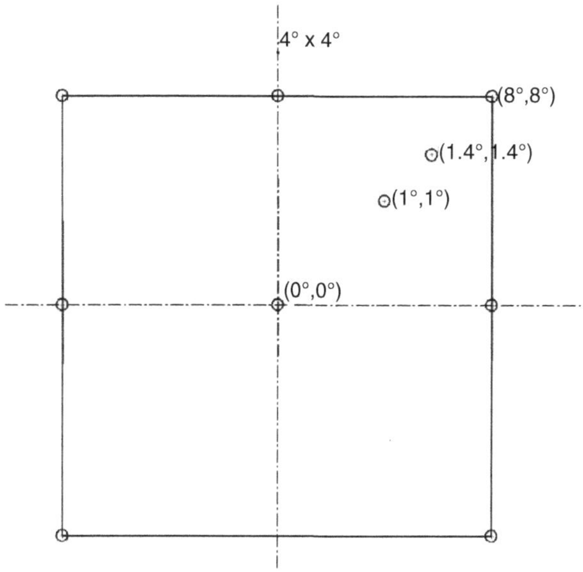


**Fig. 3** Ray tracing field of view of optical system

premise of achieving a suitable obstruct ratio, the initial structural parameters of the system are calculated by formula (1)–(9). As shown in Table 2,

## 3.2 *Architectural Design of Achromatic Mangin Mirror*

In optical design software CODEV, ten viewpoints are selected to track the whole field of view, as shown in Fig. 3. After preliminary optimization, it is found that the system has large spherical aberration, coma aberration, and distortion. A small amount of correction mirror is added to correct the aberration. It is found that the spherical aberration and coma aberration are negative values of −0.0317 mm and − 0.0613 mm, respectively, except for the positive aberration, and the sinusoidal aberration is −0.033110 − (−0.018203) = −0.014907 mm, which is much larger than the size of a single pixel. The aberration of the system must be corrected.

According to formulas (9)–(11), when the refractive index of the Mangin reflector is between 1.4 and 2, the factor formula is always positive. Therefore, when the Mangin reflector takes the negative light focus, the positive spherical and sinusoidal differences can be introduced, and the original spherical and sinusoidal differences can be corrected when the appropriate light focus is selected.

Based on the spherical and sinusoidal aberrations obtained from the initial structure calculation, the focal degrees of achromatic Mangin mirrors are calculated by formulas (9) and (10) on the premise of choosing suitable crown glass and flint glass. By combining formulas (11) and (12) of PW method, the focal power and curvature radius of the three surfaces of the achromatic Mangin mirror have been calculated by formulas (13)–(15). The calculated Mangin reflector is substituted into the optimized structure of the refractive optical system. In order to reduce the computational difficulty, the eccentricity square of the reflector is obtained by program optimization, and the refractive optical system with achromatic Mangin reflector with large relative aperture and large field of view is obtained.

It is found that aberration can be corrected by using HZF6 glass with a refractive index of 1.747905 (corresponding wavelength 650 nm) in front and DK9L mirror with a refractive index of 1.514517 (corresponding wavelength 650 nm) and a central wavelength of 650 nm in back. The surface parameters of achromatic Mangin reflector are obtained by optimizing the surface number of light rays in sequence, as shown in Table 3.

### 3.3 Optimizing Imaging Optical System

In this paper, the size of CMOS pixel is chosen to be 2 micron × 2 micron. In practical application, the cut-off frequency of optical system depends on the resolution of detector [13]:

$$f_c = \frac{1}{2\text{pixelwidth}} = 250\text{lp/mm} \tag{16}$$

Because the maximum aberration of the optical system is at 0.707 of the field of view [14], 0, 0.5, 0.7, and diagonal field of view are selected to trace light. In view of the asymmetry of image quality caused by the quadric reflector, a total of ten fields of view are traced for the system, as shown in Fig. 3.

The formulas of spherical and sinusoidal aberrations and chromatic aberrations of the achromatic Mangin reflector are derived by using PW method. The initial structure of the achromatic Mangin reflector is calculated by combining the formulas of optical focusing. The structure of the achromatic Mangin reflector is

**Table 3** Surface parameters of achromatic Mangin mirror

| Surface | Radius/mm | Quadratic coefficient | Distance/mm | Material |
|---|---|---|---|---|
| 1 | −1989.88.00 | 0 | −8 | HZF6 |
| 2 | −49828.54 | 0 | −8 | DK9L |
| 3 | −876.5298 | −9.058 | / | / |

optimized by substituting it into the structure of the refractive optical system, and the imaging optical system with the achromatic Mangin reflector with large relative aperture and large field of view is obtained. The maximum aperture of the system is 220 mm and the total length of the system is 193.5 mm. The imaging optical system with large relative aperture and large field of view is designed. The system structure is shown in Fig. 4.

The structural parameters of the imaging optical system with achromatic Mangin mirror are shown in Table 4. SPECIAL is a special glass, SCHOTT is a Schott glass, and CDGM is a bright glass in Chengdu.

## 4 Image Quality Analysis

The MTF, spot diagram, field curve, and distortion of 10 fields of view tracked by optical system are analyzed, as shown in Figs. 5 and 7. Figure 5 shows the MTF curves of ten fields of view of the imaging optical system. The MTF curves are analyzed at the Nyquist spatial frequency 250 lp/mm of the CMOS detector. The results show that the MTFs of the ten fields of view are all about 0.30 and close to the diffraction limit, which meets the requirements of imaging optical system.

The distribution of the spot diagram of the system is shown in Fig. 6. The overall approximation of the scattered speckle to the circle indicates that the coma and aberration are corrected well. The chromatic aberration of the system is also significantly improved by introducing achromatic Mangin reflector. The RMS value of the diameter of the scattered speckle reaches a maximum value of about two microns in the field of view at the diagonal edge, which is generally less than two microns in the size of a single pixel.

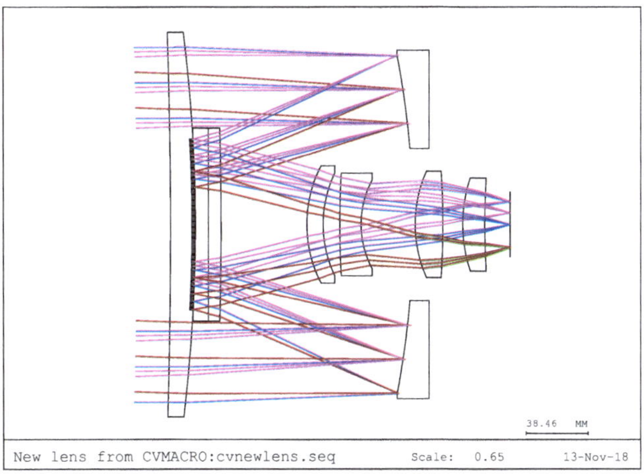


**Fig. 4** Layout of optical system

**Table 4** Initial structural parameters of optical system

| Surface | Radius/mm | Quadratic coefficient | Thickness/mm | Materials |
|---|---|---|---|---|
| 1 | −3500 | | 15.0 | SILICA_SPECIAL |
| 2 | −882.55 | −9.3316 | 138.5 | |
| STOP | −520.328 | −0.3451 | −122.5 | |
| 4 | −2100.97 | | −8 | SF4_SCHOTT |
| 5 | 83472.675 | | −8 | NFK5_SCHOTT |
| 6 | −882.55 | −9.3316 | 8 | NFK5_SCHOTT |
| 7 | 83472.675 | | 8 | SF4_SCHOTT |
| 8 | −2100.97 | | 50.8 | |
| 9 | 74.1 | 0.5664 | 10 | NSK16_SCHOTT |
| 10 | 67.12 | | 10.5 | |
| 11 | 459.756 | −8.0859 | 10 | DLAF79_CDGM |
| 12 | 53.512 | | 33.8 | |
| 13 | 73.722 | | 17.2 | HZK50_CDGM |
| 14 | −464.806 | | 12.6 | |
| 15 | 84.72 | | 13.3 | HZK9A_CDGM |
| 16 | 361.916 | | 15 | |

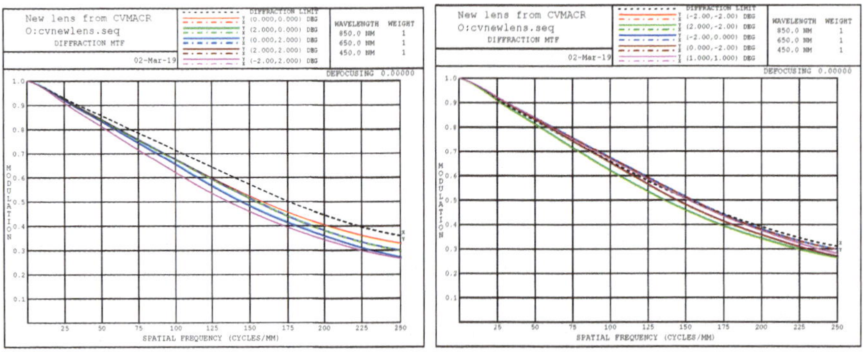


(a) (b)

**Fig. 5** MTF curves of the system. (**a**) MTF curves of field 1–5 (**b**) MTF curves of field 6–10

Figure 7 shows the astigmatism field curve and distortion curve. The left side curve shows that the meridian field curve is less than 0.4%, the arc field loss curve is less than 0.2%, the astigmatism is less than 0.3%, and the right side curve shows that the distortion of the system is less than 0.48%. In a large field of view imaging optical system, astigmatism, field curve, and distortion are well corrected.

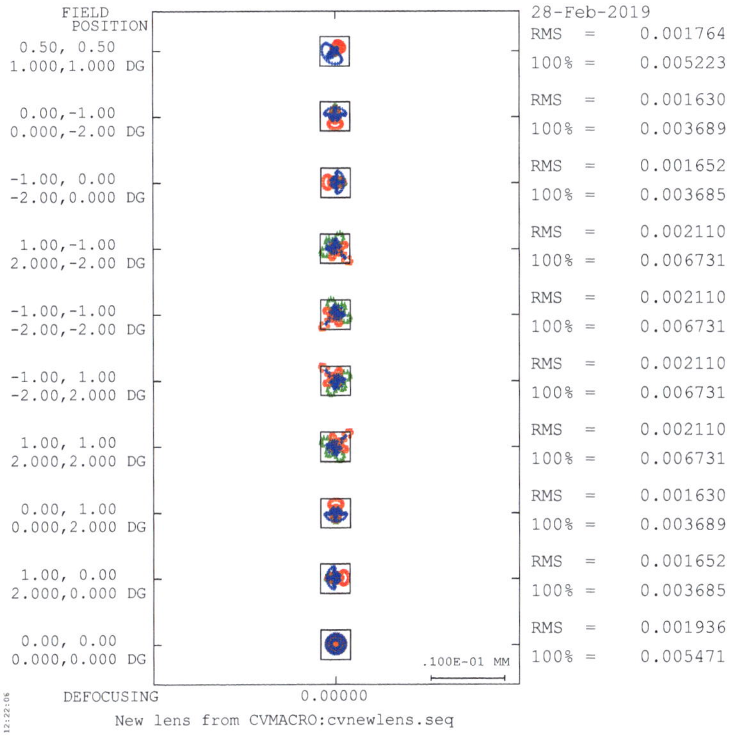


**Fig. 6** Spot diagram of the system

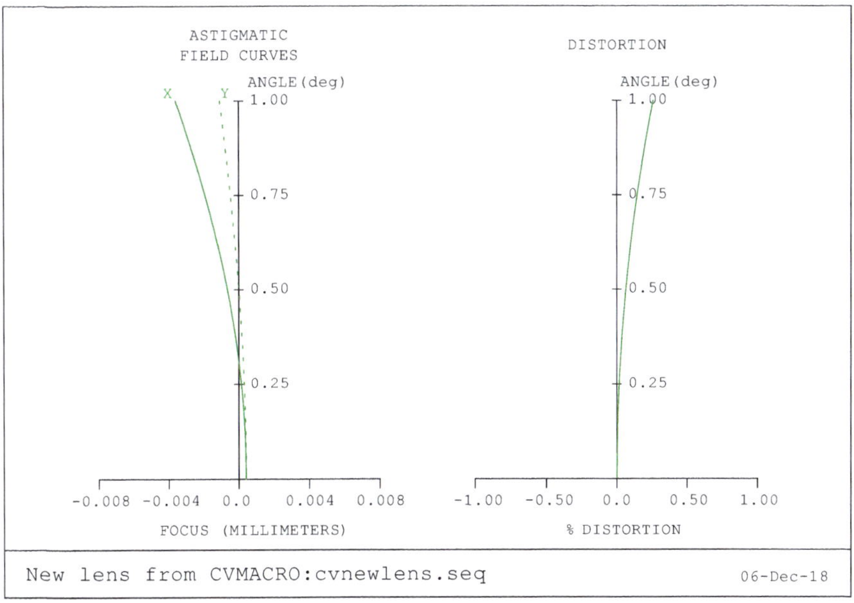


**Fig. 7** Spot diagram of the system

## 5 Conclusion

A new type of refractive imaging optical system with a large relative aperture, large field of view, and near diffraction limit is realized in this paper. At the Nyquist frequency 250lp/mm of CMOS detector, the MTF values of the main field of view of the system are all about 0.3; the maximum RMS value of the dispersion spot radius of the system is about two micron. It is smaller than the size of a single pixel of the detector as a whole; the distortion is less than 0.48%, so it can meet the requirements of the imaging optical system. The imaging optical system has the characteristics of large relative aperture, large field of view, wide spectrum, etc. It still satisfies the characteristics of near diffraction limit at high Nyquist frequency and about 0.30. The new refractive optical system with achromatic Mann-Gold reflector designed in this paper not only overcomes the difficulty of realizing large field of view in traditional refractive optical system, but also avoids the difficulty of adjusting large relative aperture, large field of view, and near diffraction limit optical system using off-axis three-reflection optical structure. At the same time, it also realizes the ultra-compact structure which is difficult to realize in off-axis optical system. It has the advantages of small spherical aberration, small sinusoidal aberration, low chromatic aberration, and small distortion. The optical structure can be widely used in military field and climate detection.

## References

1. Meng, Q.Y., Dong, J.H., Qu, H., et al.: Light optical system design with wide spectral band, large field of view for deep space exploration. Acta Photon Sin. **44**(1), 135–140 (2015)
2. Dun, X., Jin, W.Q., Wang, X.: Design of large relative aperture compact infrared optical system. Acta Opt. Sin. **34**(6), 0622002 (2014)
3. Bai, H.B., Miao, L.: Design of large aperture and long focal length zoom optical system. J. Appl. Opt. **39**(5), 644–649 (2018)
4. Chen, W., Zheng, Y.Q., Xue, Q.S.: Airborne imaging spectrometer with wide field of view and large relative-aperture. Editorial Off. Opt. Precis. Eng. **23**(1), 15–21 (2015)
5. Zhu, L.R.: Optical Design of Four-Mirror Systems with Ultra-Large Aperture. Soochow University, Suzhou (2007)
6. Hu, M.Y., Li, M.J., Zhao, Q., et al.: Design and imaging distortion correction of pin-hole objective lens with large field. Acta Opt. Sin. **37**(5), 0522002 (2017)
7. Liu, J., Liu, W.Q., Kang, Y.S., et al.: Optical design of off-axis four-mirror optical system with wide field of view. Acta Opt. Sin. **33**(10), 1022002 (2013)
8. Goldstein, N., Dressler, R.A., Shroll, R., et al.: Ground Testing of Prototype Hardware and Processing Algorithms for a Wide Area Space Surveillance System (WASSS), p. ADA591374. Air Force Research Lab Kirtland AFB NM Space Vehicles Directorate, New Mexico (2013)
9. Lv, Y., Zeng, X.F., Zhang, F.: Impact on image performance due to surface scattering for off-axis three mirror optical system. Laser Optoelectron. Progress. **55**(9), 092901 (2018)
10. Pan, J.H.: The design, manufacture and test of the aspherical optical surface, pp. 31–36. Suzhou University Press, Suzhou (2004)

11. Lin, Y.S.: The primary aberration characteristics analysis of Mangin mirror. Opt Instrum. **2**, 53–61 (1981)
12. Smith, W.J.: Modern Optical Engineering: the Design of Optical Systems, 4th edn, pp. 205–234. McGraw Hill Companies, New York (2011)
13. Wang, W.W.: Applied Optics, pp. 293–298. Huazhong University of Science and Technology Press, Wuhan (2010)
14. Yu, D.Y., Tan, H.Y.: Engineering optics, pp. 105–132. China Machine Press, Beijing (2011)

# Phase Retrieval Using Neural Networks

**Ling Li, Xiaoyong Wang, Zongwei Yu, Xiangdong Wang, Chunxiao Xu, Xin Dong, and Yun Wang**

**Abstract** The wavefront sensing and control (WFSC) is developed for NASA James Webb Space Telescope (JWST) mission (B. H. Dean, D. L. Aronstein, J. S. Smith, R. Shiri, and D. S. Acton, Proc. SPIE 6265:626511, 2006). The phase retrieval (phase retrieval: PR) is one of the key algorithms in WFSC. There are two general categories of PR algorithms (B. H. Dean, Proc. IEEE. Aerospace Conference. 2007): iterative-transform (D. L. Misell, J Phys. D6: L6–L9, 1973; R. W. Gerchberg, W. O. Saxton, OPTIK. 34: 275, 1971; W. O. Saxton, OPTIK. 35: 237–246, 1972) and parametric approach (R. A. Gonsalves, P. Considine, J. Opt. Soc. Am. 66: 961–964, 1976; W. H. Southwell, J. Opt. Soc. Am. A3: 396–399, 1977). In most scenarios, the PR algorithm takes advantage of both categories, iterative-transform for retrieval high spatial frequency and parametric for retrieval high dynamic range. However, the parametric approach sometimes returns local minimum due to the bad starting parameters set, which is very common in optimization algorithm such as nonlinear optimization algorithm and genetic algorithm. In this paper, the neural networks are used for parametric approach, which return the global minimum and cost less time. A new PR algorithm block is also proposed and verified by experiments. Unlike other algorithms, the defocus values of PSFs are not necessary, which are also difficult to be determined when dealing with large wavefront errors (wavefront error: WFE) system. Instead, the PSF images which are taken before or after the focal plane (focal plane: FP) are the inputs for this algorithm. The experiment results which include both the high spatial frequency and high dynamic range capability of PR are presented in this paper.

**Keywords** Wavefront sensing · Phase retrieval · High dynamic range · Neural network

---

L. Li (✉) · X. Wang · Z. Yu · X. Wang · C. Xu · X. Dong · Y. Wang
Beijing Key Laboratory of Advanced Optical Remote Sensing Technology, Beijing Institute of Space Mechanics and Electricity, Beijing, China



H. P. Urbach, Q. Yu (eds.), *6th International Symposium of Space Optical Instruments and Applications*, Space Technology Proceedings 7,
https://doi.org/10.1007/978-3-030-56488-9_23

# 1 Introduction

There are more engineering risks when a large-diameter (2 m and above in diameter) space telescope be launched and deployed to the space orbit such as misalignment of optical elements and drift of FP, so the secondary mirror is often designed to be capable of adjusting in six DOFs (DOF: degree of freedom). The WFE of multi-fields are the vital inputs for the optical model to solve the adjustment amounts of the secondary mirror and focal plane.

There are several methods to obtain the wavefront errors such as Shack–Hartmann wavefront sensor, 4-wave lateral shearing wavefront sensor, interferometry, and PR. But the PR is the simplest and best option for the flight missions. The PR is developed for JWST WFSC [1]. Both iterative-transform [3–5] and parametric approach [6, 7] are integrated into PR algorithm [2].The advantages are that the PR can be performed with science cameras and introduces a minimum of new error sources, which ensures maximum redundancy and reliability, also saves cost and mass as well [8].

PR is an image based wavefront sensing technique, which uses several defocus PSFs to recovery the optical phase information. PR algorithm was first used to characterize the Hubble Space Telescope primary mirror edge defect [9–11], then developed by Jet Propulsion Laboratory [12, 13], University of Rochester [14, 15], NASA Goddard Space Flight Center [1, 2, 16], and Ball Aerospace and Technologies Crop [17, 18].

In this paper, the neural networks are used for parametric approach and the iterative-transform loops are also adjusted as well to improve efficiency and reliability of PR algorithm.

# 2 Parametric Neural Networks Design

The aberration coefficients (e.g., Fringe Zernike) of wavefront can be speculated by the intensity distribution, the size and the shape of PSFs. By definition, the image sampling parameter, Q, is expressed in terms of these parameters:

$$\mathrm{Q} = \frac{\lambda \bullet f}{D \bullet d} \tag{1}$$

where $\lambda$ is the wavelength of PSF, $f$ is the focal length of optical system, $D$ is the diameter, and the d is pixel size. The greater the Q is, the more details of PSFs can be recognized. The same PSF in different Q is shown in Fig. 1.

The compressed PSF images are used as inputs of neural networks, due to the balance between efficiency and accuracy. The greater Q images contain more information of WFE, but also significantly increasing the size of neural networks which cost more time to train. Furthermore, the compressed PSF images keep most of the

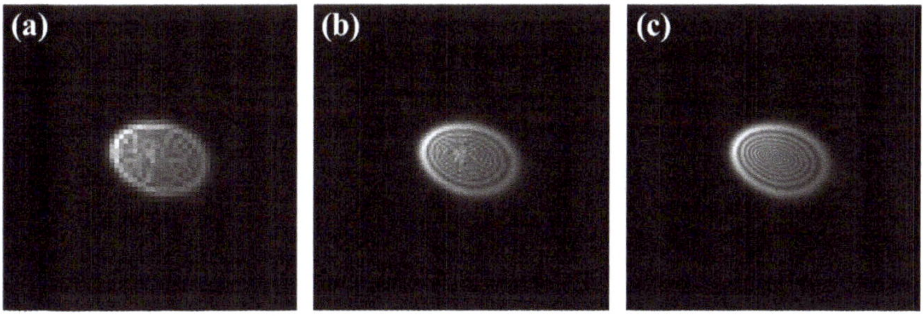


**Fig. 1** (**a**) PSF image (Q = 1/4); (**b**) PSF image (Q = 1/2); (**c**) PSF image (Q = 1)

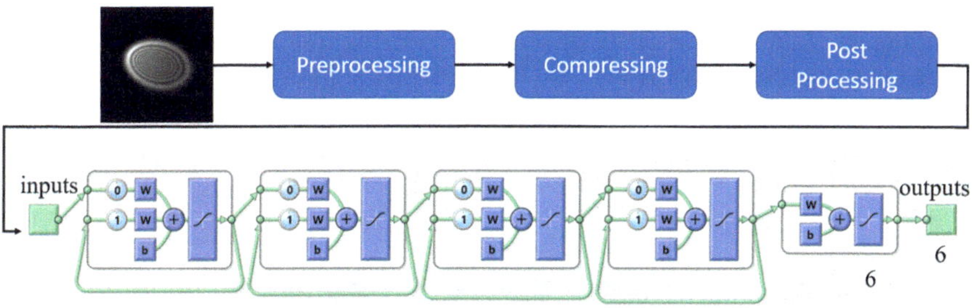


**Fig. 2** Neural network design

low spatial frequency information of WFE, which is mainly introduced by the misalignment of optical elements, the primary mirror support effect, the environment factors, and the defocus effect of telescope. Those factors which introduce the low order WFE also are the most crucial factors which affect the imaging quality during the flight mission. Training the networks with the compressed PSF images in high dynamic range ensures that the algorithm can deal with most circumstances during the flight mission. So the low order Zernike coefficients (Fringe Zernike —fourth to ninth terms) are the outputs of neural networks.

There are two neural networks which are trained to deal with different situations, one is for the PSF images which are taken before FP, and the other one is for the PSF images which are taken after FP. Each neural network is illustrated in Fig. 2. The PSF images should be checked by software to make sure that the image is well exposed. In the preprocessing step, the noise reduction and the PSF extraction will be executed. In the compressing step, the PSF image will be compressed. In the post-processing step, the compressed PSF image will be normalized and mapped to neural network inputs. The inputs information will be calculated through five hidden layers, and then the neural network will output Zernike fourth to ninth coefficients for each PSF image. 100,000 samples are used to train neural network.

## 3 Phase Retrieval Algorithm Block

Figure 3 illustrates that how the PR algorithm works with neural networks. The defocus images are divided into two groups, one is taken before the focus plane, the other is taken after the focus plane. Two groups of images are sent to different neural networks to calculate low order Zernike coefficients. The neural network also computes the fourth Zernike coefficient; therefore, the accurate defocus position is not necessary (the accurate defocus positions will be given after PR process, which helps to determine the in-focus position). The neural network parametric block will give out low order Zernike coefficients for each PSF image, which can be used to compute a priori phase for each PSF image. Then the MGS (modified Gerchberg–Saxton) algorithm [12] will be executed as inner loop to iterative-transform the phase of optical system, which will be used to calculate the PSF. The calculated PSF is compared with the PSF taken by detector by using root mean square algorithm (root mean square: RMS). If the PSF RMS meets the limit δ, the outer loop halts. Then the unwrap algorithm will be executed and the WFE will be outputted by iterative-transform block. If the RMS does not meet the limit δ, the a priori phase will be recalculated and the iterative-transform block will be executed again.

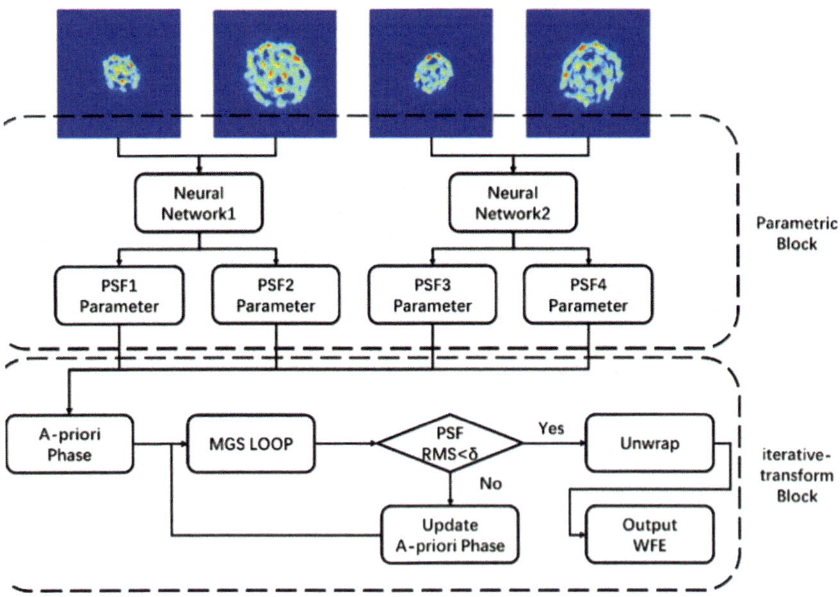


**Fig. 3** Phase retrieval algorithm block diagram

# 4 Phase Retrieval Experiments

Two experiments are designed to verify the algorithm. The first one is shown in Fig. 4. The He–Ne polarized laser, 20X beam expander, and linear polarizer are combined as light source, the output power of laser can be easily adjusted by rotating the linear polarizer. The focusing lens (f = 100 mm) and 5 μm pinhole are used as point source generator. The optical system includes refracting telescope (f = 850 mm) and refracting collimator (f = 1600 mm). Although the collimator and pinhole are used to simulate the starlight, both the collimator and the telescope introduce WFE, So they are tested by 4D interferometer and PR as one optical system. The PCO sCMOS camera with 6.5 μm pixel size is used as detector to taking defocus PSF images.

Four defocus PSF images are taken by detector, two of them are taken before FP, the rest of them are taken after FP. All of them are loaded by PR Software which embeds the neural networks. The PR result is compared with 4D interferometer test result to analyze the accuracy of algorithm (see Fig. 5 and Table.1), both results are analyzed in different spatial frequency by using ZYGO MetroPro (by removing first 4, 16, and 36 terms of fringe Zernike), the difference between them is also shown in Fig. 5, which shows that the PR result is highly reliable in high spatial frequency.

The second experiment replaces the refracting telescope (f = 850 mm) with one 2-lens optical system. The design diameter of 2-lens optical system is 100 mm, and the design RMS of WFE is 0.0256 wave (λ = 632.8 nm). In this experiment, the 40 mm diameter circle pupil is setup at pupil mount, and the 2-lens system off-axis value is 20 mm, the design RMS of WFE is 0.0222 wave (λ = 632.8 nm). The 3-axis decenter values of the second lens can be easily adjusted by linear translation stages (see Fig. 6).

Both low dynamic WFE (test1) and high dynamic WFE (test2) have been tested by 4D PhaseCam 6000 and PR, the results are also both compared with Zemax

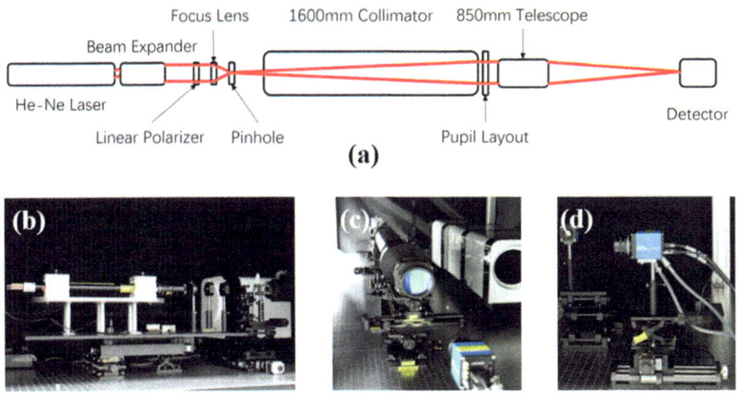


**Fig. 4** (**a**) Experiment 1 layout; (**b**) photo of laser system; (**c**) photo of 850 mm telescope; (**d**) photo of detector

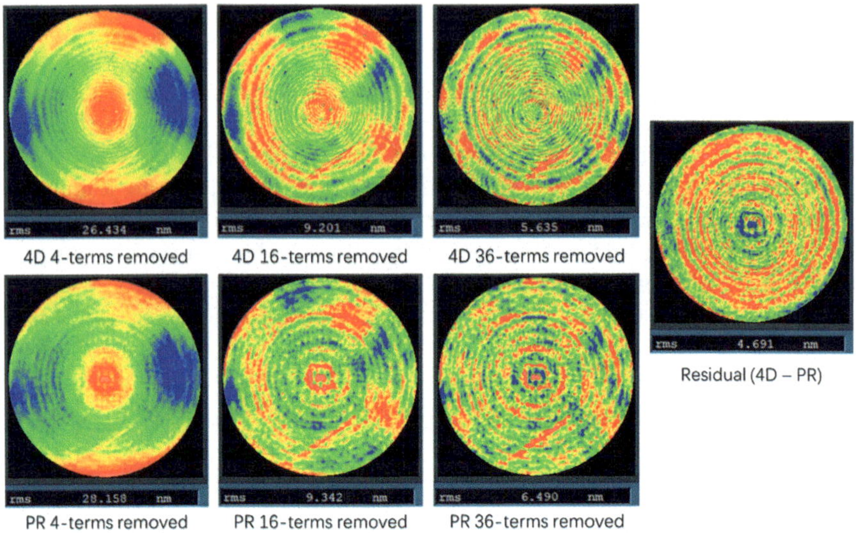


**Fig. 5** Experiment 1 results: high spatial frequency capability of PR

**Table 1** The comparison of experiment 1 results (unit: λ @632.8 nm)

| Zernike | z5 | z6 | z7 | z8 | z9 | z10 | z11 | z12 | z13 | z14 | z15 | z16 |
|---|---|---|---|---|---|---|---|---|---|---|---|---|
| PR | −0.066 | −0.009 | 0.006 | −0.002 | 0.052 | −0.017 | 0.014 | 0.005 | −0.003 | 0.009 | −0.002 | −0.021 |
| 4D | −0.073 | −0.009 | 0 | −0.005 | 0.037 | −0.015 | 0.011 | 0.01 | −0.001 | 0.006 | −0.004 | −0.02 |

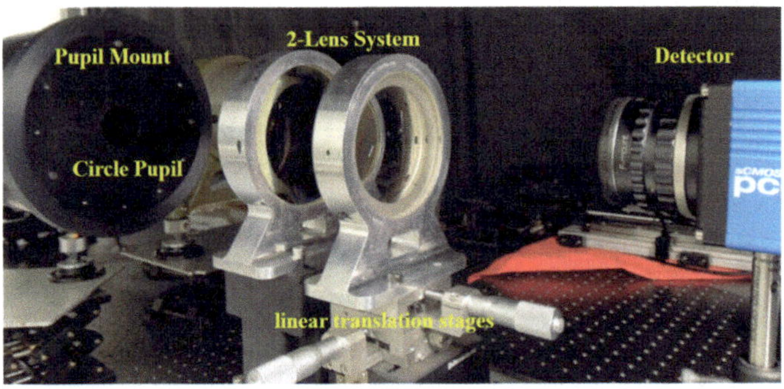


**Fig. 6** Experiment 2 layout

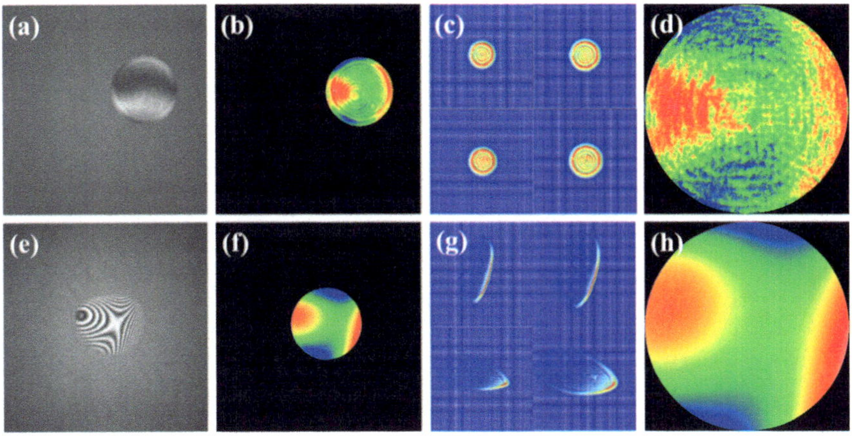


**Fig. 7** Experiment 2 results: (**a**) Test1 interferogram; (**b**) Test1 4D result; (**c**) Test1 PSF images; (**d**) Test1 PR result; (**e**) Test2 interferogram; (**f**) Test2 4D result; (**g**) Test2 PSF images; (**h**) Test2 PR result

model. With precision linear translation stages, the miss-align values of three axes decenter of second lens can be known as input of optical model, so the WFE can be computed by Zemax as comparison criterion (the Zemax data does not take account of the surface error of 2-mirror system and collimator, but it can be calibrated by the 4D test1 data which is tested in design circumstance). The test results are shown in Fig. 7 and Table 2. Both of them illustrate that the PR has the high-accuracy compared with 4D PhaseCam interferometer in low dynamic situation. As discussed above, the 4D test data in low dynamic situation also is used as calibration data for Zemax model. But PR with neural networks has better performance in high dynamic range situation than 4D PhaseCam. The 4D PhaseCam test result does not fit the Zemax model well due to the low contrast of interference fringe in the edge of the circle mask (see Fig. 7e).

## 5 Algorithm Implementation Using GPU

PR algorithm is a high intensity calculation process. As an initial design, the algorithm is developed with parallel CPU computation, which takes minutes to reach full convergence. To shorten the computation time, the nonlinear optimization algorithm is taken place by neural networks, and the MGS algorithm is redesigned to run with GPU clusters as well. Neural network costs a lot of computing resource during the training stage, but costs very few during the execution stage. The neural network is trained through NVIDIA Titan V GPU clusters, which shortens the time of parametric block from tens of seconds to less than 0.1 second. Now, with GPU helps,

**Table 2** The comparison of experiment 2 results (unit: λ @632.8 nm)

| | Test1 results | | | | Test2 results | | | |
|---|---|---|---|---|---|---|---|---|
| | 4D | PR | ZEMAX | ZEMAX Calibrated | 4D | PR | ZEMAX | ZEMAX Calibrated |
| Rms | 0.025 | 0.027 | 0.022 | – | 1.318 | 1.279 | 1.260 | |
| Z5 | 0.045 | 0.047 | 0.046 | 0.045 | 2.705 | 2.654 | 2.623 | 2.622 |
| Z6 | 0.004 | 0.000 | 0.000 | 0.004 | −1.462 | −1.375 | −1.293 | −1.289 |
| Z7 | 0.023 | 0.025 | 0.023 | 0.023 | 0.963 | 0.954 | 0.963 | 0.963 |
| Z8 | −0.006 | −0.004 | 0.000 | −0.006 | −0.434 | −0.507 | −0.477 | −0.483 |
| Z9 | −0.006 | −0.003 | −0.001 | −0.006 | 0.129 | −0.010 | 0.017 | 0.012 |
| Z10 | −0.023 | −0.019 | −0.011 | −0.023 | −0.141 | 0.007 | 0.022 | 0.010 |
| Z11 | 0.001 | 0.003 | 0.000 | 0.001 | −0.032 | 0.007 | −0.015 | −0.014 |
| Z12 | −0.015 | −0.020 | −0.008 | −0.015 | −0.252 | 0.021 | 0.009 | 0.002 |
| Z13 | 0.004 | 0.002 | 0.000 | 0.004 | 0.087 | 0.017 | −0.008 | −0.004 |
| Z14 | −0.004 | 0.000 | −0.003 | −0.004 | −0.045 | 0.004 | 0.000 | −0.001 |
| Z15 | −0.003 | 0.001 | 0.000 | −0.003 | 0.140 | −0.010 | −0.002 | −0.005 |
| Z16 | −0.005 | −0.003 | −0.002 | −0.005 | 0.083 | 0.015 | 0.000 | −0.003 |

our PR algorithm costs 10s in minimum (for 4 PSF images) to reach full convergence for the optical WFE.

## 6 Conclusion

As discussed in this paper, this PR algorithm uses neural networks to calculate the low order Fringe Zernike coefficients, which can avoid the local minimum problem of nonlinear optimization, and short the computing time of parametric block. This PR algorithm also uses MGS algorithm as an inner loop and RMS of PSF as feedback of the outer loop, which provides both high dynamic range and high spatial frequency retrieval wavefront error of the optical system. Two experiments were designed to verify the algorithm, and both of them get satisfying results. The algorithm is developed with GPU kernel, which costs 10s in minimum to process 4 PSF images to reach full convergence for the optical WFE with 512*512 pixels.

## References


1. Dean, B.H., Aronstein, D.L., Smith, J.S., Shiri, R., Acton, D.S.: Phase retrieval algorithm for JWST flight and tested telescope. Proc. SPIE. **6265**, 626511 (2006)
2. Dean, B.H.: Looking at Hubble through the eyes of JWST. In: Proc. IEEE. Aerospace Conference (2007). https://doi.org/10.1109/AERO.2007.353005
3. Misell, D.L.: A method for the solution of the phase problem in electron microscopy. J. Phys. **D6**, L6–L9 (1973)

4. Gerchberg, R.W., Saxton, W.O.: Phase determination from image and diffraction plane pictures in an Electron-microscope. Optik. **34**, 275 (1971)
5. Gerchberg, R.W., Saxton, W.O.: A practical algorithm for determination of phase from image and diffraction plane pictures. Optik. **35**, 237–246 (1972)
6. Gonsalves, R.A., Considine, P.: Phase retrieval from modulus data. J. Opt. Soc. Am. **66**, 961–964 (1976)
7. Southwell, W.H.: Wavefront analyzer using a maximum likelihood algorithm. J. Opt. Soc. Am. **A3**, 396–399 (1977)
8. Bely, P.Y.: The Design and Construction of Large Optical Telescopes, pp. 311–334. Springer, New York (2003)
9. Fienup, J.R., Marron, J.C., Schulz, T.J., Seldin, J.H.: Hubble space telescope characterized by using phase retrieval algorithms. Appl. Opt. **32**, 1747–1767 (1993)
10. Lyon, R., Miller, P.E., Grusczak, A.: Hubble space telescope phase retrieval: a parameter estimation. Proc. SPIE. **1567**, 317–326 (1991)
11. Krist, J.E., Burrows, C.J.: Phase retrieval analysis of pre and post-repair Hubble space telescope images. Appl. Opt. **34**, 4951–4964 (1995)
12. Acton, D.S., Atcheson, P., Cermak, M., Kingsbury, L., Shi, F., Redding, D.C.: James Webb space telescope wavefront sensing and control algorithms. Proc. SPIE. **5487**, 887–896 (2004)
13. Bikkannavar, S., Redding, D., Green, J., Basinger, S., Cohen, D., Lou, J., et al.: Phase retrieval methods for wavefront sensing. Proc. SPIE. **7739**, 77392X (2010)
14. Fienup, J.R.: Phase retrieval algorithms: a comparison. Appl. Opt. **21**, 2758–2769 (1982)
15. Brady, G.R., Fienup, J.R.: Effect of broadband illumination on reconstruction error of phase retrieval in optical metrology. Proc. SPIE. **6617**, 66170I (2007)
16. Scott Smith, J., Aronstein, D.L., Dean, B.H.: Phase retrieval on broadband and under-sampled images for the JWST testbed telescope. Proc. SPIE. **7436**, 74360D (2009)
17. Atcheson, P., Acton, S., Lightsey, P.: Instrument-level phase retrieval wavefront sensing and correction for astronomical telescope. Proc. SPIE. **4839**, 228–233 (2003)
18. Scott Acton, D., Scott Knight, J., Contos, A., Grimaldi, S., Terry, J., Terry, P., et al.: Wavefront sensing and controls for the James Webb space telescope. Proc. SPIE. **8442**, 84422H (2012)

# The Greenhouse Gas Instrument

**Ning An, Hedser Van Brug, Ming Li, Xiaolin Liu, Yugui Zhang, Yuxiang Liu, and Dazhou Xiao**

**Abstract** The climate change caused by the greenhouse gases in the atmosphere which are increasing year by year has become one of the most important questions on earth. To solve the climate question, we need to detect these gases in the atmosphere, obtain their concentration distribution, and analyze their sources. The Greenhouse Gas Instrument (GHGI) is a near-infrared and short-wave infrared (NIR/SWIR) nadir solar backscatter spectrometer, which is used to detect the concentration distribution of global greenhouse gases and increase the quantitative inversion precision of the gases further, gases measured include $CO_2$, $CH_4$, CO, and so on. GHGIs spatial resolution is 3 km × 3 km, spectral resolution is higher than 0.1 nm. In this paper, the instrument and its performance are discussed.

**Keywords** Greenhouse gas · Spectrometer · Grating · Polarization · Calibration

## 1 Introduction

The climate change caused by increased greenhouse gases in the atmosphere has become one of the most important global questions, but we still do not know accurately about the source and aggregation process of greenhouse gases [1, 2]. The Greenhouse Gas Instrument (GHGI) is a high resolution nadir solar backscatter spectrometer used to detect concentration distribution of global greenhouse gases and increase the quantitative inversion precision of the gases further, gases include $CO_2$, $CH_4$, CO, etc.

GHGI will work at sun-synchronous polar orbit at 836 km altitude, spatial resolution is 3 km × 3 km, swath is 100 km, wavelength ranges are near-infrared and short-wave infrared (NIR/SWIR). GHGI is a four-channel grating spectrometer, which can record spectra of the $O_2$A-band (754–770 nm, spectral resolution is 0.04 nm, denoted

N. An (✉) · M. Li · X. Liu · Y. Zhang · Y. Liu · D. Xiao
Beijing Key Laboratory of Advanced Optical Remote Sensing Technology, Beijing Institute of Space Mechanics and Electricity, Beijing, People's Republic of China

H. Van Brug
TNO – Technical Sciences, Delft, The Netherlands



H. P. Urbach, Q. Yu (eds.), *6th International Symposium of Space Optical Instruments and Applications*, Space Technology Proceedings 7,
https://doi.org/10.1007/978-3-030-56488-9_24

**Table 1** Characteristics of GHGI

| | |
|---|---|
| Orbit | Sun-synchronous polar orbit @836 km |
| Spatial resolution | 3 km × 3 km |
| Swath | 100 km |
| Field of view | ±3.6° |
| Spectra | B1: 754–770 nm<br>B2: 1591–1621 nm<br>B3: 2040–2080 nm<br>B4: 2330–2380 nm |
| Spectral resolution | B1: 0.04 nm<br>B2: 0.07 nm<br>B3: 0.09 nm<br>B4: 0.10 nm |
| Signal-to-noise (SNR) | B1: 350<br>B2: 340<br>B3: 230<br>B3: 200 |
| Observation modes | Nadir, glint, and stare |

as B1), a weak $CO_2$ band (1591–1621 nm, spectral resolution is 0.07 nm, denoted as B2), a strong $CO_2$ band (2040–2080 nm, spectral resolution is 0.09 nm, denoted as B3), and a $CH_4$ band (2330–2380 nm, spectral resolution is 0.1 nm, denoted as B4). For GHGIs four channels, the signal-to-noise values are between 200 and 350. Table 1 gives an overview of GHGIs characteristics.

Conceptually, GHGI is a passive imaging spectrometer. It comprises a scan mirror system, two telescopes, four spectrometers, four detectors, a cooler, and an on-board calibration system. A schematic diagram of the lightpath within the instrument is depicted in Fig. 1. The incoming radiation from earth enters the instrument through scan mirrors, by utilizing dichroic mirrors, the spectra are divided into four bands, every band is then dispersed by grating and imaged to the corresponding detector. Channels B2–B4 use a cooler to decrease the detector temperature to 80 K to decrease the influence of dark noise, and channel B1 utilizes a passive cooling method.

Scan mirrors comprise an along-track scan mirror (ALSM) and an across-track scan mirror (ACSM), which can make GHGI achieve various observation modes that include nadir, glint, and stare. Nadir mode is used to obtain continuous nadir scene projected to the detectors of GHGI, which can realize swath by 100 km and spatial resolution by 3 km. Glint mode is used to observe atmosphere above ocean. In this process GHGI controls scan mirrors to track the sunglint according to the parameters of orbit. Stare mode is used to observe some special targets such as volcanic eruption. In this process GHGI controls scan mirrors to compensate the relative movement between instrument and targets.

The on-board calibration system can achieve various calibration modes, which include achieving absolute calibration by diffuse reflector solar calibration, achieving pixel inconsistency measurement by lamp calibration, and achieving instrument line shape (ILS) measurement by laser diode calibration.

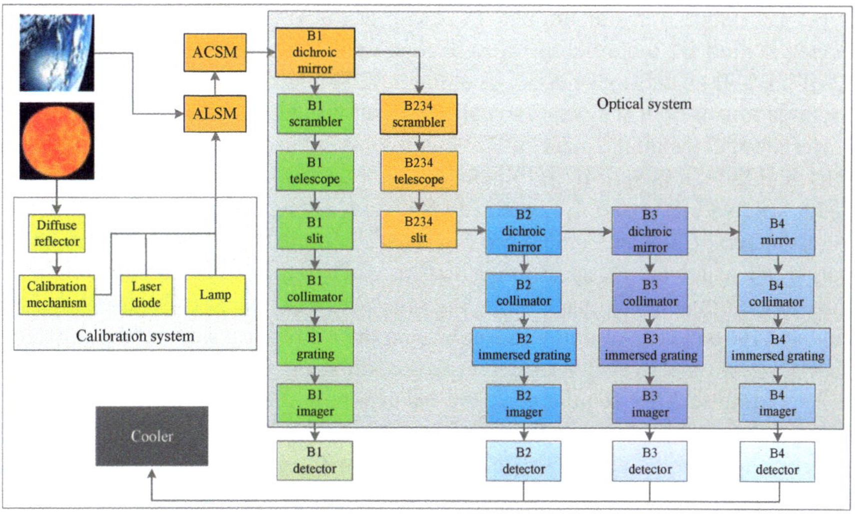


**Fig. 1** System configuration of GHGI

**Fig. 2** Optical system of GHGI

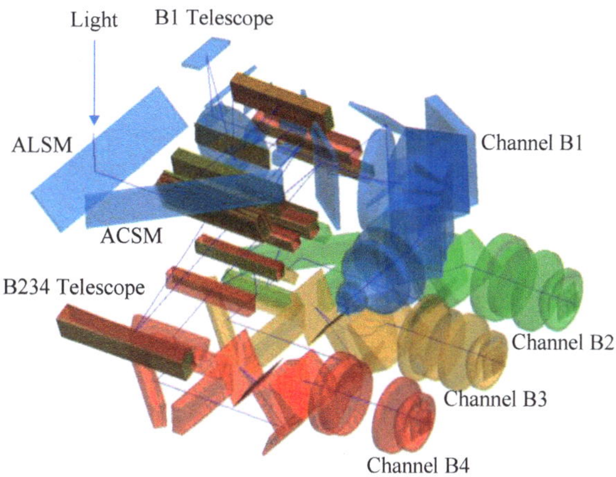


## 2 Optical System

The optical system is depicted in Fig. 2, which includes four channels. The incoming radiation from earth enters GHGI through scan mirrors. After the scan mirrors, B1 band is completely split off from B2–B4 bands by a reflection at B1 dichroic mirror. B1 scrambler is used to decrease polarization sensitivity of channel B1. Then B1 beams convergence at the image plane of B1 telescope, which is the entrance port of B1 slit. B1 telescope is a two mirror optical system. Exit light of B1 slit is collimated by B1 collimator and enters B1 grating, by which the B1 spectrum is dispersed. Finally the dispersed light is imaged to B1 detector through B1 imager.

B2–B4 bands transmit through B1 dichroic mirror, and a B234 scrambler is located behind B1 dichroic mirror to decrease polarization sensitivity of channels B2–B4. B2–B4 bands make use of a common B234 telescope, because the telescope is a three mirror optical system, so color aberrations can be ignored. B2–B4 beams convergence at the image plane of B234 telescope and use a common B234 slit. Exit light of B234 slit is split by B2 dichroic mirror and B3 dichroic mirror into three bands which are B2 band, B3 band, and B4 band. Three bands enter three different spectrometers. B2 beams are collimated by B2 collimator and enter B2 immersed grating, by which the B2 spectrum is dispersed. Then the dispersed light is imaged to B2 detector using B2 imager. Imaging processes and optical configuration of B3 band and B4 band are similar to B2 band, and their differences are different parameters of spectrometer, such as different radius, different grating grooves, and so on.

According to SNR requirement, the results show that the etendue of channels B2–B4 is about 6 times that of channel B1, channels B2–B4 require a much larger telescope aperture, so we use two telescopes and will solve the co-registration of B1 and B2–B4 through assembly. For channel B1, the telescope comprises two cylindrical mirrors, the aperture is F/9.9 × F/7 and the slit is 64.9 mm × 0.12 mm. For channels B2–B4, the telescope comprises a sphere mirror and two cylindrical mirrors, the aperture is F/14.4 × F/6.8 and the slit is 169.55 mm × 0.36 mm.

Grating formula is expressed as:

$$nd\left(\sin\alpha + \sin\beta\right) = m\lambda \tag{1}$$

where $n$ is the refractive index of the medium, $d$ is the groove distance of grating, $\alpha$ is the incidence angle, $\beta$ is the diffraction angle, $m$ is the diffraction order, $\lambda$ is the wavelength. So the resolving power of grating can be expressed as:

$$R = mN = Nd\frac{n\left(\sin\alpha + \sin\beta\right)}{\lambda} \tag{2}$$

where $N$ is the total number of grooves. If the grating substrate is planar and the groove distance is uniform, the term $Nd$ can be regarded as the physical size of the grating. From the resolving power formula we know that if the grating works in a high refractive index medium, the resolving power $R$ will scale with $n$. If we keep the resolving power unchanged, we can scale down the physical size of grating $Nd$ by a factor of $n$. When an equal F-number is used for the optics, the focal lengths of collimator and imager also scale with $n$, thus a reduction in volume of the spectrometer can be obtained up to $n^3$, also a reduction in mass can be obtained [3].

According to the weight and volume requirements of GHGI, channel B1 uses an ordinary plane grating, while channel B2–B4 use three immersed gratings to make their volume and mass smaller. The glass material of immersed grating is silicon by refractive index 3.4, which can yield a volume reduction factor of up to 40. For channel B1, the size of grating is 176 mm × 173 mm. For channels B2–B4, the size of gratings is 125 mm × 125 mm.

For channel B1, a single resolution element on the detector consists of a rectangle having the size of 3 detector pixels in spectral direction (3 × 22.5 μm = 67.5 μm) and 14 pixels in spatial direction (14 × 22.5 μm = 315 μm), the design results show that the RMS size of spots ranges from 5 to 20 μm in spectral direction and 10 to 65 μm in spatial direction, the MTF is higher than 0.87@7.4 cycles/mm in spectral direction and higher than 0.85@1.59 cycles/mm in spatial direction. For channels B2–B4, a single resolution element on the detector consists of a rectangle having the size of 3 detector pixels in spectral direction (3 × 24 μm = 72 μm) and 28 pixels in spatial direction (28 × 32 μm = 896 μm), the design results show that the RMS size of spots ranges from 8 to 30 μm in spectral direction and 20 to 110 μm in spatial direction, the MTF is higher than 0.8@6.9 cycles/mm in spectral direction and higher than 0.93@0.558 cycles/mm in spatial direction. From the results of spots size and MTF we know that the imaging performance of the spectrometer is very good.

## 3 Polarization

Due to Rayleigh scattering of atmospheric molecules and aerosols, the light reflected by the Earth atmosphere is generally polarized. GHGI is sensitive to the state of polarization of the light that enters the instrument, due to polarization dependent properties of mirrors, dichroics, and gratings. So we need to take some measures to make the instrument insensitive to the state of polarization of the light.

For GHGI, we use scrambler to make the instrument insensitive to polarization, which is a Dual Babinet Spatial Pseudo-depolarizer [4]. The scrambler comprises two pairs of H-V depolarizers; the wedge angle, center thickness, and glass material are identical, and the second pair has an optical axis orientation of 45° with respect to the optical axis of the first pair. Such a configuration can cover all incident polarization states. The sketch diagram of scrambler is depicted in Fig. 3.

The glass material of B1 scrambler and B234 scrambler is birefringent quartz, the wedge angles are 0.53° for B1 and 0.69° for B2–B4 [5]. When a linearly polarized light enters the scrambler, the polarization degree of exit light is depicted in Fig. 4. The results show that the polarization degree of exit light for all four bands is less than 2%.

The polarization effect of scan mirrors and B1 dichroic element should also be considered since they are before the scramblers. We made an orthogonal arrangement of ALSM and ACSM; thus, polarization effect for nadir mode can be avoided. With regard to coats for scan mirrors and B1 dichroic mirror, we also considered the design of coats to make the polarization sensitivities as small as possible according to their ray angles. At last, we analyzed the polarization sensitivity of the system, the results show that polarization sensitivity for four channels is less than 1.5%.

Because of the birefringent effect of quartz, the beam separation could not be overlooked. If the maximum separation is less than one sample pixel, then the effect could be ignored. We can use ZEMAX software to simulate the beam separation, and the results of beam separation of B1–B4 are depicted in Fig. 5. For

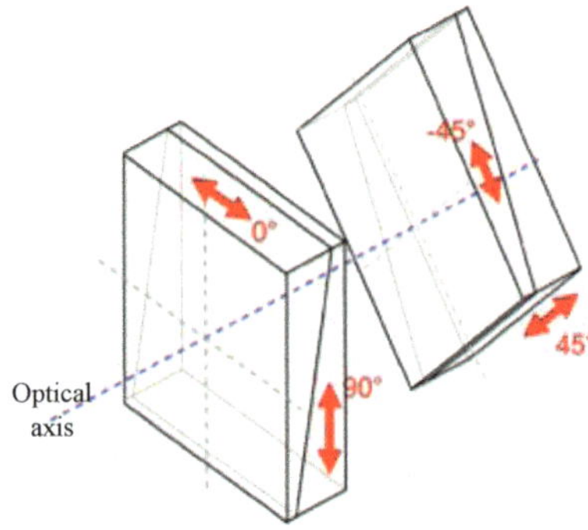


**Fig. 3** Sketch diagram for the scrambler. The second pair has an optical axis orientation of 45° with respect to the optical axis of the first pair

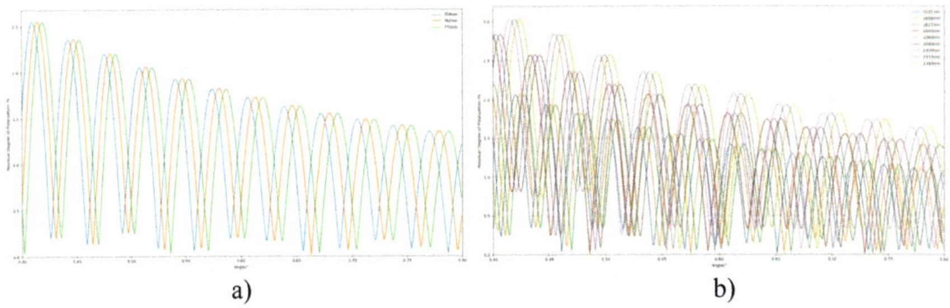


**Fig. 4** (**a**) Polarization degree of exit light of B1 scrambler; (**b**) polarization degree of exit light of B234 scrambler

channel B1, the maximum separation is about 25 μm, which is less than one sample pixel by 315 μm (co-adding number is 14 pixels and pixel size is 22.5 μm). For channel B2–B4, the maximum separations are almost identical by 91 μm, also less than one sample pixel by 896 μm (co-adding number is 28 pixels and pixel size is 32 μm). From the results we can conclude that the effect of beam separation can be ignored.

## 4 Calibration System

The on-board calibration system can achieve various calibration modes that include solar calibration, lamp calibration, and laser diode calibration. Solar calibration can achieve absolute calibration by diffuse reflector, lamp calibration can achieve pixel inconsistency measurement, and laser diode calibration can achieve instrument line shape (ILS) measurement.

The on-board calibration system comprises calibration mechanism, lamp assembly, and laser diode assembly. Calibration mechanism comprises a wheel and a rotating shaft. The diffuse reflectors are installed on the wheel. When solar calibration mode works, the wheel is rotated to let the sunlight shine on the diffuse reflector,

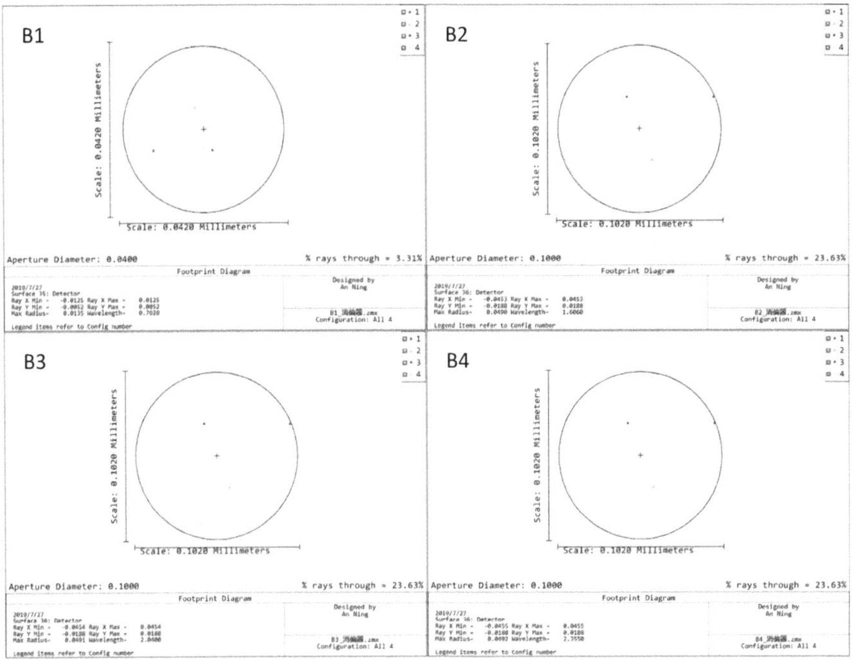


**Fig. 5** Beam separation of channel B1–B4. For channel B1, the maximum separation is about 25 μm; for channel B2–B4, the maximum separations are almost identical by 91 μm

and then the diffused light enters ALSM. When lamp and laser diode calibration work, the wheel is rotated to let the light from lamp and laser diode assembly go through and then enter ALSM. The working modes are depicted in Fig. 6.

Lamp assembly comprises a lamp, and laser diode assembly comprises 8 laser diodes, 8 collimating lens, dichroic mirrors, and a beam splitter. Lamp assembly and laser diode assembly use a common integrating sphere and collimator. The schematic diagram of optical system is depicted in Fig. 7. Every band for channel B1–B4 utilizes two laser diodes to achieve high-precision on-board ILS measurement, so there are eight laser diodes of the laser diode assembly, which needs eight collimating lens to aggregate the energy. For each laser diode there is an attenuator to avoid the energy too strong. With dichroic mirrors and beam splitter, collimating light from eight laser diodes is relayed to the integrating sphere together. The lamp is installed on the integrating sphere. The integrating sphere is used to make the light from lamp and laser diodes become uniform. Exit light from integrating sphere is then collimated by collimator, which is a two mirror optical system. Collimated light by collimator can cover full stop pupil and full field angle and provide very uniform illumination. The illumination distribution of on-axis field and 3.6° field at stop pupil position is depicted in Fig. 8, which shows that the nonuniformity of illumination distribution is very small and can satisfy the calibration requirements.

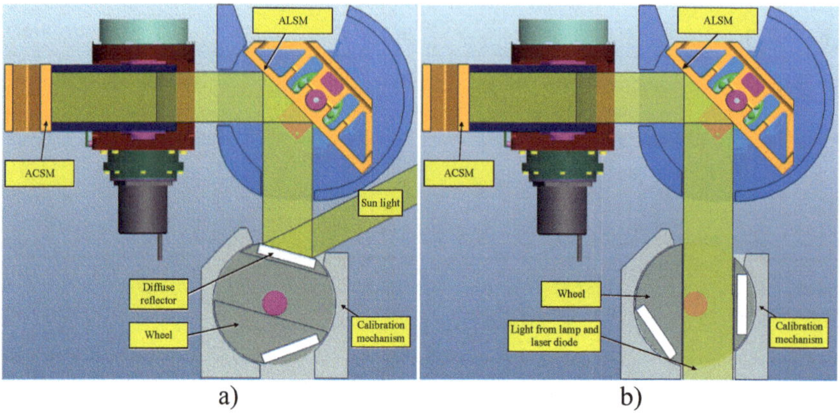


**Fig. 6** (**a**) Solar calibration mode; (**b**) lamp and laser diode calibration mode

**Fig. 7** Schematic diagram of optical system of lamp and laser diode assembly

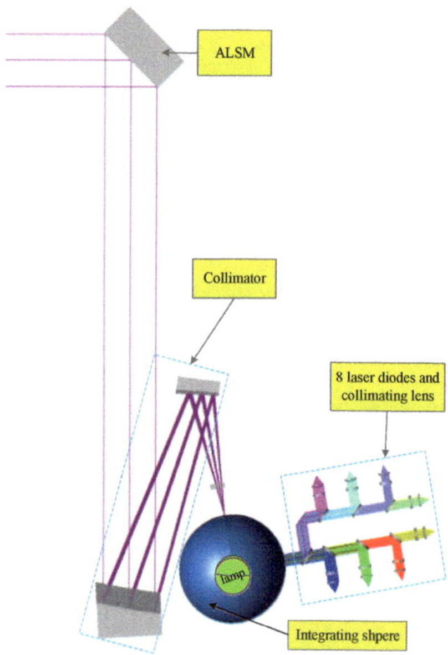


## 5 Performance Testing

Through using tunable diode laser measurements to achieve instrument test and fit test data with Gaussian function, ILS of channel B1 and B2 is obtained. ILS of channel B1 is depicted in Fig. 9, and ILS of channel B2 is depicted in Fig. 10. The results show that full width at half maximum (FWHM) is 0.04037 nm by wavelength 762 nm and 0.0637 nm by wavelength 1606 nm, which means that the spectral resolution can meet the requirement. Channel B3 and channel B4 are still in the assembly phase and will be completed by the end of 2019.

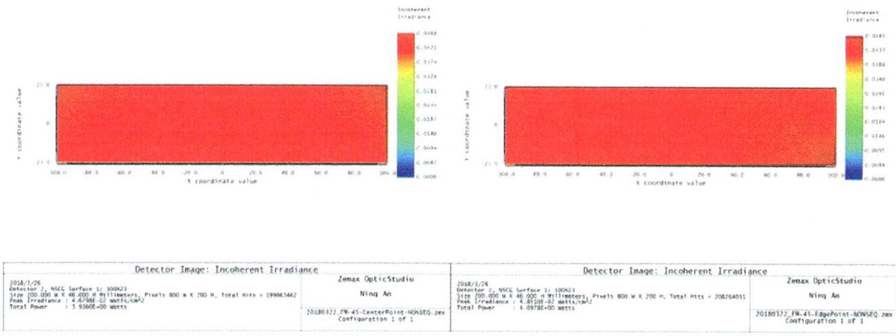


a) b)

**Fig. 8** (**a**) Illumination distribution of on-axis field at stop pupil position; (**b**) illumination distribution of 3.6° field at stop pupil position

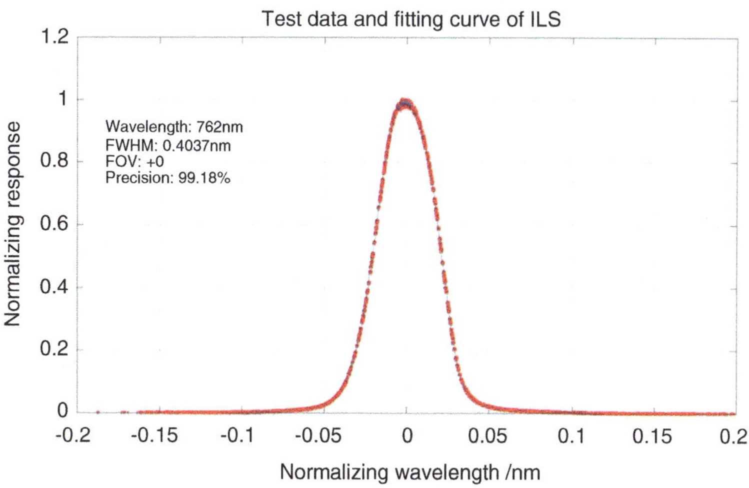


**Fig. 9** ILS of channel B1. The FWHM is 0.04037 nm@762 nm

## 6 Conclusion

GHGI is used to detect the concentration distribution of global greenhouse gases and provides a high-precision quantitative inversion of greenhouse gases. In this paper, we discussed the function and design of the instrument. Currently, the instrument is not completed as channel B3 and channel B4 are still in assembly phase. However, from the test results of channel B1 and channel B2 we can see that the instrument can achieve its technical specifications; GHGI will play an important role in the detection of greenhouse gases.

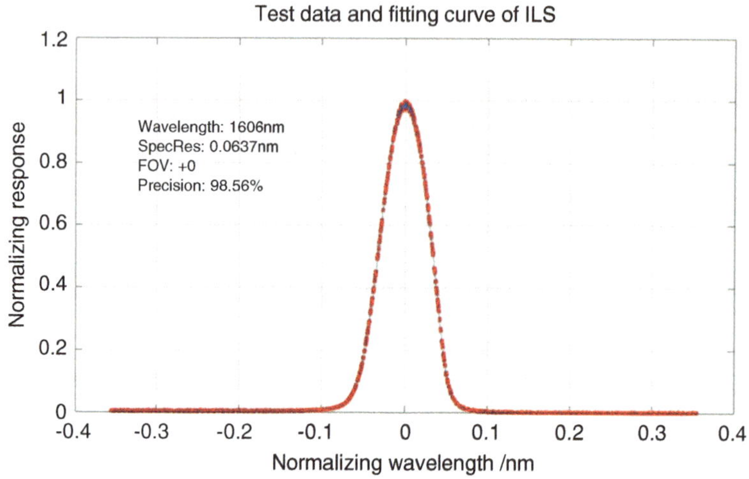


**Fig. 10** ILS of channel B2. The FWHM is 0.0637 nm@1606 nm

# References


1. Bovensmann, H., Burrows, J.P., Buchwitz, M., et al.: SCIAMACHY: mission objectives and measurement modes. J. Atmos. Sci. **56**(2), 127–150 (1999)
2. Levelt, P.F., van den Oord, G.H.J., Dobber, M.: TROPOMI and TROPI: UV/VIS/NIR/SWIR instruments. Proc. SPIE. **6296**, 629619 (2006)
3. van Amerongen, A.H., Visser, H., Vink, R.J.P., Coppens, T., Hoogeveen, R.W.M.: Development of immersed diffraction grating for the TROPOMI-SWIR spectrometer. Proc. SPIE. **7826**, 8 (2010)
4. McGuire Jr, J.P., Chipman, R.A.: Analysis of spatial pseudodepolarizers in imaging systems. Opt. Eng. **29**(12), 1478–1484 (1990)
5. Xiaolin, L., Ming, L., Ning, A., Shaofan, T., Qian, S.: Design and analysis of dual babinet depolarizer applied to rectangular pupils. Laser Optoelectr. Progr. **55**, 082601 (2018)

# Lightweight Design of a Kind of Primary Mirror Component for Space Telescope

**Yang Song, Wenyi Chai, Yongming Hu, Wei Xin, Yaoke Xue, and Chenjie Wang**

**Abstract** With the improvement of space optics technology, lightweight design is becoming more and more important in the process of designing a space telescope. The primary mirror component is a main part in the space telescope system, the deadweight of it affects the weight of the whole telescope. The paper makes a lightweight design for a kind of primary mirror component. The primary mirror component has three parts. These three parts are primary mirror, supports, and support plate. Special hexagonal holes are dug on the back of primary mirror. Furthermore, the back of mirror has a particular curvy surface. The purpose is to make the primary mirror lighter. The supports are three bipod supports, which have some flexible links. The support plate adopts the symmetrical design, and most areas of it have been hollowed out with unique method for weight loss. What's more, some region of support plate has stiffeners and ribs for high reliability. Structural analysis, which is based on the finite element method, is conducted under different kinds of load cases. The results of simulations show that the structure is reliable with a large safety margin. The structure has enough strength, stiffness, and stability. It indicates that lightweight design in this paper is feasible.

**Keywords** Lightweight design · Space telescope · Primary mirror component · Structure analysis

---

Y. Song · W. Chai (✉) · Y. Hu · W. Xin · Y. Xue
Xi'an Institute of Optics and Precision Mechanics of Chinese Academy of Sciences, Xi'an, China
e-mail: songyang@opt.ac.cn; huyongming@opt.ac.cn; xinwei@opt.ac.cn

C. Wang
Xi'an Institute of Optics and Precision Mechanics of Chinese Academy of Sciences, Xi'an, China

Xi'an Jiaotong University, Xi'an, China
e-mail: wangchenjie@opt.ac.cn



H. P. Urbach, Q. Yu (eds.), *6th International Symposium of Space Optical Instruments and Applications*, Space Technology Proceedings 7,
https://doi.org/10.1007/978-3-030-56488-9_25

# 1 Introduction

The number of space telescopes is increasing in recent years for the development of space optics technology. How to design and manufacture the space telescope is becoming an important issue for designers and engineers. Due to the payload of remote sensing or observation satellite, the performance, volume, and weight of space telescope are the key factors affecting the performance of satellite. Besides, the extensive use of small and light satellites puts forward more stringent requirements for the weight and performance of space telescopes. Therefore, the development trend of space remote sensing or observation is to design a space telescope with lighter weight, smaller volume, and better performance. The primary mirror component is the main part in the space optical system. Therefore, more and more engineers put their efforts in making a lightweight design of primary mirror component. This is a matter of great significance, which will lower the launch cost.

The paper makes a design of a primary mirror component, the aperture of which is 720 mm. The paper adopts a special method to make the weight reduction holes in the structure. What's more, the flexible supports are designed particularly. To test the reliability of structure component, a series of simulations based on finite element method is conducted. The analysis indicates that the design scheme is safe and reliable.

# 2 Present Status on Lightweight Design of Primary Mirror

The primary mirror component plays a vital role in space optical system. The design scheme includes the specific structural style of primary mirror and the way of supporting. Some typical designs of primary mirror component in telescope are shown in the figures below (Fig. 1).

Pleiades [1], Snap [2], and Kepler [3] are three typical designs about primary mirror component. A brief introduction is made here about the primary mirror in these three kinds of design. In Pleiades, the diameter of primary mirror is 650 mm. It is made of glass ceramics. And there are some round holes in the bottom of mirror.

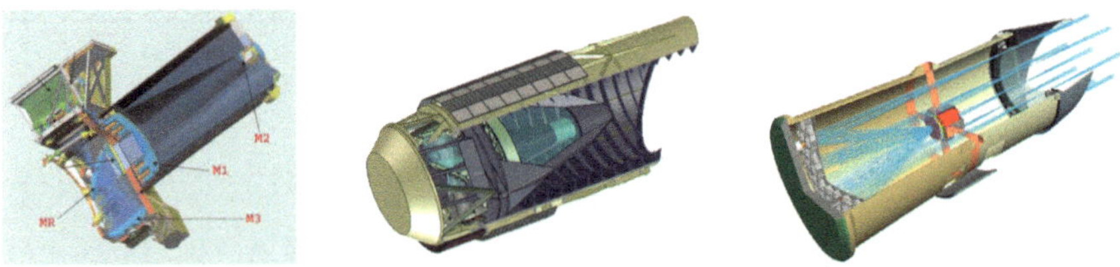

(a) (b) (c)

**Fig. 1** Some typical designs of primary mirror component in telescope: (**a**) Pleiades, (**b**) Snap, (**c**) Kepler

In Snap, the diameter of primary mirror is 2050 mm. It is made of ultra-low expansion glass (ULE). In Kepler, the primary mirror, which is also made of ultra-low expansion glass (ULE), is 1400 mm. The primary mirror of these telescopes all adopted the lightweight design with holes in the back of mirror for weight loss. The shape of weight loss hole is mainly round, triangle, hexagon, and so on [4]. However, strength, stiffness, and stability should be ensured for safe operation. What's more, the ways of supporting for primary mirror cannot be ignored in the process of structure designing. Generally speaking, there are three kinds of supporting. First kind is that the mirror is supported on the outer edge. Second kind is that the mirror is supported at through-hole on the central axis. Third kind is that the mirror is supported by three-point flexible supports. Bipods are often used in the three-point supports. Pleiades and Snap both adopt bipod as support. A bipod is a typical two-degree freedom element. It can give the mirror statically determinate constraints. Two common kinds of bipods are normal bipod and inverted bipod according to different connection styles with reflective mirror. It means that there are three connection points as we use normal bipod style. Besides, it means that there are six connection points as we use inverted bipod style. Though inverted bipod style can provide mirror with more stiffness, it increases the complexity of assembly.

Moreover, the material for the primary mirror in space telescope should be considered carefully. The common materials for mirror are zerodur, ultra-low expansion glass (ULE), silicon carbide (SiC), and carbon fiber composite [5, 6]. The parameters of the materials can be seen in the following table.

# 3 Design Scheme of Primary Mirror Component

## 3.1 The Whole Component

There are three parts in primary mirror component, including primary mirror, supports, and support plate. The component in this paper is called special lightweight primary mirror component. The whole structure is shown in the following figures.

The aperture of the primary mirror is 720 mm. To some extent, it is a telescope with large aperture, so the stiffness should be considered first. SiC is selected as the material for primary mirror because of the high specific stiffness. The material parameters of SiC can be seen in Table 1. Invar steel is selected for the supports by engineering experience. At last, titanium alloy is chosen for support plate because of high strength. The material parameters can be seen in Table 2. From Fig. 2, it can be seen that the three supports are connected to support plate by screws. Three supports are connected to the side face of mirror by epoxy adhesive. The mass of whole assembly body is 25.33 kg.

**Table 1** Common material parameters for primary mirror

| Material | Density (kg/m³) | Young's modulus (GPa) | Poisson's ratio | Specific stiffness | Coefficient of thermal expansion ($10^{-6}$/K) |
|---|---|---|---|---|---|
| Zerodur | $2.5 \times 10^3$ | 92.9 | 0.24 | 36.9 | 0.05 |
| ULE | $2.2 \times 10^3$ | 67 | 0.17 | 30.4 | 0.03 |
| SiC | $3.1 \times 10^3$ | 310 | 0.25 | 100 | 2.2 |
| Carbon fiber composite | $1.6 \times 10^3$ | 81.4 | 0.28 | 50 | According to layers |

**Table 2** Parameters of material for supports and support plate

| Material | Young's modulus (GPa) | Poisson's ratio | Density (kg/m³) | Coefficient of thermal expansion ($10^{-6}$/K) |
|---|---|---|---|---|
| Invar steel (4 J32) | 145 | 0.23 | $8.15 \times 10^3$ | 2.1 |
| Titanium alloy (TC4) | 109 | 0.34 | $4.45 \times 10^3$ | 8.8 |

(a) (b)

**Fig. 2** The whole component: (**a**) the explosive view, (**b**) assembly body

## 3.2 *The Design of Primary Mirror*

The aperture of primary mirror is 720 mm. The diameter of through-hole on the central axis is 150 mm for light path. The curvature radius of mirror surface is 4313 mm. The detailed characteristics of mirror can be seen in the figures below.

On the side surface of the mirror body, there are several ribs with a triangle shape to improve the stiffness of mirror. What's more, some hexagonal holes are dug on the back of the mirror. The diameter of inscribed circle where the hexagon is located is 54 mm. The wall thickness of these holes is 3 mm. The depth of these holes should meet the case that the minimum thickness of mirror is 4 mm. Generally speaking, the back of the mirror is a platform, although there are holes on the back. However, in order to get a lighter mirror, a big region with sphere shape is removed from the mirror body. The radius of this sphere is about 420 mm, which is about

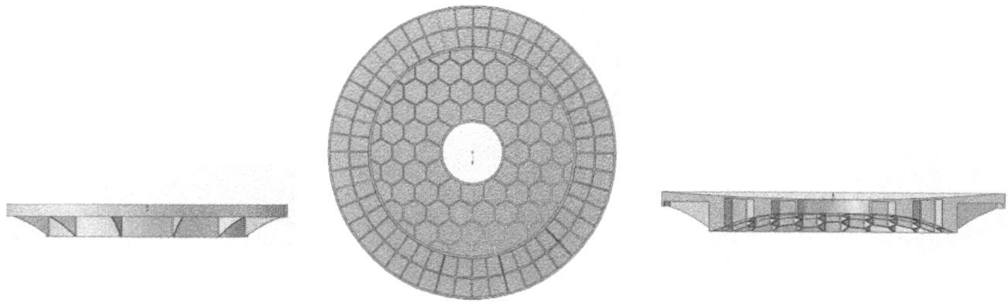

(a) (b) (c)

**Fig. 3** The primary mirror with different views: (**a**) the side view, (**b**) the bottom view, (**c**) the profile map view

1/10 of the curvature radius of mirror surface. That is to say, there is a big curvy surface at the bottom of the mirror. By now, such structural design has not been reported in the open literature. The profile map of the mirror body is shown in Fig. 3c. And the characteristic of the sphere can be seen in this figure. The purpose of this design is to make the mirror lighter and lighter with enough strength and stiffness. The lightweight rate [7] of the primary mirror is about 80%.

## 3.3 *The Design of Support Plate*

Strictly speaking, the support plate in this paper should be called as support frame, because much material is removed from the support plate. Some detailed features of support plate can be seen in the figures below (Fig. 4).

The support plate is a combination of two circles which are connected with five cross girders. Cross girder is similar to I-beam for its high modulus of flexural section. The areas where the support will be assembled are strengthened with ribs. And there are several holes which are distributed evenly in the outer circle part, the whole component is connected to other structure through these holes by screws.

## 3.4 *The Design of Supports*

There are three supports for supporting the primary mirror. The positions of the supports in the assembly body are on a circle. The reason why three supports are adopted is that this method can provide a statically determinate constraint. As is known to all, the structure will generate thermal stress and thermal deformation under the changing temperature [8]. Therefore, support with flexible links is designed, which is shown in the figures below.

This kind of supports has much elasticity. They can release the thermal stress when the temperature changes. The flexible links are in the normal bipod style. The

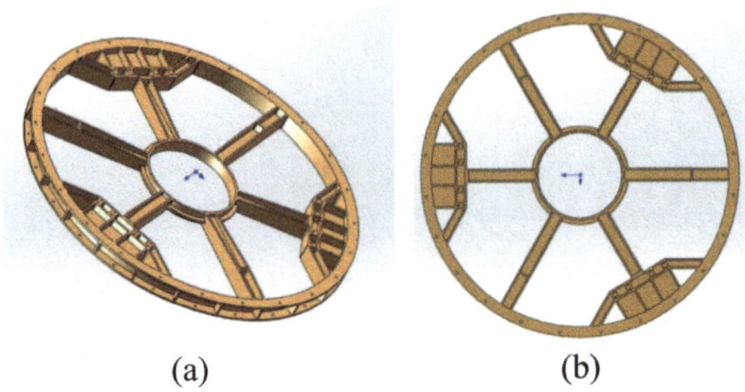

Fig. 4 The support plate with different views: (a) three-dimensional view, (b) the bottom or top view

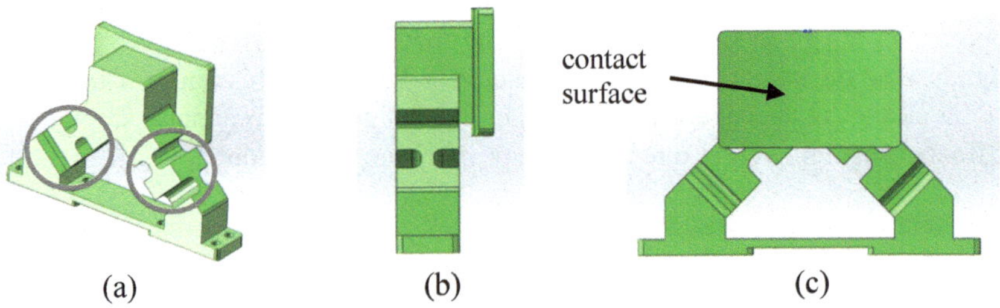


Fig. 5 The support with different views: (a) three-dimensional view, (b) side view 1, (c) side view 2

implementation method is that the section area of bipod section changes, which can be seen in circles in Fig. 5a. Moreover, the whole structure will not generate excessive assembly stress because of flexible links [9]. The contact surface of support will be glued to side surface of mirror by epoxy adhesive.

## 4 Structural Analysis

After completing the design scheme of the primary mirror component, structural analysis should be conducted immediately to verify the feasibility and reliability of the design scheme. Simulation software which is based on finite element method is used in the process of structural analysis. Finite element model is constructed according to the geometry model. Some thin-wall part in the model is meshed by shell elements, other part is meshed by hexahedral elements. There are 116,221 nodes and 137,751 elements in the whole model. The finite element model is shown in Fig. 6. The boundary condition is stated here. The holes on the outer circle of support plate are connected to other structure, so the six degrees of freedom of the nodes at these positions are constrained. The supports are connected to support plate

Fig. 6 The finite element model

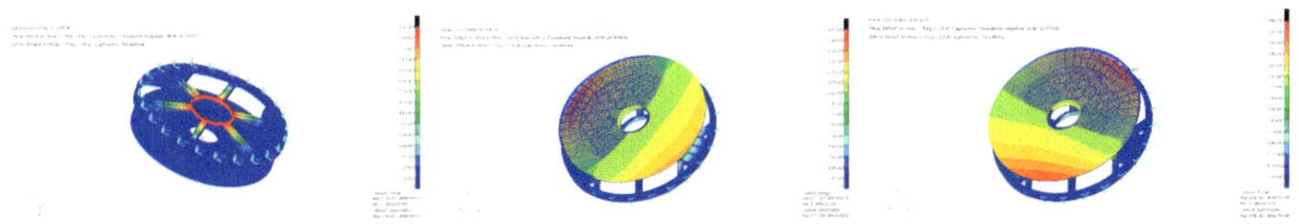


Fig. 7 The first three modals: (a) first order modal, (b) second order modal, (c) third order modal

by screws. So the contact elements are used in these contact areas. The supports are glued to the side surface of mirror, nodal degree of freedom coupling is used at the contact areas between the supports and mirror.

As a main part of the space telescope, the primary mirror component will be launched in space by rocket. Therefore, the fundamental frequency of the whole structure should be far away from the excitation frequency during the rocket launching. Therefore, modal analysis is done first. The result is shown in Fig. 7 and Table 3.

The frequency of first-order modal is 190 Hz, which occurs at the support plate. Second and subsequent modals occur on mirror, which are larger than 228 Hz. The fundamental frequency of primary mirror component in Snap, which is mentioned in Sect. 2, is only 111 Hz. The fundamental frequency of design scheme in this paper is more than that in Snap. Form the result of modal analysis, it can be seen that the structure design is feasible.

The primary mirror component is assembled and adjusted on the ground, so the whole structure is under 1 g gravity. The deformation is calculated by software. The deformation cloud charts of mirror under different directions of gravity are shown in Fig. 8. The PV values and RMS values are provided in Table 4.

From the result, the case of maximum PV and RMS is under 1 g gravity in X direction. The maximum PV is 52.54 nm, the maximum RMS is 9.29 nm. In other words, when $\lambda = 632.8$ nm, the maximum PV is $\lambda/12$, the maximum RMS is $\lambda/68$. This is acceptable for engineering practice.

**Table 3** The part result of modal analysis

| Order of modal | 1 | 2 | 3 | 4 | 5 | 6 | 7 |
|---|---|---|---|---|---|---|---|
| Frequency (Hz) | 190.21 | 228.46 | 228.47 | 256.57 | 256.59 | 288.85 | 290.86 |

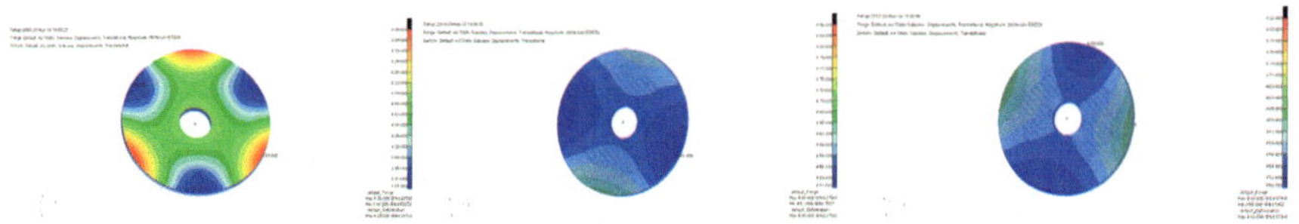

(a) (b) (c)

**Fig. 8** The deformation under different directions of 1 g gravity: (**a**) gravity in *X* direction, (**b**) gravity in *Y* direction, (**c**) gravity in *Z* direction

**Table 4** The PV and RMS under 1 g gravity

| | gravity in *X* direction | gravity in *Y* direction | gravity in *Z* direction |
|---|---|---|---|
| PV (nm) | 52.49 | 17.79 | 23.27 |
| RMS (nm) | 9.29 | 3.33 | 3.42 |

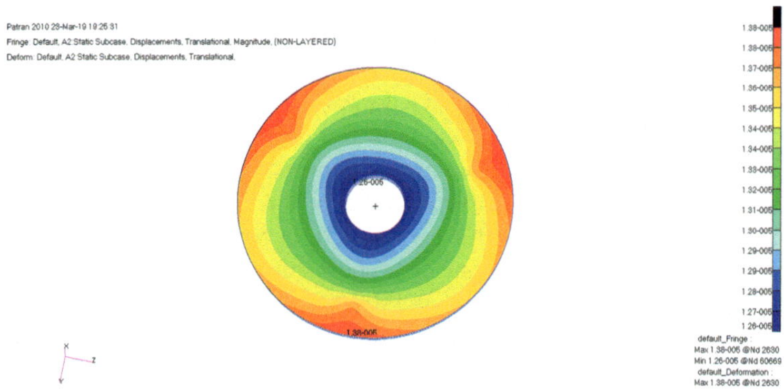


**Fig. 9** The deformation under temperature change (20–22 °C)

The whole structure will undergo temperature change in practical use in space. The temperature range is between 18 and 22 °C. The temperature of assembly on ground is 20 °C. So real temperature change in structure is 2 °C. The thermal analysis of structure, which is under 2 °C temperature change, is completed. The deformation cloud charts of mirror are shown in Fig. 9.

The PV and RMS value can be calculated according to the surface deformation of mirror. The PV is less than $\lambda/10$, the RMS is less than $\lambda/65$; the $\lambda$ is 632.8 nm. From this result, the small deformation of mirror is acceptable for engineering.

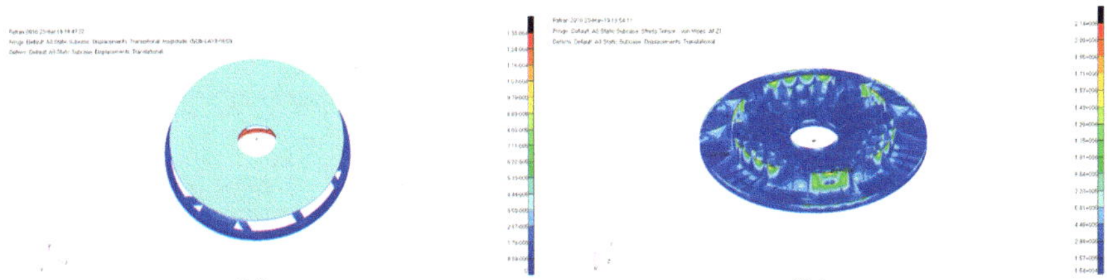

(a) (b)

**Fig. 10** The deformation and stress under overload in *X* direction: (**a**) the deformation cloud chart, (**b**) the Mises stress cloud chart

**Table 5** The deformation and Mises stress under overload in three directions

| | *X* direction | *Y* direction | *Z* direction |
|---|---|---|---|
| Max displacement (m) | $1.33 \times 10^{-4}$ | $4.86 \times 10^{-5}$ | $4.82 \times 10^{-5}$ |
| Max Mises stress (MPa) | 2.14 | 3.7 | 3.46 |

**Table 6** The input condition of random vibration

| | | | | |
|---|---|---|---|---|
| Range of frequency (Hz) | 10–100 | 200–600 | 600–2000 | Grms |
| Power spectral density | +3 dB/oct | $0.12g^2$/Hz | −3 dB/oct | 12.35 |

The whole structure will undergo the overload during the rocket launching. The analysis is done under 10 g static overload in three directions, respectively. The deformation and stress cloud charts under overload in X direction are shown in Fig. 10, for the space limit of paper. The maximum displacement and stress of three directions are shown in Table. 5.

It can be seen in Table 5 that the maximum of displacement is $1.33 \times 10^{-4}$ m, the maximum of Mises stress is 3.7 MPa. It indicates that the structure has enough strength and stiffness.

The situation of vibration during rocket launching is complex. So the random vibration analysis should be done. The characteristic of random vibration is that the phase and amplitude of vibration are unknown at any given moment. Random vibration describes the vibration process through its statistical feature of amplitude [10]. The input condition of random vibration is provided in Table 6.

The response curves in three different directions are in Fig. 11. The red curve is input curve. The response curves are in other colors. The response points are selected on the surface of inner and outer side on mirror. It can be seen that the maximum of dynamic response is 24 g. The maximum dynamic amplification coefficient is 2. It indicates that the primary mirror component is safe by previous analysis and engineering experience.

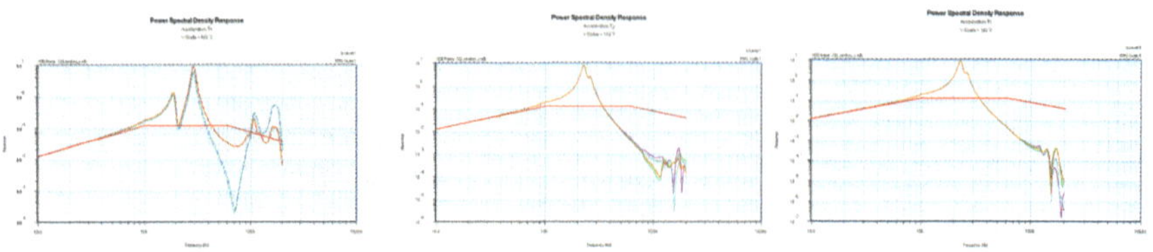

(a) (b) (c)

**Fig. 11** The input and response curves in random vibration analysis: (**a**) *X* direction, (**b**) *Y* direction, (**c**) *Z* direction

# 5 Conclusion

The paper makes a lightweight design of primary mirror component with an aperture of 720 mm. In order to get a lightweight design, weight loss holes are designed on the back of mirror. An innovative point in the paper is that the back surface of the mirror is sphere surface for further weight loss. The lightweight rate of it is about 80%. The support plate is made up of two circle parts with flanging structure and several connection cross girders. The mirror is supported by three bipod supports with flexible links, which can decrease the thermal deformation and thermal stress. Moreover, proper material is selected for the whole component to ensure that the structure has enough strength, stiffness, and stability. In order to test the feasibility and reliability of the design scheme, a series of simulations are conducted, which are based on finite element method. All the simulation results indicate that the whole structure is reliable with large safety margin. In other words, the lightweight design is reasonable. What's more, the similar size of primary mirror component can adopt the design scheme proposed in this paper.

# References


1. Gleyzes, M.A., Perret, L., Kubik, P.: Pleiades system architecture and main performances. International Archives of the Photogrammetry, Remote Sensing and Spatial Information Sciences. **39**(1), 537–542 (2012)
2. Lampton, M.L., Sholl, M.J., Krim, M.H.: SNAP telescope: an update. In: Uv/Optical/IR Space Telescopes: Innovative Technologies & ConceptsSPIE, Bellingham (2004)
3. Koch David, G., William, J., Borucki, L.W., Dunham, E.W.: Kepler: a space mission to detect earth-class exoplanets. In: Space Telescopes and InstrumentsInternational Society for Optics and Photonics, Bellingham (1998)
4. Yanjun, Q., Yanru, J., Liangjie, F., Xupeng, L., Bei, L., Wei, W.: Lightweight design of multi-objective topology for a large-aperture space mirror. Appl. Sci. **8**(11), 2259 (2018)
5. Jian, Y., Jianyue, R.: Improvement and optimization of lightweight Structure for SiC reflective mirror. Acta Photonica Sinica. **44**(8), 2259 (2015)
6. Bao, Q., Wei, S., Changzheng, C., Jianyue, R.: Ultra-lightweight design of 610 mm circular primary mirror supported in centre. Acta Photonica Sinica. **45**, 9 (2016)

7. Guo, S., Zhang, G., Li, L., Wang, W., Zhao, X.: Effect of materials and modelling on the design of the space-based lightweight mirror. Mater. Des. **30**(1), 9–14 (2009)
8. Shenhua, L., Yingjun, G., Hongwei, X., Ye, Y., Zhilai, L.: Lightweight design and flexible support of large diameter mirror in space camera. Laser & Infrared. **47**, 11 (2017)
9. Zongxuan, L., Xue, C., Lei, Z., Guang, J., Yuan, Z., Zhixue, J., et al.: Design of cartwheel flexural support for a large aperture space mirror. Acta Optical Sinica. **34**, 6 (2014)
10. Newland, D.E.: Random vibration. In: Handbook of Noise and Vibration ControlWiley, Hoboken (2007)

# SHE: Design and Development of 60 K Telescope Downsize Prototype

**Ruicong He and Zhimin Liu**

**Abstract** SHE (Search for Habitable Exoplanets) is the first large infrared astronomical telescope project in China dedicated to searching for habitable exoplanets. The project goals are to complete the satellite and payload technologies research and system design and to test key performances of the telescope by developing a downsize prototype. The SHE downsize prototype is an optical space telescope of 50 cm clear aperture with three channels of spectrum ranging from 0.4 to 5 micron. The telescope is passively cooled and thermally controlled to a cryogenic temperature of around 60 K. This paper gives an overview of the SHE instrument optical, mechanical, and thermal design.

**Keywords** Exoplanets · Space telescope · Cryogenic temperature

## 1 Introduction

The SHE (Search for Habitable Exoplanets) project is a state-supported mission with the aim of satisfying the high demand from scientists of academic and research institutes in the field of space science at home and across the world. The idea behind the payload design is to be pulled by scientific needs, including atmospheric transmission spectrum detection of exoplanets, study of the physical and chemical characteristics of earth-like exoplanets, further study of the origin and evolution of planets and life, and exploration of extraterrestrial life and so on.

The project research task consists of two parts. The first one is to complete the satellite and payload technologies research and system design of a 2 m clear aperture telescope, and the second one is to test key performances of the telescope by developing a 50 cm clear aperture downsize prototype. The whole telescope works

R. He (✉) · Z. Liu
Beijing Key Laboratory of Advanced Optical Remote Sensing Technology, Beijing Institute of Space Mechanics and Electricity, Beijing, People's Republic of China
e-mail: herc_bisme@spacechina.com



H. P. Urbach, Q. Yu (eds.), *6th International Symposium of Space Optical Instruments and Applications*, Space Technology Proceedings 7,
https://doi.org/10.1007/978-3-030-56488-9_26

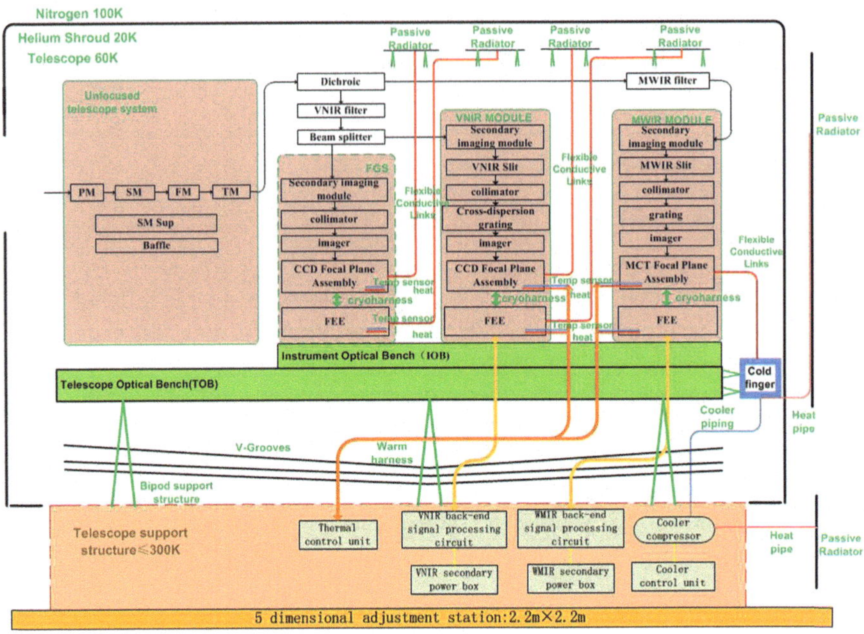


**Fig. 1** SHE prototype functional block diagram

at cryogenic temperature. The current prototype baseline architecture is shown in Fig. 1 as a functional block diagram.

## 2 Optical Design

Figure 2 shows the layout of the telescope and three channels. The optical system is mainly composed of the telescope module, common optics, filters, visible near-infrared (VNIR) module, medium wavelength infrared (MWIR) module, and fine guidance system (FGS) module.

The telescope design has an obscured central field which is a kind of annular field Korsch telescope. In this design we set an off-set field of ~0.35° to use an unobscured FOV to avoid any beam clashes in the optical path with the other mirrors. To get a compact layout, a fold mirror (FM) is added just behind the primary mirror (PM) which allows the system to be folded within the required space. There is an intermediate image plane between the FM and the tertiary mirror (TM). The entrance aperture is at the PM. After prime focus, TM, an off-axis parabola (OAP), produces a collimated beam.

After that, the beam is split into the two channels 0.4–0.9 μm and 3.2–5 μm by the dichroic [3]. The dichroic reflects the VNIR wavelengths to the VNIR and FGS channels, and the reflected beam passes to a beam splitter, which reflects to the

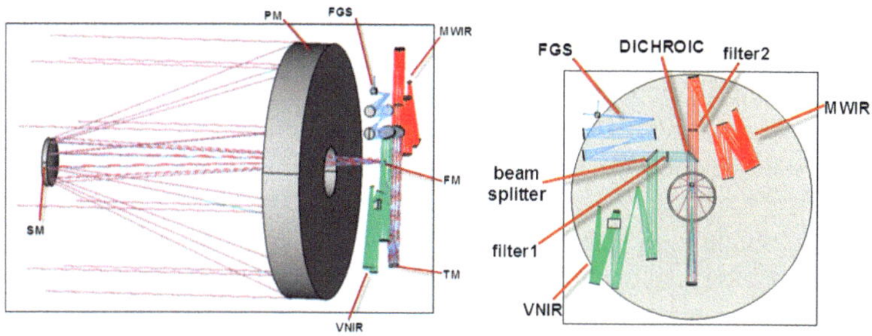


**Fig. 2** The layout of the telescope and three channels

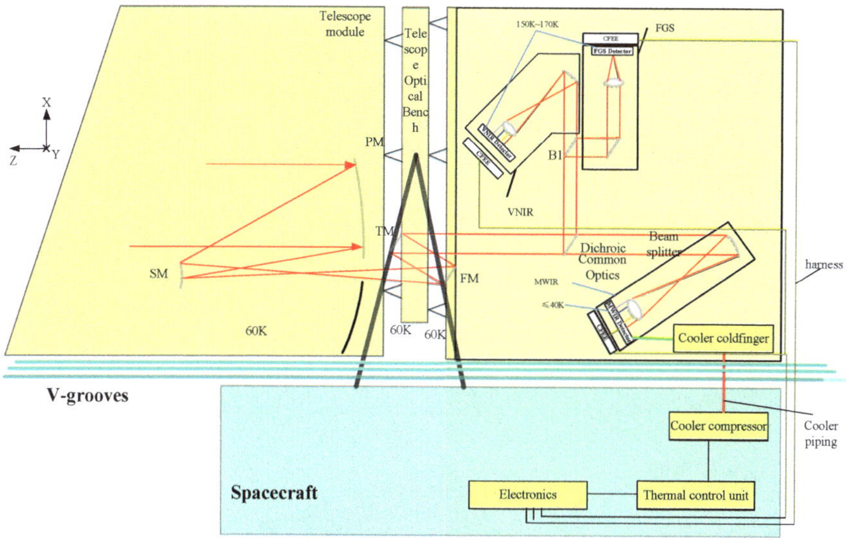


**Fig. 3** Layout of the SHE prototype optical design, which is divided into the telescope module, the instrument modules (including VNIR, MWIR, and FGS module)

VNIR and transmits to the FGS. At the same time, the dichroic transmits the MWIR wavelengths to the MWIR channel. The layout of the SHE prototype optical design is shown in Fig. 3.

## 3 Mechanical Design

In order to satisfy the image quality of the optical system at cryogenic temperature, athermal design philosophy is adopted as the most possible way to success, and thus the same material for both optics and structures. Aluminum alloy 6061-T6 is chosen for the main material for the prototype. The homogeneity of the material should be considered, too. It is better to apply the same piece of material to manufacture the optics and structures, at the same time the monolithic structures are designed to make sure the shrinkage is as uniform as possible [1]. Another key procedure to maintain the size stability of the material from room temperature to cryogenic temperature is a complex heat treatment in the process to minimize the machining residual stress.

In order to obtain high stability of the instrument, a "no adjustment" strategy is chosen. In this case, design method, tolerance analysis, material properties, and manufacturing process will be needed to ensure optical alignment accuracy and instrument optical performance [2]. SPDT (single point diamond turning) technique is applied to realize the precision requirement of the optical surface profile and high flatness of the mounting surface of the structures. To get more stable instrument performance, mechanical interfaces are designed to be minimized and thus structures are more complex, too. This strategy will not only increase the stability of the system, but also improve the efficiency of assembly significantly.

Despite the chosen design philosophy, it is impossible to ensure the prototype entirely without alignment. Figure 4 gives an overview of the mechanical design of the prototype. Overall, the prototype consists of a TOB, on which all are mounted, PM, SM, FM, TM, baffle, common optics, and instrument optical bench (IOB). VNIR, MWIR, and FGS modules are mounted on the IOB. All the optics are made of RSA-6061 and the structures are made of 6061-T6, except some common optics and prisms which are transmission components. Based on the design concept, most optics are mounted on the open space of the supporting benches, making the assembly easier.

**Fig. 4** 3D view of SHE prototype

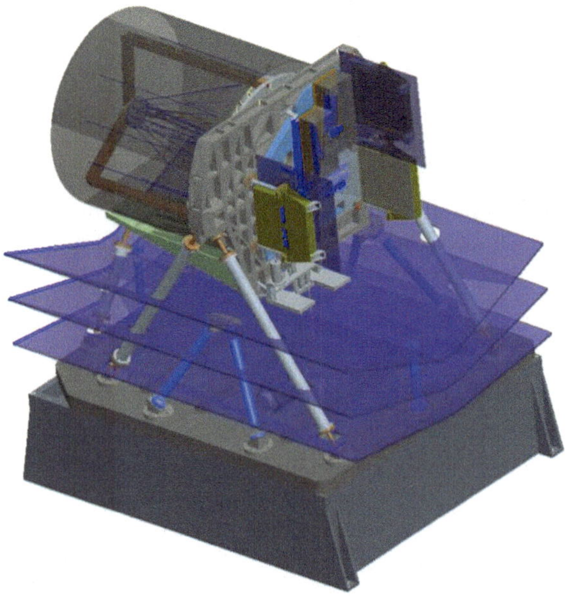

The TOB is supported on the platform via three bipods. There are three points of connection on the bench. The bipod members will be hollow cylinders made of CFRP (carbon fiber reinforced plastics), a low conductive material at cryogenic temperature with excellent mechanical properties. The two ends of the bipod are connected by a titanium alloy joint to the structures. These bipods have been simulated by finite element model and optimized for the best performance of both stiffness and thermal insulation.

## 4 Thermal Design

The thermal design of the SHE payload is based on a strategy of passive and active cooling systems (Fig. 5). For the passive cooling part, the three cold temperature stages consist of V-Grooves passive radiators that, exploiting the stable L2 thermal environment on orbit, will provide 40 K cryogenic temperature environment for the modules in the telescope. It will be simulated by a helium shroud for the prototype testing on the ground. Three channel detectors will be cooled around 60 K by means of the V-Grooves.

For the active cooling part, the MWIR detector is cooled down to 40 K by the 20 K Stirling-type pulse tube cryo-cooler. The cooler is composed of the compressor, first coldhead, secondary coldhead, pulse tube, and the flexible conductive links

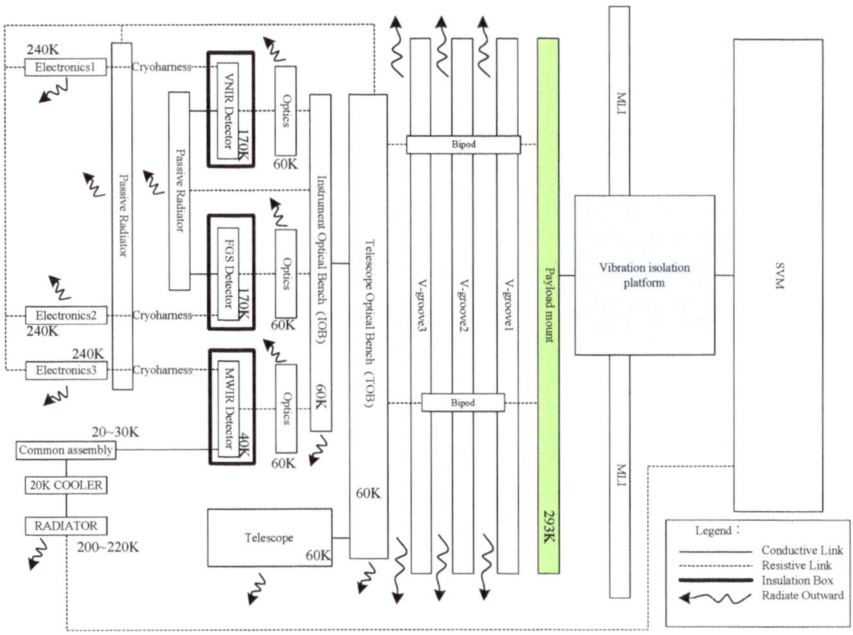


**Fig. 5** SHE thermal scheme with main thermal IFs

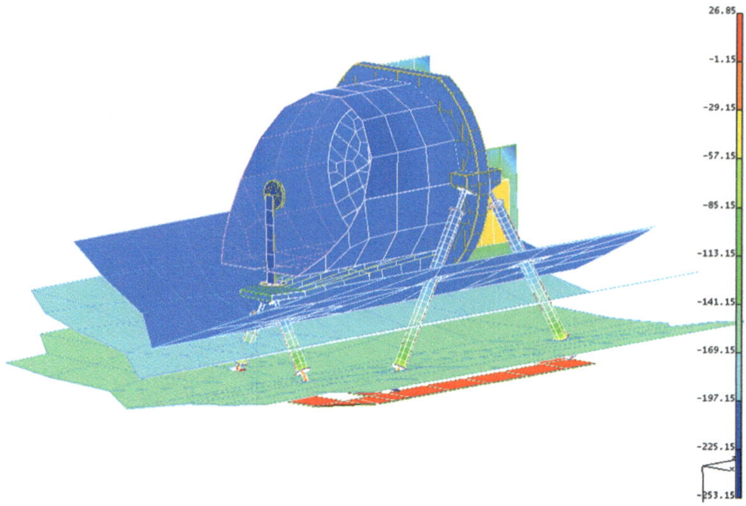


**Fig. 6** SHE prototype average temperatures

which link to the MWIR detector. The cooling capacity is above 0.3 watts at 20 K, when the power is below 400watts. The thermal stability is not more than 0.1 K every 8 h. Life is more than 35,000 h. The weight of the cooler is below 13 kg.

Each channel module of the instrument thermally consists of an optical module (OM) and a focal plane assembly (FPA). In order to ensure electrical performances, the cryo-harness connecting the electronics to the detectors cannot be longer than a few tens of cm (around 20 cm max). The electronic box should be mounted nearby the detectors. For the VNIR and FGS channels, the electronics are installed near the detectors on the instrument optical bench (IOB). For the MWIR module, in order not to overload the cooler cold end, the electronic box is mounted on the telescope optical bench (TOB).

According to the thermal scheme of active and passive thermal control, the unit temperatures of the prototype at steady-state conditions are graphically shown in Fig. 6 which shows that the first V-Groove temperature is 130 K, the second is 100 K, and the third is 60 K.

## 5 Testing

The test phase of the SHE prototype has been split up into three parts:

1. component level test: Φ500mm primary mirror WFE test;
2. subsystem tests: telescope module, VNIR module, MWIR module, and FGS module;
3. integral SHE prototype test.

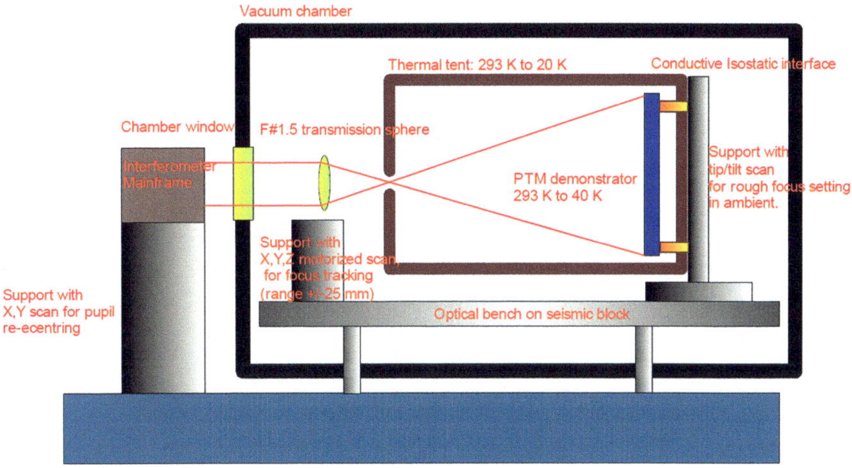


**Fig. 7** Primary mirror test set-up overall view

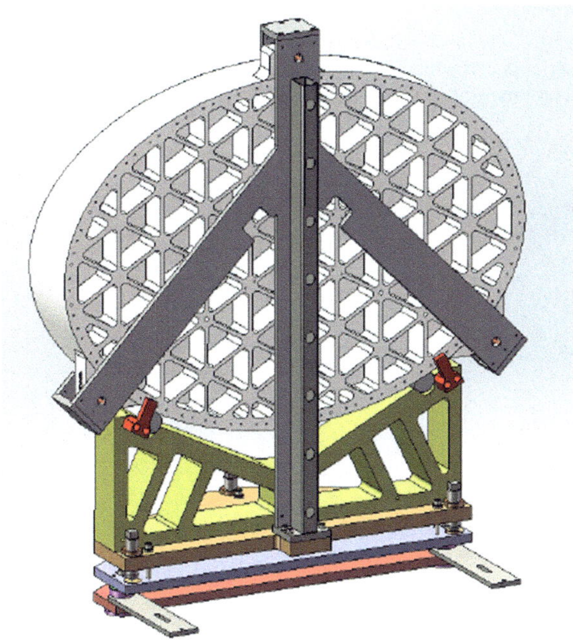

**Fig. 8** Primary mirror supporting structure in the test set-up

All the tests are planned to be conducted both at ambient and cryogenic working temperature. To satisfy the 60 K testing temperature, a helium shroud is being designed and manufactured.

Figure 7 is a primary mirror test set-up overall view and an isostatic support of the primary mirror is shown in Fig. 8. CSL will provide to us a conceptual design of the optical performance of the primary mirror test under cryogenic environment in a way of cooperation. Then we finish the test by ourselves. The subsystem tests and the integral prototype test design are still in progress.

## 6 Summary

We have provided an overview of SHE project background and science objectives, as well as a description of the main features of the optical, mechanical, thermal design, and test.

## References

1. Felix, C.M., Bettonvil, G., Kroes, T., Agoćsa, A., van Duinc, E., Elswijka, M., de Haana, R., et al.: Manufacturing, integration and test results of the MATISSE cold optics bench. Proc. SPIE. **9147**, 91477Q (2014)
2. Kroes, G., Kragta, J., Navarroa, R., Elswijka, E., Hanenburga, H.: Opto-mechanical design for transmission optics in cryogenic IR instrumentation. Proc. SPIE. **7018**, 70182D (2008)
3. Ning, A.: SHE prototype 50 cm aperture telescope optical system design report. Unpublished

# Payload Characteristics and Performance Verification of Atmospheric Infrared Ultra-Spectral Spectrometer

**Zhang Yugui, Hou Lizhou, Xu Pengmei, Lu Zhijun, Jiang Cheng, and Li Lijin**

**Abstract** Atmospheric infrared ultra-spectral spectrometer (AIUS), which is onboard GF-5 satellite, is the first Chinese solar occultation infrared spectrometer payload. It can realize vertical distribution information detection of atmosphere gases by solar occultation for the first time in China. This payload was launched on May 9, 2018 and is fully operational on-orbit at present. In this paper, the detection principle, system composition, characteristics, and engineering realization of the payload are introduced, and the characteristic parameters are analyzed and verified by using the on-orbit data. The results show that the characteristic parameters fully meet the design specifications, and the detection data can give important support for the application of atmospheric science research, environmental protection, numerical weather prediction, and so on.

**Keywords** Fourier transform spectrometer · AIUS · Payload characteristics Trace gases · GF-5

## 1 Introduction

GF-5 is the first hyperspectral satellite in the world to achieve comprehensive observation of the atmosphere and land and is also an important scientific research satellite in CHEOS (Chinese High-resolution Earth Observation System) Project. AIUS is one of the important payloads of GF-5.

Based on time-modulated Fourier transform spectroscopy [1–5], as the highest spectral resolution infrared hyperspectral space payload in China so far, AIUS has realized the vertical distribution detection of atmospheric components by employ-

Z. Yugui (✉) · H. Lizhou · X. Pengmei · L. Zhijun · J. Cheng · L. Lijin
Beijing Institute of Space Mechanical and Electricity, Beijing, China

Beijing Key Laboratory of Advanced Optical Remote Sensing Technology, Beijing, China
e-mail: jingzhou8112@126.com



H. P. Urbach, Q. Yu (eds.), *6th International Symposium of Space Optical Instruments and Applications*, Space Technology Proceedings 7,
https://doi.org/10.1007/978-3-030-56488-9_27

ing solar occultation for the first time in China. During on-orbit sunrise, AIUS can measure atmospheric transmittance spectra at different altitudes of 8–100 km by occultation and obtain vertical distribution information of trace gas concentration by data processing and retrieval algorithms according to the depths of specific spectral absorption lines of different gases. As AIUS has a wide spectral range and super high spectral resolution, more than 40 atmospheric gases can be retrieved [6–8], which provides a scientific basis for climate change research and atmospheric environmental monitoring and helps us understand the atmospheric environmental changes we live in.

AIUS is successfully launched into orbit by GF-5 on May 9, 2018, and is now fully operational on orbit. On-orbit test results show that the performance parameters meet the requirements and the preliminary gas retrieval results are good.

## 2 Detection Principle and Brief Payload Description [9]

### *2.1 Solar Occultation Detection Principle for Atmosphere*

The solar occultation operation principle of AIUS is shown in Fig. 1. During on-orbit sunrise, AIUS automatically seeks, captures, and locks the mass center of solar radiation by the solar tracker to achieve continuous solar occultation observation. During the tracking process, the solar observation begins from the beginning of sunrise which is tangent to the horizon and continues working until the sun goes beyond the atmosphere. Meanwhile, AIUS also continuously works to measure a series of solar transmission spectrum data which are corresponding to different atmospheric height, forming an atmospheric detection data package. A series of spectral curves of atmospheric transmittance varying with tangent height can be obtained by dividing the solar transmittance spectra by those outside the atmosphere in the same data package.

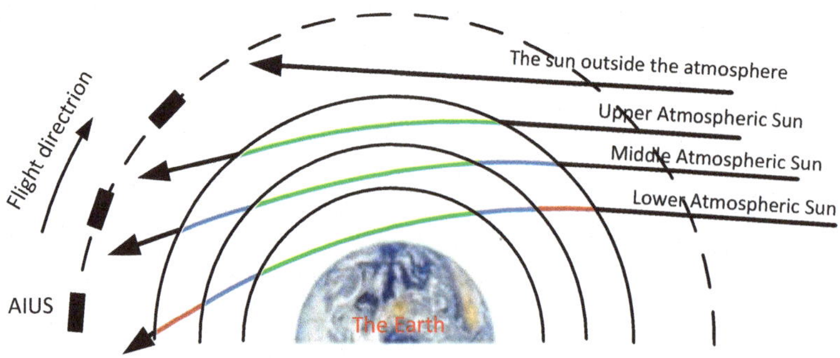


**Fig. 1** AIUS solar occultation observation

According to series depths of specific spectral absorption lines of different gases, the corresponding gas concentrations at different tangent heights can be retrieved [10–12], and the gas profiles with vertical distribution information can be obtained.

The flow chart for obtaining vertical distribution information of gas concentration is shown in Fig. 2. The treatment process is as follows:

Firstly, the atmospheric detection data package is decoded to obtain interference data and auxiliary data.

Secondly, the interferometric data are pre-processed and Fourier transformed to obtain the spectral sequence of the whole process of the sun from sunrise to leaving the atmosphere. The spectral sequence is divided by the solar spectrum outside the atmosphere to obtain the spectral transmittance sequence and is followed by the spectral calibration. Spectral transmittance curves of the sequence correspond to different atmospheric altitudes.

At the same time, the parameters of payload state, such as geographic position information and flight attitude information, can be obtained by processing the auxiliary data. Thus, according to the Earth's atmospheric refraction model, the atmospheric tangent height corresponding to each curve of the spectral sequence can be estimated, which provides the necessary input for gas retrieval.

Finally, according to the sequence of solar spectral transmittance, the information of payload location, and atmospheric tangent heights, the products of atmospheric gas vertical profiles can be obtained by retrieving target gases with appropriate algorithms.

## 2.2 System Specifications and Its Configuration

The prominent features of AIUS are its wide spectral range, high spectral resolution, and high accuracy and stability requirements for solar tracking. Its main technical specifications are shown in Table 1.

According to the technical specifications, the design and engineering implementation of AIUS is more complex. The system configuration is shown in Fig. 3. AIUS system consists of optical-mechanical main body, management controller, temperature controller, and refrigerator controller. Its optical and mechanical main body mainly includes: sun tracking module and its controller, front optical module, inter-

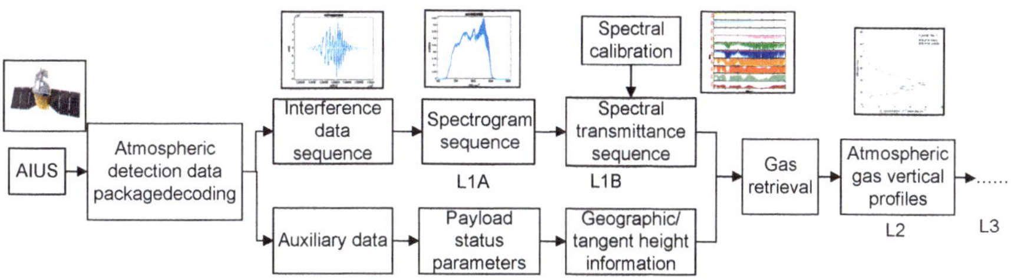


Fig. 2 The flow chart for obtaining atmospheric gas vertical profiles

**Table 1** Main specifications of AIUS

| Items | Technical specifications |
|---|---|
| Spectral range | 750–4166 $cm^{-1}$ (2.4–13.3 μm) |
| Spectral resolution | 0.03 $cm^{-1}$ |
| Dynamic range | 800– 5800 K |
| Sun tracking precision | 0.1 mrad |
| Sun tracking stability | 25 μrad |
| Spectral SNR | ≥100 @5800 K |

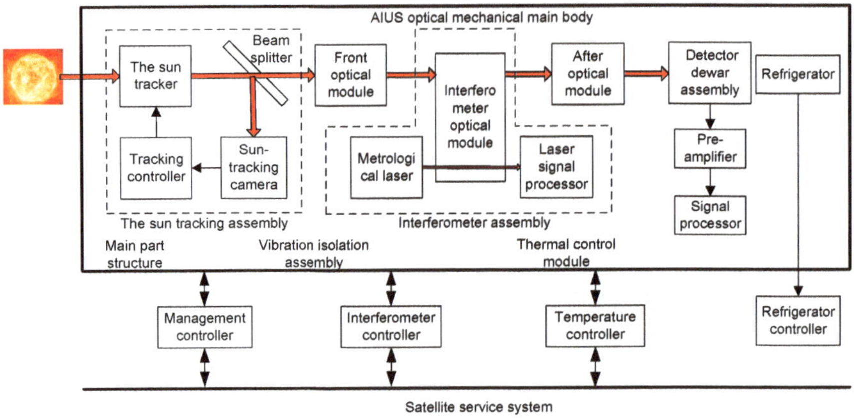


**Fig. 3** AIUS system configuration

ferometer assembly, after optical module, detector dewar assembly and signal processor, etc.

The sun tracking module completes automatic sun tracking in the visible spectrum and introduces stable solar infrared radiation into the interferometer. The interferometer is responsible for interferometric modulation, in which the monochrome frequency stabilized laser emitted by the metrological laser is in common optical path with the infrared beam. The laser interferometric signal is used for scanning control and providing sampling signal. Dewar assembly and signal processing of the detector are responsible for the signal acquisition of the interferograms.

## 3 Payload Characteristics and Its Realization [13]

The characteristics of AIUS include wide spectral range, very high spectral resolution, and extremely high sun tracking accuracy. The following is an analysis and description of these features and their implementation.

## 3.1 Wide Spectral Range

According to the technical specifications in Table 1, AIUS is required to realize 2.4–13.3 μm wide-spectrum detection. The difficulties of instrument design are mainly in two aspects: optical system and detector assembly.

### 3.1.1 Optical System

Considering that the optical system needs to realize wide spectrum energy collection, in order to minimize the energy loss of input radiation, the gold-plated reflective mirrors are used in all the optical system design except the beam splitter and compensator. Both beam splitter and compensator use ZnSe material as the substrate. The former is coated with broadband beam splitter film and the latter with broadband antireflective film to realize broadband spectral energy transmission and collection.

### 3.1.2 Detector Assembly

It is difficult for a single detector to meet the performance requirements of wide spectrum and high precision at the same time. The detector assembly employs an integrated dual-band intra band-division cell-based detector assembly, which are tightly coupled with dewar. InSb detector is used in mid-wave band and MCT detector in long-wave band. The two detectors are integrated in sealed dewar. They are divided into two spectral bands by dichroic splitter, and the wavelength demarcation point is 5.5 μm. InSb detector detects signals in 2.4–5.5 μm band, while MCT detector detects signals in 5.5–13.3 μm band. The dewar assembly model with two detectors is shown in Fig. 4.

## 3.2 High Spectral Resolution

According to the Fourier transform spectrum detection theory, the spectral resolution is mainly determined by the maximum optical path difference (OPD) of the interferometer and is also limited by the field of view (FOV) of the interferometer. The relationship between the non-apodization spectral resolution $\delta_\upsilon$ and the maximum OPD L of the interferometer is shown in (1).

$$\delta_\upsilon = \frac{1}{2L} \tag{1}$$

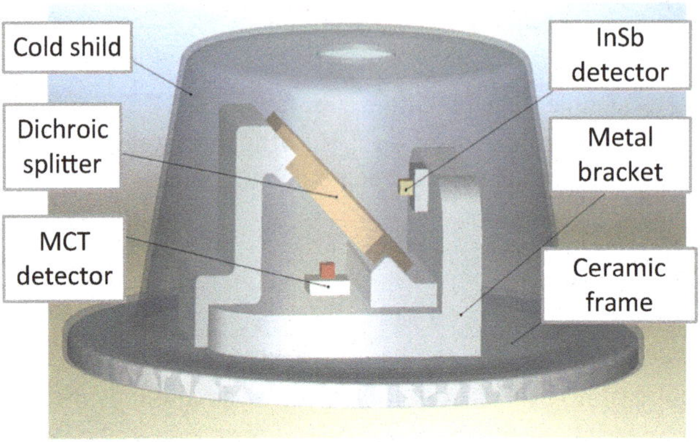


**Fig. 4** Detector dewar assembly model

The true spectral resolution is determined by the full width at half maximum (FWHM) of the instrument line shape (ILS), which is theoretically 1.207 times $\delta_\upsilon$. Therefore, for the requirement of spectral resolution in Table 1, the maximum $L$ of the interferometer is 25 cm according to (1), which satisfies the requirement of 0.03 $cm^{-1}$ with a suitable margin.

It is not easy to realize the maximum OPD of 25 cm. In the design of the interferometer, a large-range scan-arm cube corner interferometer based on eight-fold OPD technology is used to realize the maximum OPD of ±25 cm with scan angle of ±15°. The principle of AIUS optical system is shown in Fig. 5. The incident beam is divided into two beams by the beam splitter and goes to the cube corners and the end mirrors, respectively, and then returns to the original OPD. With the difference between the two arms, the OPD is magnified eight times of the mechanical movement. Finally, the interference signal is generated on the detector.

The model of AIUS interferometer assembly is shown in Fig. 6. The system consists of three parts: the optical mechanical module of the interferometer, the metrological laser signal processor, and the interferometer controller. Optical mechanical module realizes high-efficiency interference modulation of the measured beam, the measurement laser signal processor is responsible for real-time feedback of OPD signal, and the interferometer controller is responsible for the motion control of the scan-arm to achieve high stability OPD scanning.

### 3.3 *High Sun Tracking Precision*

Accurate and stable sun tracking is the premise of normal work of AIUS. In the process of sun tracking, as AIUS FOV is 20 mrad, which is about twice the view angle of the sun, the sun is required to be in the middle of the image with a precision of 0.1 mrad to avoid significant errors due to the sun image "jitter" introduced in the sun tracking process. The stability of the sun tracer is required no more than 25 μrad.

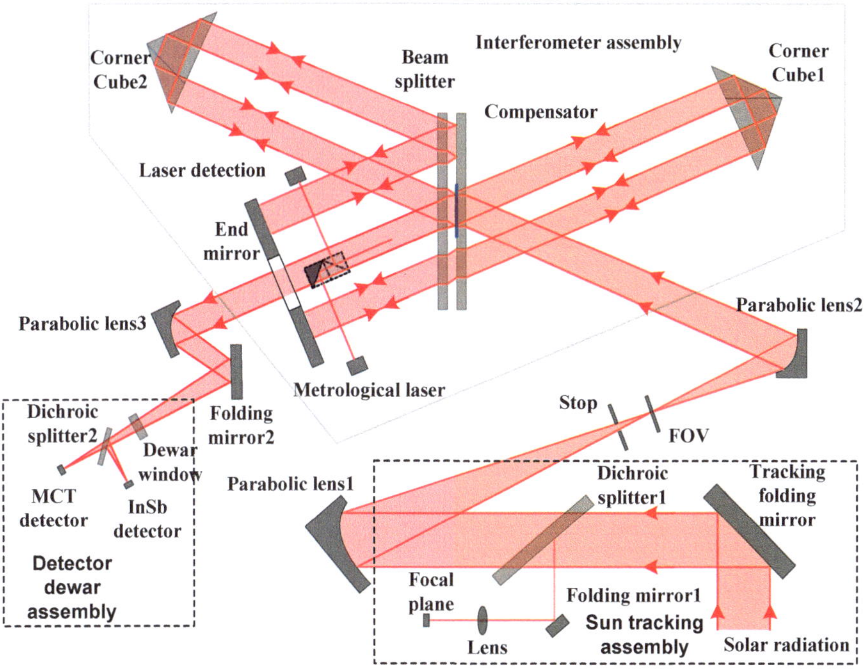


**Fig. 5** AIUS schematic optical layout

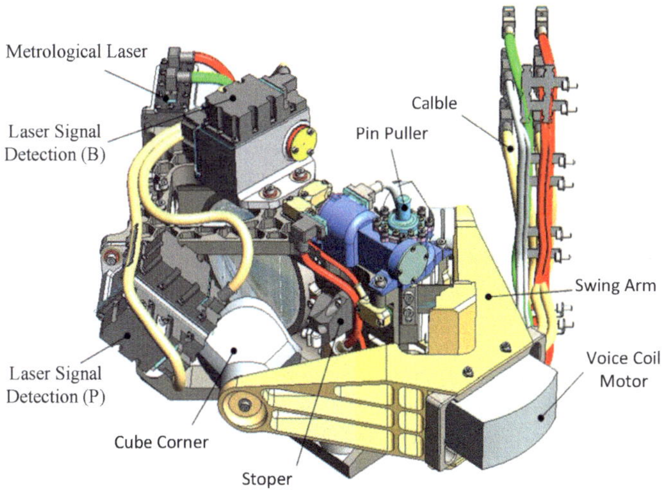


**Fig. 6** AIUS interferometer model

Especially in the early beginning of sunrise, solar images often appear deformations or even separation due to atmospheric refraction and cloud occlusion. At this moment, the sun is more difficult to capture and track.

In order to solve these problems, AIUS adopts high precision image feedback sun tracking technology, including:

### 3.3.1 Design of Sun Tracking System

The solar tracking system is composed of a sun tracking mechanism and a sun tracking camera. The former uses a "double-loop" two-dimensional pointing mechanism and a reflective pointing mirror to capture and track the sun with high sensitivity. The latter acquires sun image data and achieves autonomous tracking through closed-loop feedback between the image and the mechanism.

### 3.3.2 Sun Tracking Strategy

AIUS automatically seeks, captures, and tracks the sun during orbiting sunrise to achieve occultation observation. The sun tracking strategy is designed as follows:

1. Before on-orbit sunrise: according to the payload state parameters and atmospheric refraction model, the sun tracker calculates to point to the position where the sun will appear above the horizon;
2. After on-orbit sunrise: with the refraction angle of the atmosphere corrected in real time, the system works in an open loop and always points to the calculated sun position; once the sun tracking camera captures the sun image, it enters the image feedback tracking mode and calculates the pointing angle synchronously as a reference, so as to track the direction when the sun is occluded;
3. The sun beyond the atmosphere: until the occultation path exceeds 100 km above the atmosphere, the sun tracker continues to track and observe solar radiation for a short period of time, then stops tracking the sun and observes deep space for a short period of time.

### 3.3.3 Atmospheric Refraction Correction and Pointing Angle Calculation

The coordinate definition of the sun tracker in the optical-mechanical main body, as well as the relationship between the sun vector s and the pitch angle $\beta$ of the inner loop and the deflection angle α of the outer loop of the sun tracker, is shown in Fig. 7. The process of pointing angle calculation is to calculate the angle of $\alpha$ and $\beta$ according to the s change, so that AIUS always points to the center of the solar image mass.

The equations for calculating the angles of $\alpha$ and $\beta$ are as follows:

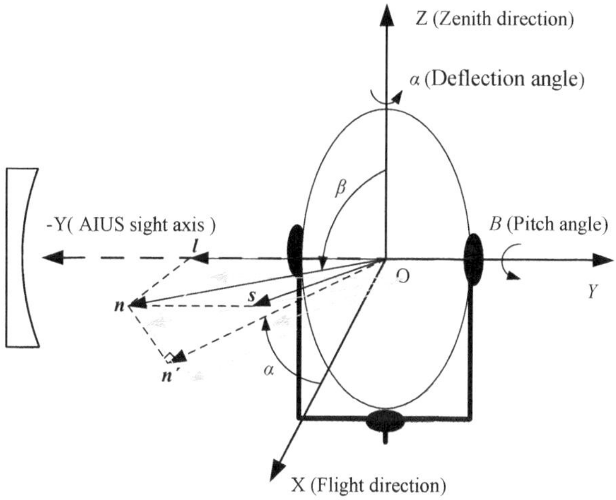


**Fig. 7** Coordinate definition of sun tracker

$$\alpha = \arccos\left(\frac{\frac{\cos\theta_x}{L}}{\sqrt{\left(\frac{\cos\theta_x}{L}\right)^2+\left(\frac{\cos\theta_y}{L}-1\right)^2}}\right) \tag{2}$$

$$\beta = \arccos\left(\frac{\cos\left(\theta_z+\Delta i\right)}{\left|\left\langle \frac{\cos\theta_x}{L}\ \frac{\cos\theta_y}{L}-1\ \cos\left(\theta_z+\Delta i\right)\right\rangle\right|}\right) \tag{3}$$

In Eqs. (2) and (3), $L=\sqrt{\cos^2\theta_x+\cos^2\theta_y+\sin^2\theta_z\text{ctan}^2\left(\theta_z+\Delta_i\right)}$, and $\theta_x$, $\theta_y$, $\theta_z$ are the angles between ***s*** and the *X*, *Y*, and *Z* axes, respectively. $\Delta_i$ is the correction angle of atmospheric refraction, which can be obtained by looking up tables, as shown in Table 2.

The calculation method of $\Delta_i$ is as follows:

According to the standard atmospheric parameters and satellite orbit parameters, an optical model of atmospheric refractive index gradient for solar occultation observation can be established. The atmospheric refractive angle at any altitude is calculated accurately by ZEMAX (optical design software). In this paper, we calculate the atmospheric refraction compensation angle table for the solar vector sequence from satellite broadcasting. As shown in Table 2, the sequence number in the table corresponds to the solar vector broadcasting sequence after the on-orbit sunrise.

**Table 2** The atmosphere refraction angle

| S.N. | Angle (°) | S.N. | Angle (°) | S.N. | Angle (°) | S.N. | Angle (°) |
|---|---|---|---|---|---|---|---|
| 1 | 1.11806 | 6 | 0.608843 | 11 | 0.199164 | 16 | 0.015389 |
| 2 | 1.012218 | 7 | 0.51561 | 12 | 0.140748 | 17 | 0.008397 |
| 3 | 0.908033 | 8 | 0.42683 | 13 | 0.093245 | 18 | 0 |
| 4 | 0.805813 | 9 | 0.343566 | 14 | 0.05701 | 19 | 0 |
| 5 | 0.705953 | 10 | 0.267332 | 15 | 0.031386 | 20 | 0 |

In the process of sun occultation, according to Table 2 and Eqs. (2) and (3), the sun tracker calculates the angle of the pointing mirror $\alpha$ and $\beta$ to track the sun in real time. Before the on-orbit sunrise, the value of $\theta_z + \Delta_i$ is 64.7°, then $\alpha$ and $\beta$ can be calculated by Eqs. (2) and (3). The sight axis of AIUS can be fixed to the position of the sunrise beginning, and AIUS start to wait for sunrise.

## 4 Tests and Performance Verification

On-orbit data are selected to test AIUS for performance verification. Firstly, through the processing of on-orbit data, an example of data transformation from interferogram sequence to gas profile products is given to show the validity of the data. Then, the payload characteristics, including spectral response range, spectral resolution, sun tracking accuracy, spectral signal-to-noise ratio, and spectral long-term stability, are verified. The test results meet the performance requirements.

### 4.1 *An Example of Data Transformation*

Taking MCT channel as an example, one orbit of the on-orbit data is processed and analyzed to the sequence of interferograms, spectrograms, and transmittance, as shown in Figs. 8, 9, and 10, respectively.

According to the spectral transmittance sequence, geographic location, and tangent height information, the profile information of vertical distribution of gas concentration can be obtained by using appropriate gas retrieval algorithms, as shown in Fig. 11.

### 4.2 *Performance Verification*

#### 4.2.1 Spectral Range Verification

By analyzing the spectrum of MCT and InSb channels, the spectral response range of AIUS is larger than 750–4550 cm$^{-1}$, which meets the system requirements, as shown in Fig. 12.

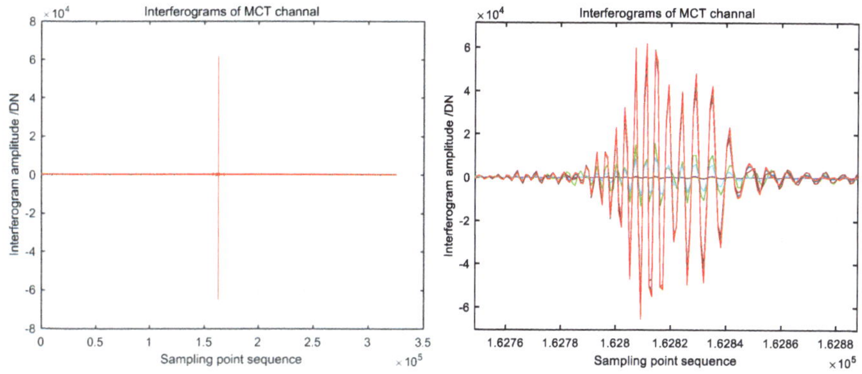


Fig. 8 Interferogram sequence and its detailed view

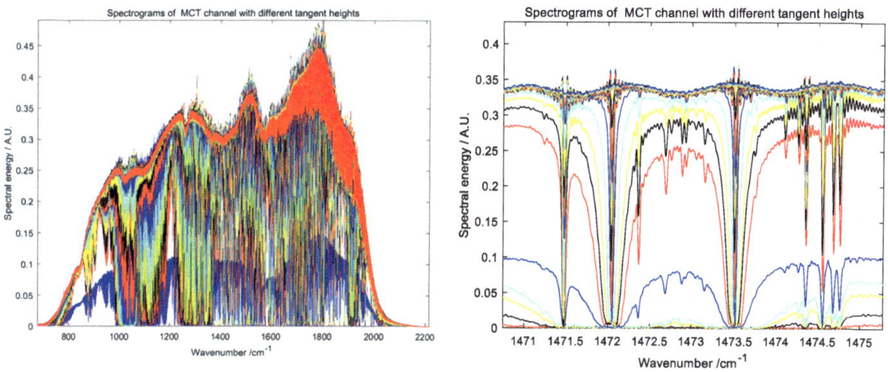


Fig. 9 Spectrogram sequence and its detailed view

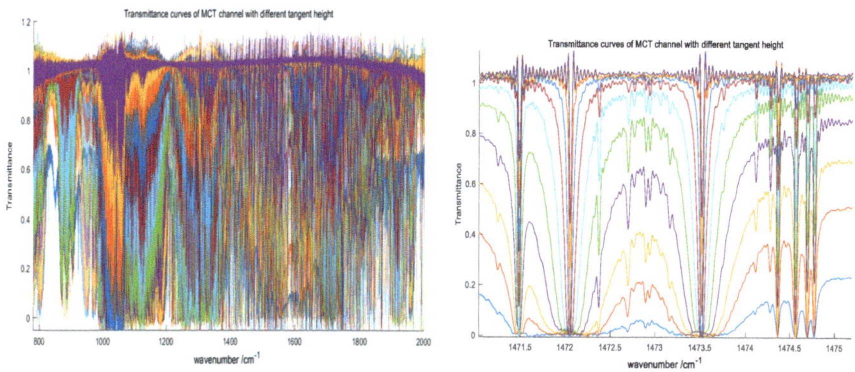


Fig. 10 Spectral transmittance sequence and its detailed view

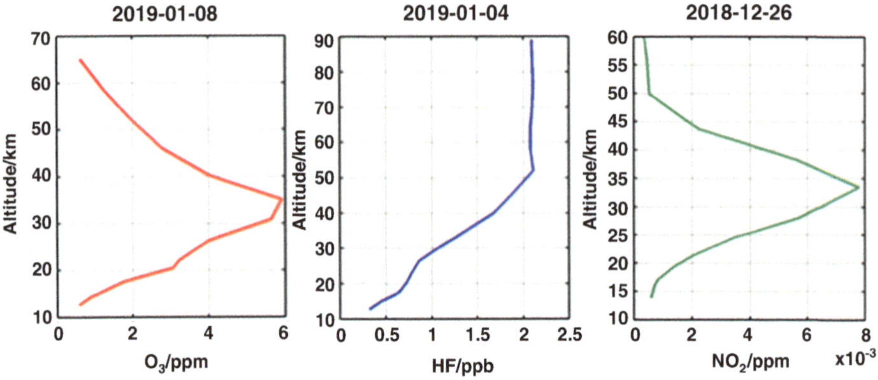


Retrieved $O_3$ profile | Retrieved HF profile | Retrieved $NO_2$ profile

**Fig. 11** Profile retrieval results of several typical gases by AIUS

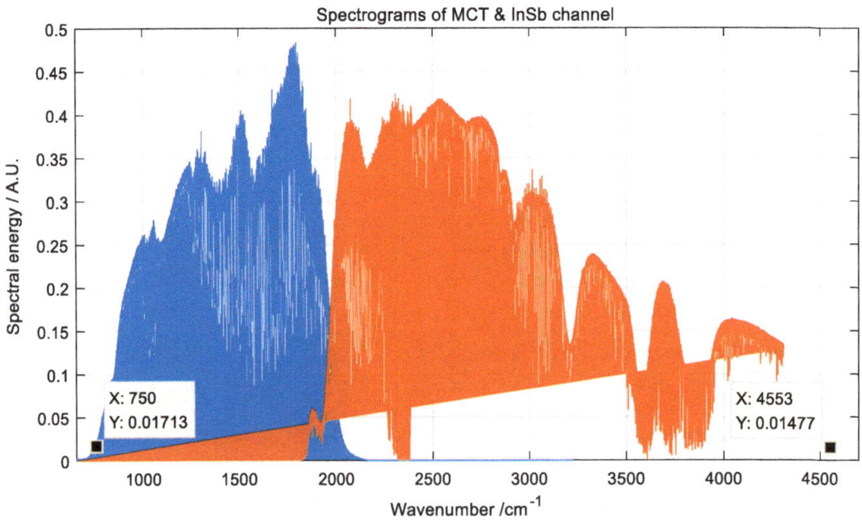


**Fig. 12** Spectral response range of AIUS

### 4.2.2 Spectral Resolution Verification

The narrow linewidth absorption lines of some specific gases are selected to evaluate the ILS and spectral resolution of AIUS, as shown in Fig. 13. An absorption line near 1808 $cm^{-1}$ is selected and its half-width is 0.0242 $cm^{-1}$ by *sinc* function fitting algorithm, which indicates that the spectral resolution of AIUS meets the requirement of 0.03 $cm^{-1}$.

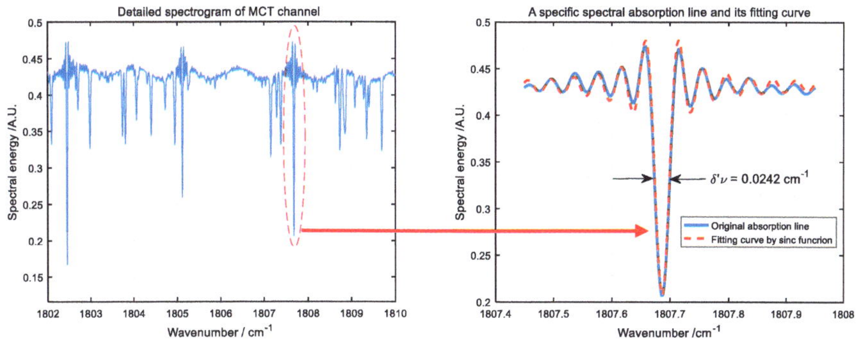


**Fig. 13** In-orbit verification of ILS and spectral resolution

### 4.2.3 Sun Tracking Precision and Stability Verification

Figure 14 (left) is a sequence of on-orbit sun tracking images, representing the solar tracking during a complete solar occultation observation (about 18 μrad/pixel). The tracking accuracy and stability of AIUS can be evaluated by calculating the distribution standard deviation of the sun image mass center. The results are shown in Fig. 14 (right). The tracking accuracy is 0.0734 mrad and the tracking stability is 24.47 μrad, which meet the requirements of AIUS.

### 4.2.4 Spectral SNR Verification

More than 30 interferograms among the on-orbit data of the sun after leaving the atmosphere (which is considered to be stable at this time) are selected for processing, and corresponding spectrograms are obtained. Mean value and standard deviation of each spectral position are calculated, and SNRs of all spectral positions are obtained. Spectral SNR curves can be drawn as shown in Fig. 15. The results show that SNR performance meets the system requirements (SNR = 100 at dotted line).

### 4.2.5 Long-Term Spectral Stability

The long-term spectral stability is an important parameter for quantitative gas retrieval. In this paper, the spectral drift after a spectral calibration from December 1, 2018 to July 21, 2019 is taken into investigation, as shown in Fig. 16. For more than half a year, the maximum spectral drift is 3.1 ppm, which has not yet affected the gas retrieval. When it is found that spectral drift affects gas retrieval, spectral re-calibration can be used to solve the problem.

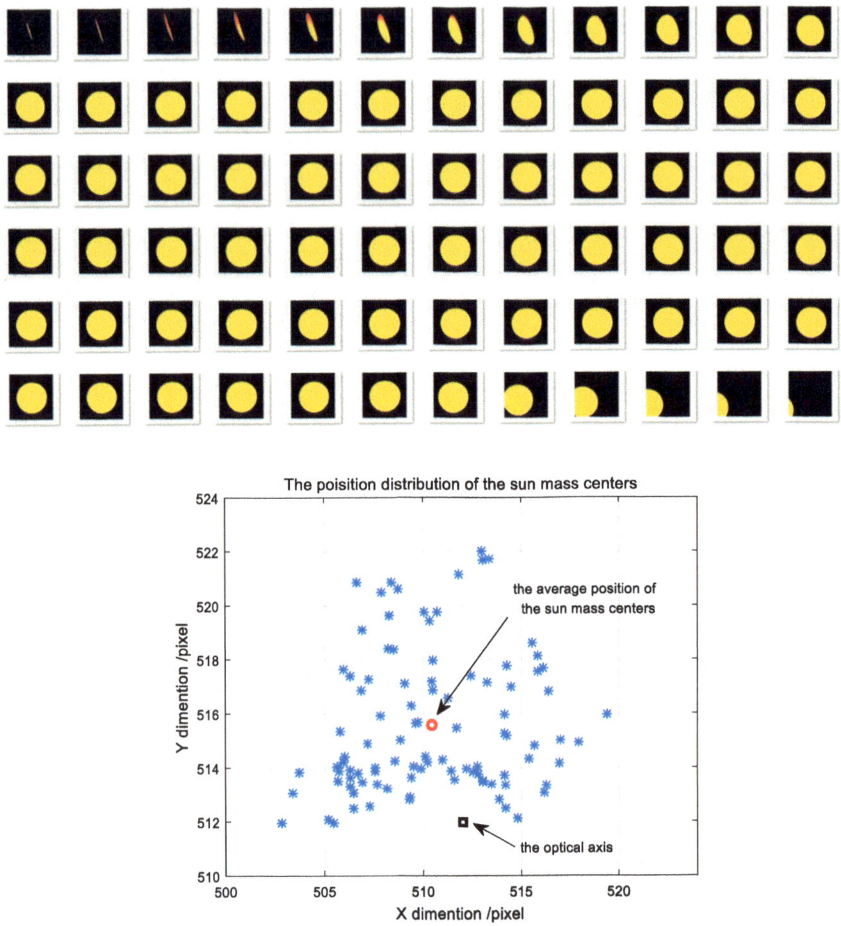


**Fig. 14** Sun image series during solar occultation and sun tracking accuracy and stability

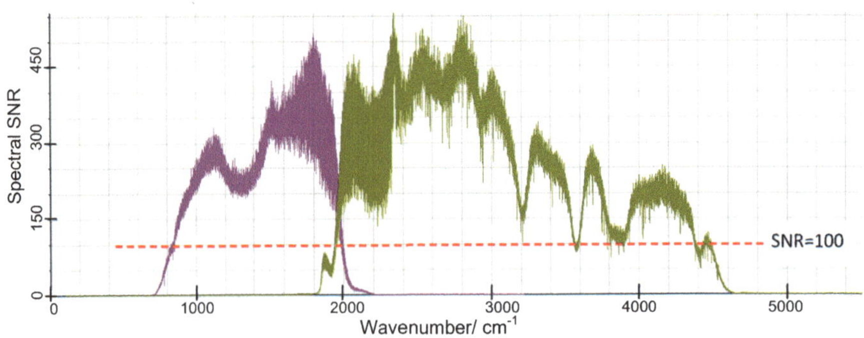


**Fig. 15** Spectral SNR testing results of AIUS

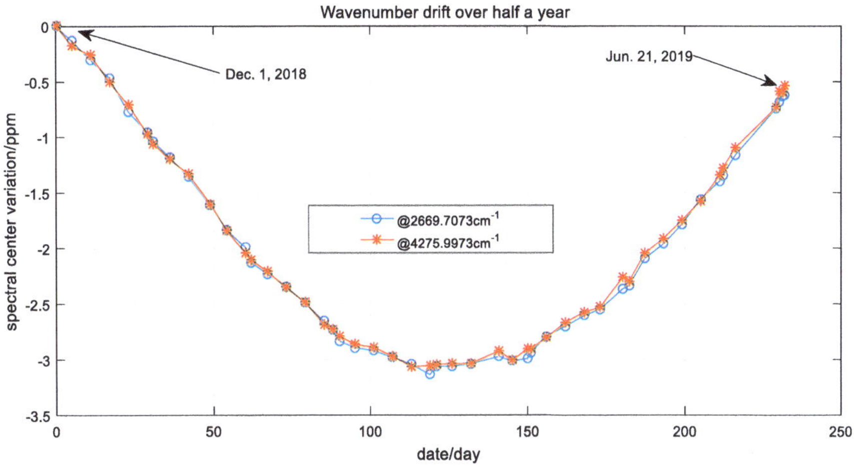


**Fig. 16** Changes of AIUS spectral drift with time

## 5 Conclusions

AIUS is the first satellite-borne hyperspectral payload and also the highest spectral resolution infrared detection payload in china. This paper describes the detection principle, payload characteristics, and performance verification of AIUS. All terms of performance meet the system requirements and preliminary gas retrieval results are good, which shows that AIUS can meet the anticipated on-orbit application requirements.

**Acknowledgement** Thank Ms. Li Xiaoying from RADI (The Institute of Remote Sensing and Digital Earth) of CAS (Chinese Academy of Sciences) for her work in atmospheric gas retrieval of AIUS and thank all reviewers for their valuable suggestions. This work is supported by National Key R&D plan project "Hyperspectral Detection Technology of Atmospheric Radiation" (Project No. 2016YFB0500700).

## References

1. Wu, X., Fan, D., Wang, P.: Fourier-transform infrared spectrometer for space atmospheric component detecting. Spacecr. Recov. Remote Sens. **28**(2), 15–20 (2007) (in Chinese)
2. Griffiths, P.R., de Haseth, J.A.: Fourier transform infrared spectrometry, pp. 161–175. Wiley, New Jersey (2007)
3. Poulin, R., Dutil, Y., et al.: Characterization of the ACE-FTS instrument line shape. Proc. SPIE. **5151**, 166–172 (2003)
4. Soucy, M.-A., Chateauneuf, F., Deutsch, C., et al.: ACE-FTS instrument detailed design. Proc. SPIE. **4814**, 70–81 (2002)

5. Chateauneuf, F., Fortin, S., Frigon, C., Soucy, M.-A., ABB-Bomem Inc: ACE-FTS test results and performances. Proc. SPIE. **4814**, 82–90 (2002)
6. Puckrin, E., Evans, W., et al.: Test measurements with the ACE FTS instrument using gases in a cell. Proc. SPIE. **5151**, 192–200 (2003)
7. Boone, C., Nassar, R., et al.: SciSat-1 retrieval results. Proc. SPIE. **5542**, 184–194 (2003)
8. Gilbert, K.L., Turnbull, D.N., et al.: The onboard imagers for the Canadian ACE SCISAT-1 Mission. J. Geophys. Res. **112**(D12207), 1–12 (2007)
9. Dong, X., Xu, P., Lizhou, H.: Design and implementation of atmospheric infrared ultra-spectral sounder. Spacecr. Recov. Remote Sens. **39**(3), 29–37 (2018) (in Chinese)
10. Burrows, J.P., Platt, U., Borrell, P.: The remote sensing of tropospheric composition from space, pp. 15–18. Springer, Heidelberg (2011)
11. Hase, F., Blumenstock, T., Paton-Walsh, C.: Analysis of instrumental line shape of high-resolution FTIR spectrometers using gas cell measurements and a new retrieval software. Appl. Optics. **38**, 3417–3422 (1999)
12. Boone, C., Bernath, P.: Scisat-1 Mission overview and status. Proc. SPIE. **5151**, 133–142 (2003)
13. Hou, L., Xu, P., Wang, C.: Design and implementation of scan-arm corner cube interferometer with large opd and high robustness. Airiti. Lib. **39**(3), 51–59 (2018) (in Chinese)

# A Spaceborne Calibration Spectrometer

**Jie Wang, Yue Ma, and Shaofan Tang**

**Abstract** The spaceborne calibration spectrometer is a specialized cross-calibration payload in orbit. It was equipped on the HY-1C satellite and had been launched in 2018 (Yang, Chin. Space Sci. Technol. 5:2, 2011). It achieves a high signal-to-noise ratio and the capability of large dynamic detection through the grating spectrometer and high full-well detector. Using the sun and diffuser as the light source, the spectrometer achieves high radiometric and wavelength calibration accuracy through the full aperture and all lightpath solar calibration. The method of spectrum reconstruction is adopted to achieve the high accuracy of cross-calibration when the spectrometer obtains the same ground-image which is detected by a calibrated remote sensor at the same time. It will help to improve the quantitative level of the calibrated remote sensor. The on-orbit test results show that the accuracy of radiation calibration is 2%, the accuracy of wavelength calibration is less than 0.5 nm, and the accuracy of cross-calibration is less than 5%.

**Keywords** Spaceborne calibration spectrometer · Grating spectrometer · High dynamic detection · High radiometric calibration accuracy

## 1 Introduction

Hyperspectral imaging sensor can obtain 3-dimensional data cube which includes 2-dimensional image and 1-dimensional spectral data. It has the ability to detect geometric, radiative, and spectral characteristics of the target. Because of this, hyperspectral imaging technology became an advanced method of earth observation. In order to improve the quantification level of HY-1C [1], the hyperspectral imaging sensor can be used to achieve the function of cross-calibration. Recently, the main spectrometer which is classified by light-splitting elements includes diffraction grating imaging spectrometer and prism imaging spectrometer [2, 3].

J. Wang (✉) · Y. Ma · S. Tang
Beijing Key Laboratory of Advanced Optical Remote Sensing Technology, Beijing Institute of Space Mechanics and Electricity, Beijing, People's Republic of China



H. P. Urbach, Q. Yu (eds.), *6th International Symposium of Space Optical Instruments and Applications*, Space Technology Proceedings 7,
https://doi.org/10.1007/978-3-030-56488-9_28

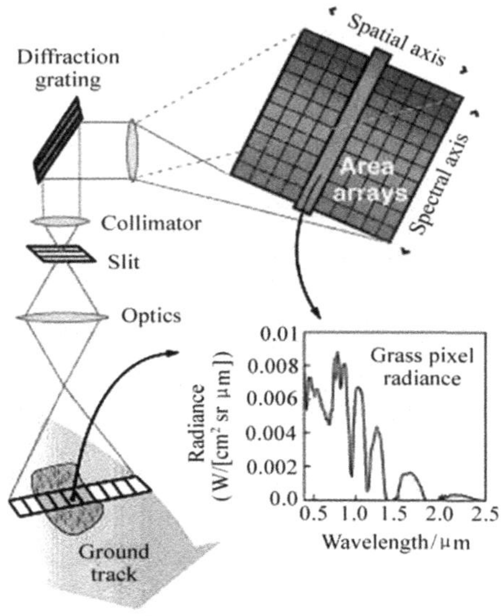


**Fig. 1** Principle diagram of hyperspectral imaging

Along with the development of manufacture capability of light-splitting element, high-speed array CCD or CMOS, large volume data processing, we have the ability to design and manufacture the spectrometer which has the characteristics of high precision, high stability, high spatial and spectral resolution, and high signal-to-noise ratio [4] (Fig. 1).

Nowadays, the spaceborne cameras are developed in the direction of large aperture, high resolution, and high quantification. To achieve the high quantification, the calibration equipment must be set in the camera, but the calibration equipment which is too large does not have enough space. Some payload which doesn't have calibration device need the camera, such as MODIS which is high stability, to achieve cross calibration and correct the calibration coefficient. In this way, the calibration accuracy will be affected, because of the difference of the observation time, the target and spectral response function between the camera and the spectrometer (Fig. 2).

## 2 System Function, Performance, and Composition

The purpose of the spectrometer is to calibrate another camera, so it must be high precision, high signal-to-noise ratio, and stability. As a result, the index of the equipment is shown in Table 1.

The spaceborne calibration spectrometer has two main functions: Firstly, it can obtain the image and spectral data of the earth to cross calibrate with another camera. Secondly, it can get the radiant energy from the diffuser which can reflect the

**Fig. 2** The spaceborne calibration spectrometer

**Table 1** The index of the spectrometer

| Serial number | Name of the index | Target value |
|---|---|---|
| 1 | Wavelength range | 0.4–0.9 μm |
| 2 | Swath | 12.1 km |
| 3 | Spectral resolution | 5.2 nm |
| 4 | Observation modes | Nadir |
| 5 | SNR@typical | >1000 |
| 6 | Radiometric accuracy | 2% |
| 7 | Accuracy of wavelength | <0.5 nm |
| 8 | Accuracy of cross-calibration | 5% |

energy of the solar. Especially, the lightpath of solar calibration includes sun, screen, and diffusing panel.

Because of these functions, the spectrometer has two channels. The two channels can be switched by a rotating mechanism which includes two mirrors. When one mirror turns an angle, the other mirror turns half to keep the direction of the light. The depolarizer is before the rotating mechanism, it can eliminate the polarization of incoming light. The fore optics and Offner spectrometer split the light and CCD detects the 2-dimensional signal, one-dimensional is the image and the other is the spectrum. The system composition is shown in Fig. 3.

# 3 Characterization

## *3.1 Large Dynamic Range and High SNR*

The spectrometer is used as standard equipment to supply data which is used for the calibrated camera. Because of this, the subsystem must own high dynamic range and high SNR. On the other hand, the target of the spectrometer is the ocean, so the

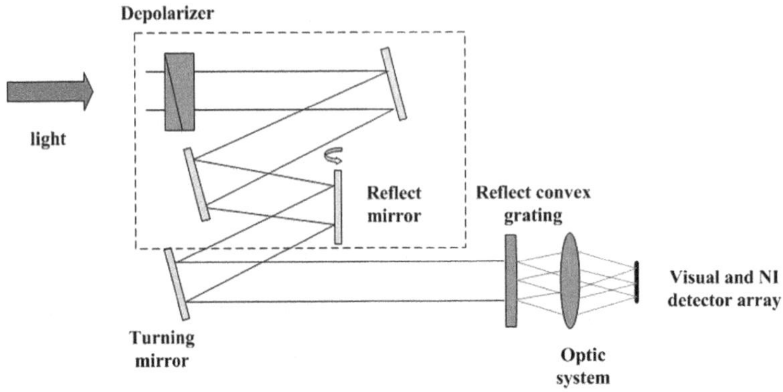


**Fig. 3** The system composition

payload also needs a high dynamic range and SNR. When the equipment was designed, we used the below formulation to compute the signal-to-noise ratio of the system, the formulation is shown below:

$$\frac{S}{N} = \frac{\pi}{4} \cdot L(\lambda) \frac{\tau(\lambda) \cdot T \cdot A \cdot \sigma}{\frac{\text{hc}}{\lambda} \cdot N \cdot F^2} \cdot \eta(\lambda) \cdot \text{CVF} \cdot \Delta\lambda \tag{1}$$

From the formulation, there are six main elements which can affect the result of the system.

1. F# of the optics;
2. Diffraction efficiency of grating;
3. Optical transmittance;
4. Quantum efficiency of the detector;
5. The noise of the detector;
6. The noise of the circuit;
7. Full well of the detector.

Decreasing the F# of the subsystem can get more incoming energy; however, the volume and weight of the spectrometer will get larger. The diffraction of grating, optical transmittance, quantum efficiency, noise of the system, and the full well are limited by the level of the technology. When the subsystem is designed, we need to consider all these factors. According to the design, the test result is shown in Fig. 4.

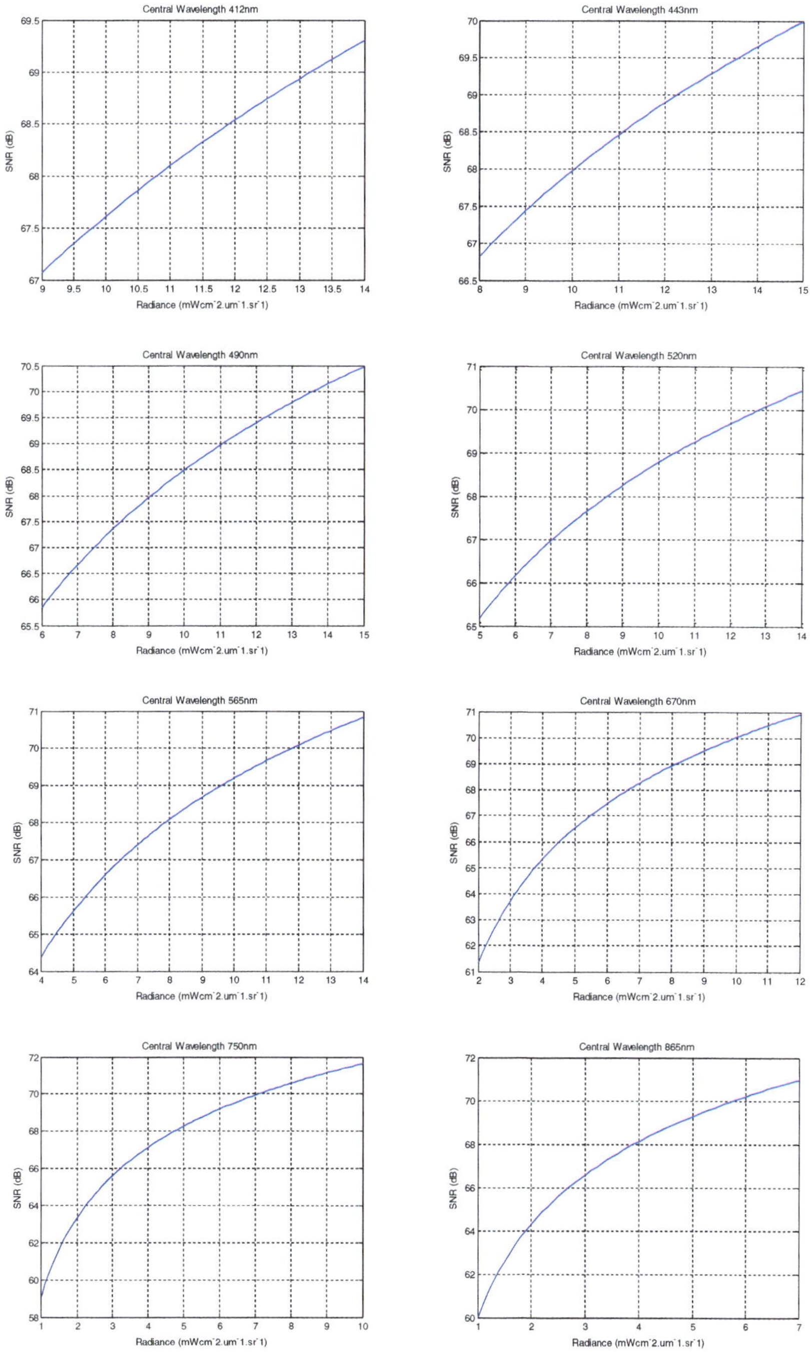


**Fig. 4** The SNR results

## 3.2 High Precision On-board Calibration

The method of solar calibration is adopted to achieve high accuracy radiation calibration and wavelength calibration. The light source consists of solar, screen, and diffuser. The diffuser includes radiation diffuser and wavelength diffuser to achieve different functions. The screen and diffuser are shown in Fig. 5.

### 3.2.1 Calibration Timing

In order to avoid the stray light from the earth, the calibration timing is set at the moment when the satellite can detect the solar energy and beneath the satellite the areas are still in the darkness. The simulated result in STK is shown in Fig. 6.

The design of the sun baffle is needed to concern, because the zenith and azimuth of incoming light are various in every on-board calibration test. The variation of the altitude and azimuth is shown in Fig. 7.

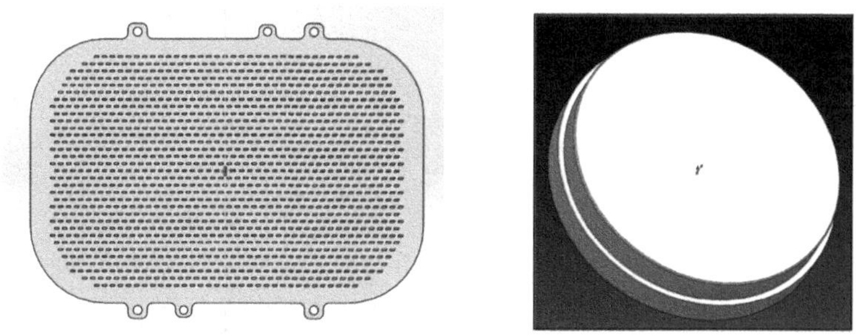

**Fig. 5** The screen and diffuser

**Fig. 6** The simulated result in STK

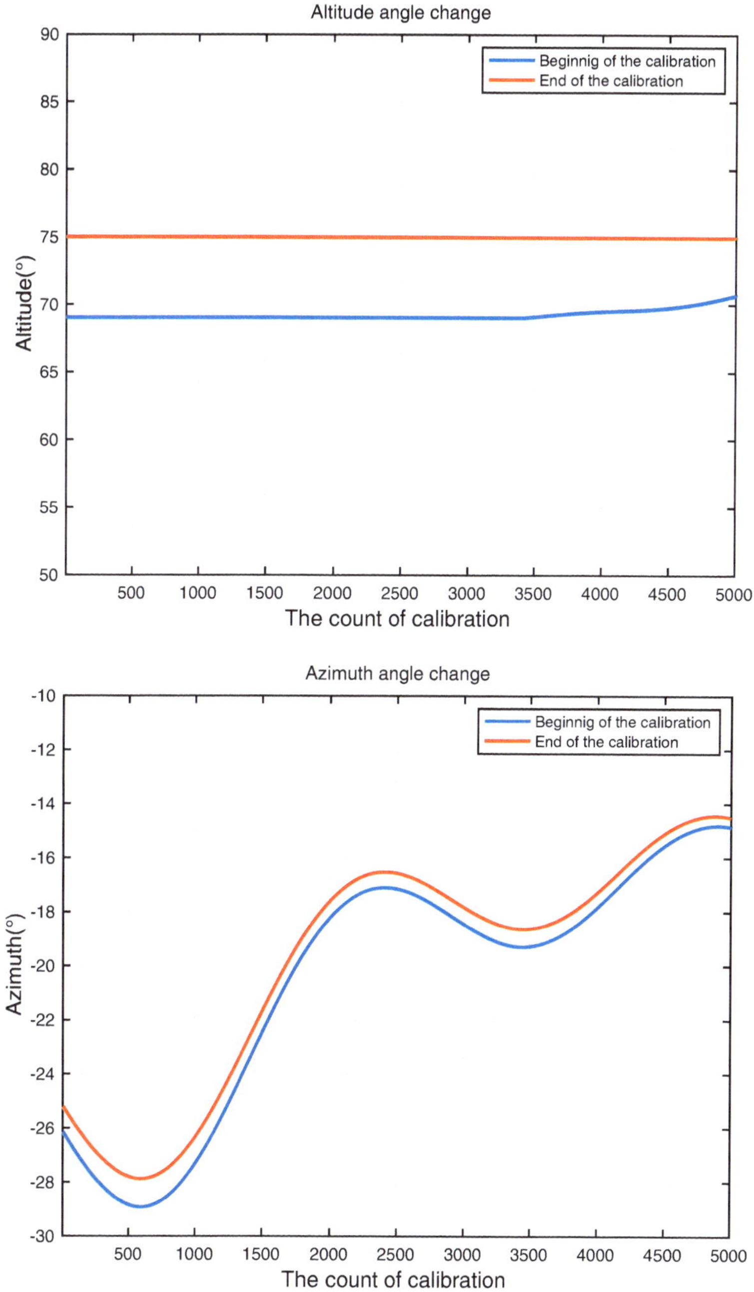


**Fig. 7** The variation of the altitude and azimuth

### 3.2.2 In-Orbit Radiation Calibration Process

In the lab, the model of the radiometric diffuser plane should be built, it includes the BRDF of the diffusers, the solar input angle, and the solar irradiance. Meanwhile, the spectrometer requires accurate absolute and inter-band radiometric calibration to get the calibration coefficient. The formulation is shown below:

$$L_{\text{CAL}} = \frac{E_s(\lambda) \times \cos\theta}{R(t)^2} \cdot f_{\text{cal}} \cdot \tau \tag{2}$$

In-orbit, according to solar input angle, satellite attitude, and solar irradiance data, the irradiance of the sun on the calibration plate is obtained. The formulation is shown below:

$$F_{i,j} = \frac{L_{\text{cal}}}{L_{\text{lab}}} \tag{3}$$

Compared with the irradiance which is computed by using the calibration equation and the theoretical model result, the irradiance correction factor can be obtained, it is shown below:

$$L_{\text{TG}} = F_{i,j} \times L_{\text{lab}} \tag{4}$$

Finally, the calibration equation of the spectrometer is shown below:

$$\text{DN}_i = F \times K_i \times L_i + B_i \tag{5}$$

Radiometric calibration accuracy of the spectrometer is less than 2%.

### 3.2.3 In-Orbit Wavelength Calibration Process

The spaceborne calibration spectrometer achieves the high precision wavelength calibration based on the wavelength diffuser which contains rare earth. There are several characteristic absorption peaks in the whole spectrum [5].

In the lab, the spectral reflectivity of the radiometric diffuser and wavelength should be measured. At the same time, the spectrum response function should be accurately detected in a small step (<1 nm), the result is shown in Fig. 8.

Construct the theoretical ratio of the radiometric diffuser and wavelength diffuser, the ratio is shown below:

$$V(i) = \int_{\infty}^{-\infty} \frac{\rho(\lambda)}{\gamma(\lambda)} \cdot R\left(\lambda - \lambda_c^i\right) d\lambda \tag{6}$$

According to the data of spectrometer, the actual ratio can be obtained. According to construct a pricing function, we could get the central wavelength and wavelength

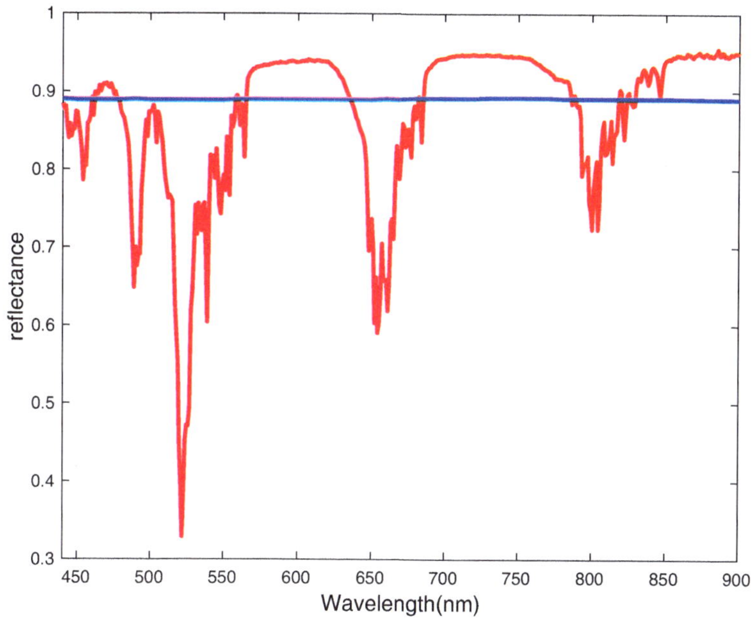


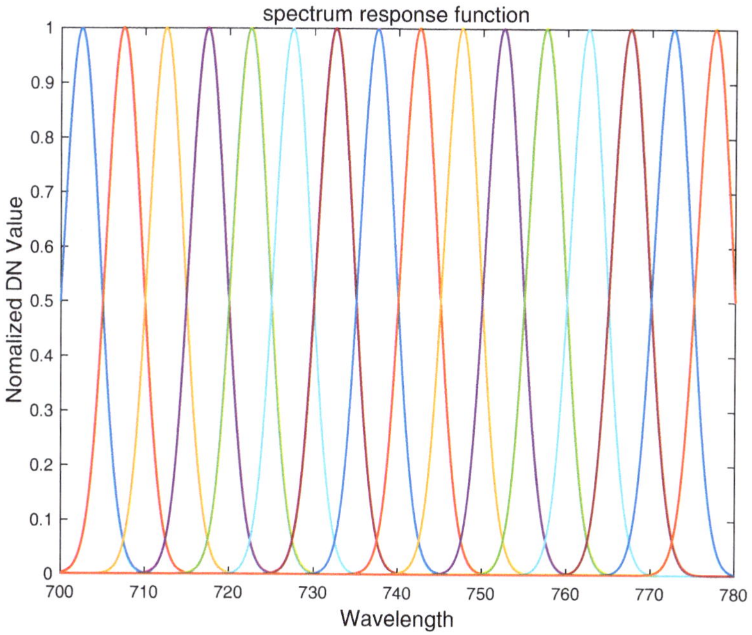


**Fig. 8** Spectral reflectivity and spectrum response function

drift when the value of the pricing function is minimal, the central wavelength and wavelength drift can be confirm. The formulation of the function is shown below:

$$\Delta_i = \frac{\sum_{j=-N}^{j=N}\left(M(i+j)-\bar{M}\right)\times\left(V(i+j)-\bar{V}\right)}{\sqrt{\sum_{j=-N}^{j=N}\left(M(i+j)-\bar{M}\right)\times\left(M(i+j)-\bar{M}\right)}\sqrt{\sum_{j=-N}^{j=N}\sum_{j=-N}^{j=N}\left(V(i+j)-\bar{V}\right)\times\left(V(i+j)-\bar{V}\right)}} \quad (7)$$

According to the dispersion model, the relationship between the number of the pixel and central wavelength is shown below:

$$\lambda_c^i = a \cdot i + b \quad (8)$$

So the central wavelength drift is shown below:

$$\lambda_{c\_\mathrm{cor}}^i = \lambda_c^i + \Delta\lambda \quad (9)$$

According to the on-board test, the wavelength calibration accuracy is 0.5 nm.

### 3.3 Cross-Calibration

The influence factors of cross-calibration accuracy include the calibration uncertainty of the spectrometer, spectral matching factor uncertainty, the calibration uncertainty of the target load, and another unknown uncertainty. The spectral matching factor uncertainty consists of earth object matching uncertainty, the time matching uncertainty, the geometric matching uncertainty, and spectral response matching uncertainty. In order to improve the accuracy of the cross-calibration, the method based on the spectrum recovery is adapted (Fig. 9).

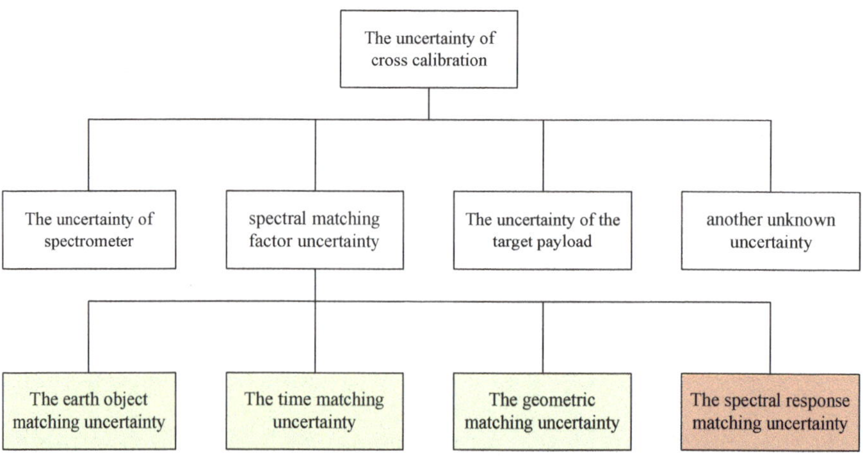


**Fig. 9** The influence factors of cross-calibration

The same target is imaged simultaneously with the spectrometer and the target load, which have completed high precision radiation calibration and wavelength calibration. The discrete spectral radiance can be obtained by the calibration equation of the spectrometer after on-orbit correction. Then the continuous spectral radiance is obtained by interpolation and iteration. Through the continuous spectral radiance, the target load is cross-calibrated. The key point is how to get continuous spectral radiance. According to the absolute calibration equation, we can get the discrete spectral radiance, it is shown below:

$$DN_i = K_i \times L_i + B_i \tag{10}$$

Through data interpolation, we can get the initial continuous spectral radiance.

$$L^0 = L = \{L_i \mid (i = 1, 2, 3, \ldots, n)\} \tag{11}$$

$$L^0(\lambda) = \text{spline\_interp}(L^0) \tag{12}$$

The error is reduced by iteration, and the continuous spectrum is obtained when the error judgment condition is satisfied. The iteration formula is as follows:

$$L_i^k = \Sigma L^k(\lambda) S(\lambda) / \Sigma S(\lambda) \tag{13}$$

$$L^k = \{L_i^k \mid (i = 1, 2, 3, \ldots, n)\} \tag{14}$$

$$L^{k+1} = L_I^k + a(L - L^k) \tag{15}$$

$$L^{k+1} = \text{spline\_interp}(L^k) \tag{16}$$

Error judgment condition is shown below:

$$L^k - L_2 \ll 10^{-6} \tag{17}$$

After a comprehensive analysis of the indicators, the accuracy of cross-calibration is less than 5%.

## 4 Application

The spectrometer has large dynamic detection capability, high signal-to-noise ratio, and high stability. It can be used as a special calibration instrument for large diameter and high-resolution payload. These instruments are usually limited by volume and weight, so it is difficult to design calibration lightpath and calibration device [6].

On the other hand, the load is capable of graph and spectrum detection with high signal-to-noise ratio, and its spectrum can be extended to 2.5 um. The data can be used for atmospheric correction of the calibrated payloads.

## 5 Conclusion

The on-board calibration spectrometer is the first imaging spectrometer dedicated to calibration on the same platform. The spectrometer owns the large dynamic range, high SNR, and high accurate radiometric calibration and wavelength calibration. It provides a solution for quantitative increase of large camera on the same platform.

## References

1. Yang, B.: Constructing China's ocean satellite system to enhance the capability of ocean environment and disaster monitoring. Chin. Space Sci. Technol. **5**, 2 (2011)
2. Yang, Z.-p., Tang, Y.-g., Bayanheshig, Ji-cheng, C., Yang, J.: Research on small-type and high-spectral-resolution grating monochromator. Spectrosc. Spectr. Anal. **36**, 273 (2016)
3. Yan, L.-w.: Study and design on Dyson imaging spectrometer in spectral broadband with high resolution. Spectrosc. Spectr. Anal. **34**, 1135 (2014)
4. Sun, J.-m., Guo, J., Shao, M.-d., Jin-song, Y., Zhu, L., Gong, D.-p., Qi, H.-y.: Precise focusing for TDICCD camera with wide field of view. Optics Precis. Eng. **22**, 3 (2014)
5. Chen, H.-y., Zhang, L.-m.: Spectral calibration for dispersive hyperspectral sensor based on doped reflectance standard panel. Optics Precis. Eng. **12**, 2643 (2010)
6. Du, X.-w., Shen, Y.-c., Li, C.-y., An, N., Shi, Y.-j., Wang, Q.-p.: EUV flat field grating spectrometer and performance measurement. Spectrosc. Spectr. Anal. **8**, 2272–2273 (2012)

# Visible and Near-Infrared Spectral Calibration of Greenhouse Gas Monitor

**Du Guojun, Zhang Yugui, Lu Zhijun, Ou Zongyao, Li Ming, and Dong Xin**

**Abstract** The detection target of greenhouse gas monitor is the concentration distribution of global atmospheric greenhouse gases (CO2, CH4, CO, etc.). In order to achieve the spectral calibration data of the monitor, a spectral calibration device was established, including tunable laser, wavelength meter, rotating engineering scatter, and other equipment, and the spectral response of the monitor was tested by spectral calibration device. In order to obtain the ILS function of a single pixel, a tunable laser is used to scan the monochrome light. The relative spectral response curve is fitted by the Gauss function. Nearly 30 spectral response curves are merged to obtain the accurate distribution of ILS. The spectral resolution and the measurement accuracy of ILS are calculated. The spectral calibration equation is obtained by choosing 10 wavelength points in the whole spectral range and fitting with cubic polynomial. Spectral calibration results show that the spectral resolution of O2A is 0.036–0.039 nm, the measurement accuracy of ILS is better than 1%, the spectral calibration accuracy is better than 0.9 pm, and the spectral resolution of weak CO2 Absorption band is 0.061–0.068 nm, and the measurement accuracy of ILS is better than 0.8%. The accuracy of spectral calibration is better than 2.5 pm. It meets the spectral calibration requirements of greenhouse gas monitor.

**Keywords** Spectrometer · Grating · Greenhouse gas · Spectral resolution · ILS Spectral calibration

## 1 Introduction

The detection target of greenhouse gas monitor is the concentration distribution of global atmospheric greenhouse gases (CO2, CH4, CO, etc.) [1]. Its detection accuracy can describe the spatial and temporal changes of greenhouse gases at regional and global scales, through high precision quantitative inversion. The hyperspectral

D. Guojun (✉) · Z. Yugui · L. Zhijun · O. Zongyao · L. Ming · D. Xin
Beijing Key Laboratory of Advanced Optical Remote Sensing Technology, Beijing Institute of Space Mechanics and Electricity, Beijing, People's Republic of China



H. P. Urbach, Q. Yu (eds.), *6th International Symposium of Space Optical Instruments and Applications*, Space Technology Proceedings 7,
https://doi.org/10.1007/978-3-030-56488-9_29

greenhouse gas monitor mainly obtains the following secondary products: average column CO2 dry air mixing ratio XCO2A at regional scale (>1000 km) and column average CH4 dry air mixing ratio XCH4. These data will be further assimilated into numerical models to improve the quantitative estimation of surface greenhouse gas fluxes at regional scale and to analyze and monitor global carbon sources and sinks [2]. The spectral resolution of the greenhouse gas monitor is 0.04 nm in the O2 band and 0.07 nm in the weak CO2 Absorption band. The high spectral resolution brings big challenges to the spectral calibration. This paper mainly describes the spectral calibration method, calibration device, data processing, and calibration results of the monitor.

## 2 Overview

### *2.1 Spectrum Performance Requirements*

Four spectrographic systems with similar structure were used to detect hyperspectral bands of 0.76, 1.61, 2.06, and 2.36 μm, respectively. Table 1 lists the specification of the monitor's O2A band and weak CO2 Absorption band. Extremely high spectral resolution improves the sensitivity of inversion. The spectral sampling rate of 3 guarantees the requirement of signal-to-noise ratio under typical observation conditions and reduces the influence of spectral resolution on inversion accuracy. The spatial dimension system design uses multipixel combination to achieve 3 km × 3 km ground resolution.

### *2.2 Principle and System Composition of Monitor*

The monitor adopts a four-channel grating spectrometer scheme. Each spectral band corresponds to a spectrometer separately. The telescope system is designed with two optical paths. Among them, the O2A band corresponds to one telescope and the weak CO2 Absorption band corresponds to another telescope. Figure 1

**Table 1** Design results of the monitor system

| Specification | O2A band | Weak CO2 Absorption band |
|---|---|---|
| Spectral range | 754–770 nm | 1591–1621 nm |
| Focal length (space) | 88.26 mm | 251.05 mm |
| Focal length (spectral) | 146.52 mm | 109.05 mm |
| Field of view (space) | ±3.604° | ±3.604° |
| Field of view (spectral) | ±0.014° | ±0.019° |
| Slit dimension | 64.9 mm × 0.1204 mm | 169.55 mm × 0.36 mm |
| Spectral resolution | 0.04 nm | 0.07 nm |

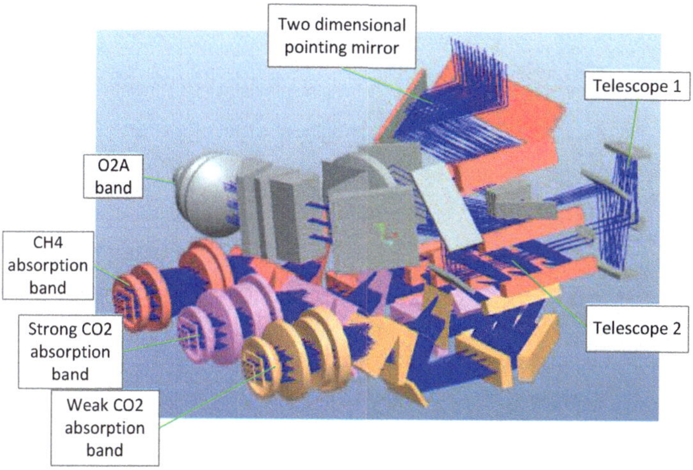


**Fig. 1** Principle diagram of optical system

shows the optical system layout of the monitor. The light carrying the information of the detected target enters the telescope system and illuminates on the slit. The light emitted from the slit passes through the collimator and then incident on the diffraction grating. Dispersed and, imagines on the detector by the imaging system. The spectral characteristics are obtained by data processing. The slit size of the detector's O2A band is 66.4 × 0.1204 mm, and the slit size of the weak CO2 Absorption band is 171.55 × 0.36 mm. Planar grating is used in the O2A band, and immersion grating is used in the weak CO2 Absorption band. The system corrects the spectral line bending caused by grating dispersion by the combination of grating and prism.

The spectrometer part is assembled on the main frame with similar optical–mechanical structure to ensure the overall stiffness and thermal stability of the spectrometer system. Figure 2 shows the structure diagram of greenhouse gas monitor. The main body of the optical machine is composed of a base plate component, a ground view shield, an on-board calibration device, an along-orbit pointing mechanism, a through-orbit pointing mechanism, a telescope component, a collimation component, an imaging component, a detector component, a detector Dewar refrigeration unit, a data processing unit, a light source control unit, a thermal control component, and a direct subordinate component.

## 3 Calibration Method

Tunable laser can provide a continuous monochrome light source with a linewidth less than 1 pm, which is the optimal scheme to achieve ultrahigh spectral resolution spectral calibration. The system adopts a similar calibration scheme as OCO, uses

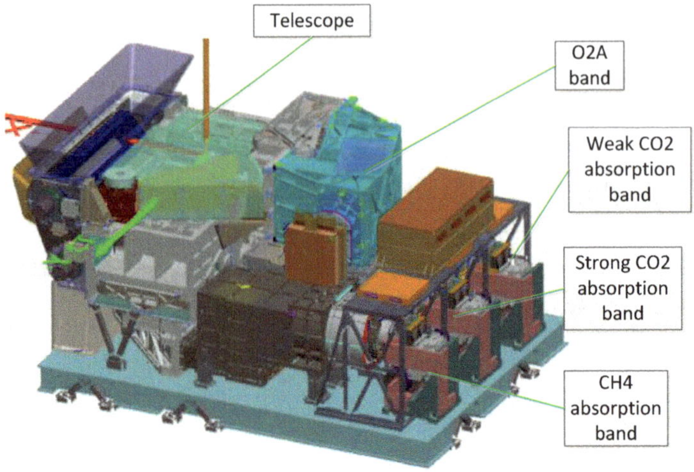


**Fig. 2** Structure diagram of greenhouse gas monitor

monochrome light covering the whole spectrum to scan ILS data in the O2A band and weak CO2 Absorption band, and places a rotating engineering scatter in front of the integrating sphere. The influence of laser speckle is eliminated by a rotating engineering scatter, and high precision spectral calibration is achieved by a data processing system.

## 3.1 Calibration System Design

The spectral calibration system of the monitor is shown in Fig. 3. The system uses tunable laser as the calibration light source to meet the response requirements of the detector while ensuring the output of narrow linewidth [3]. The tunable laser is divided into two beams by a beam splitter. One beam enters the integrating sphere as a calibration light source. The other beam enters the wavelength meter to detect the change of wavelength. The ratio of beam splitter is 9:1. The output wavelength of the laser can be real-time controlled by a data control system. Integrator sphere is homogenized and collimated by collimator, filling the whole aperture of monitor. Rotating engineering scatter is added between integrator sphere and parallel light tube to eliminate the influence of laser speckle.

Tunable lasers with different spectral bands are switched through a mirror. When using tunable lasers with O2A band, the mirror is translated to the corresponding tunable lasers with O2A band by translating the guide rail. The integrating sphere is

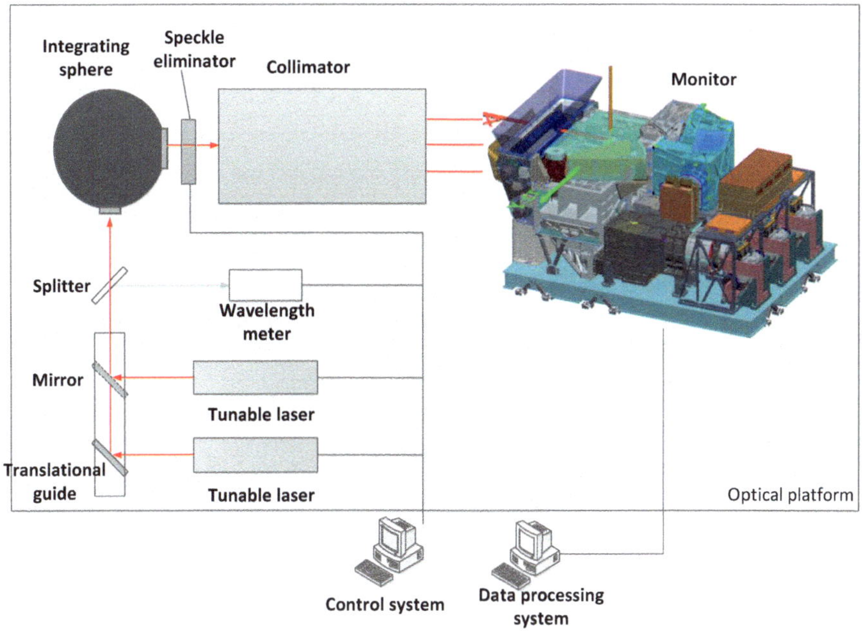


**Fig. 3** Spectral calibration system

filled with monochrome light of O2A band, and the integrating sphere is filled with tunable lasers corresponding to weak CO2 Absorption band when translating the mirror to the corresponding tunable lasers with weak CO2 Absorption band.

After illuminated by uniform monochrome light source, the slit function of monochrome light is formed on the focal plane of the detector, and the single frame ILS function is obtained. By controlling the output of the tunable laser at a certain wavelength interval, a series of ILS functions are formed on the focal plane. After dark pixel correction, energy normalization, and centroid translation, 30-frame ILS function in a certain spectral range is merged to obtain an accurate ILS function. The sampling interval of wavelength scanning is set to 0.1 FWHM. The average of 10 frames is used to improve the data stability and obtain accurate spectral calibration data.

## *3.2 Key Calibration Equipment*

Two tunable lasers are come from Sacher Lasertechnik, Germany, are used for the corresponding O2A band and weak CO2 Absorption band. The specific parameters are shown in Table 2. The wavelength stability of the tunable laser is less than 0.2 pm within 10 s.

**Table 2** Tunable laser parameters

| Type | TEC-520-765-030 | TEC-520-1590-030 |
|---|---|---|
| Band | O2A band | Weak CO2 Absorption band |
| Wavelength range (nm) | 754–780 nm | 1520–1640 nm |
| Minimum power (mW) | 30 | 30 |
| Linewidth (kHz) | 100 | 100 |
| Stability (HZ/°C) | 100 M | 100 M |

**Table 3** Wavelength meter parameters

| Type | Range of wavelength (nm) | Accuracy of wavelength ($10^{-6}$) | Sampling rate (Hz) | Input power (uW) |
|---|---|---|---|---|
| 621B | 600–1800 | ±0.2 | 1 | ≥20 |

The Bristol wavelength meter is used. The error of the wavelength is less than 0.3 pm. The specification is shown in Table 3.

## 3.3 *Data Processing Method*

The data processing flow of ILS is shown in Fig. 4. After scanning the ILS data of a single frame, the background noise is reduced by dark background correction, and the influence of random noise is reduced by taking an average of 10 frames. The simulation results show that the ILS of the system changes slowly in the whole spectral range, so the ILS profile changes a little in a small spectral range, which can be approximately considered to be consistent. Because the designed spectral sampling rate is 3 and the effective data points of single frame spectral response are limited, the detailed profile of ILS cannot be obtained. In order to improve the sampling rate of ILS [4], the accurate ILS is obtained by combining multiframe spectral response. The central wavelength of each ILS function is obtained by Gauss fitting, and the central wavelength of each ILS is set to 0, 30-frame ILS is merged into one ILS. An ILS with 900 data points is obtained, and the cut-off range of single channel ILS is ±5 FWHM [5].

The accuracy of ILS measurement is determined by analyzing 1% wavelength points A and A′ of peak response on both sides of the central wavelength. The percentage of energy in A–A′ region is $x$, $(1 - x)$ is calculated as ILS measurement accuracy, and ILS measurement accuracy must be less than 1%. The formula is as follows:

$$0.99\int_{-\infty}^{\infty}\varnothing(\lambda_0,\lambda)\,d\lambda \le \int_{A}^{A'}\varnothing(\lambda_0,\lambda)\,d\lambda \le 1. \quad (1)$$

Wavelength calibration is achieved by using the corresponding relationship of 10 wavelengths and pixels in the spectral range. The calibration equation is obtained by

**Fig. 4** ILS data processing flow

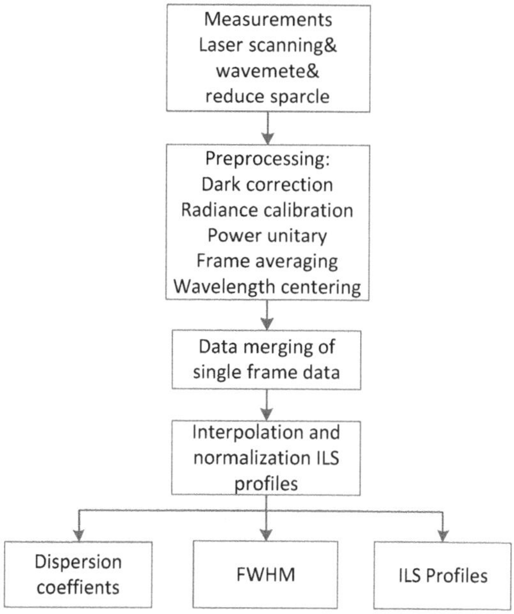


fitting the calibration curve with polynomial. The corresponding model of wavelength lambda and detector pixel number x is established. The formula is as follows:

$$\lambda = \sum_{i=0}^{n} a_i x^i. \tag{2}$$

# 4 Calibration Results

## *4.1 Wavelength Calibration Results*

The wavelength calibration is based on cubic polynomial. When speckle is not eliminated, the fitting residual is less than $3 \times 10^{-3}$ nm. After laser speckle is eliminated by rotating engineering scatter, the fitting residual is less than $0.9 \times 10^{-3}$ nm (Figs. 5 and 6).

Ten typical wavelengths are selected in the spectrum range of the O2A band and the weak CO2 Absorption band, 10 frames of data are collected at each wavelength for average, and the selected wavelength positions are evenly distributed. The tunable laser output a wavelength to measure an ILS, get the corresponding pixel position of the wavelength through Gauss fitting, use cubic polynomial fitting for 10 typical wavelengths, get the wavelength calibration equation of the wavelength and pixel position in the whole spectral band, and verify the fitting accuracy by selecting

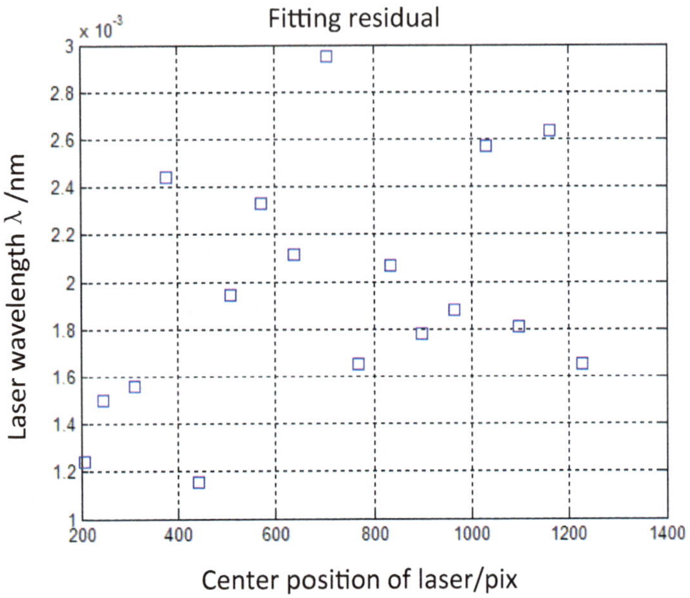


**Fig. 5** Residual of unremoved speckle

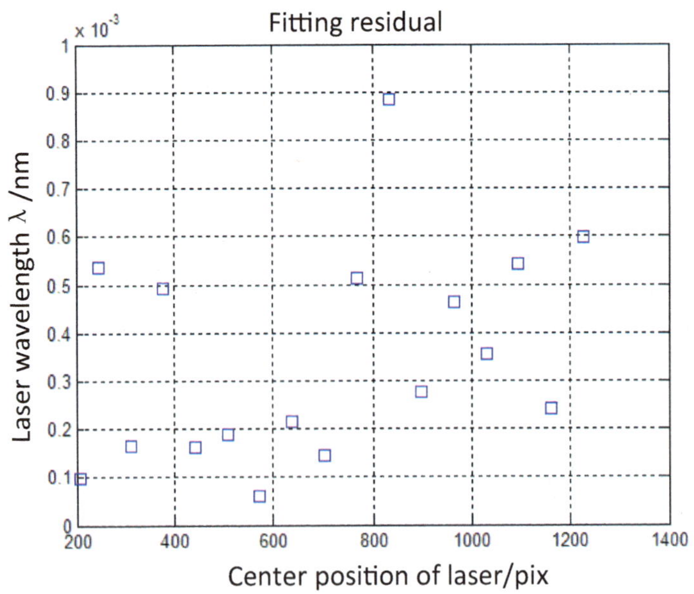


**Fig. 6** Residual of removed speckle

another seven characteristic wavelengths. Calibration results show that the O2A band is fitted by cubic polynomial and the fitting residual is less than 0.9 pm. The weak CO2 Absorption band is also fitted by cubic polynomial with a fitting accuracy of less than 2.5 pm (Figs. 7, 8, 9, 10, and 11; Tables 4, 5, 6, and 7).

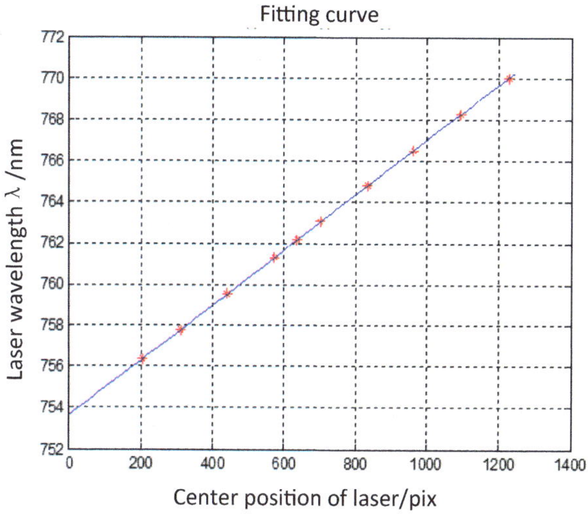


**Fig. 7** Spectral calibration fitting curve of O2A band

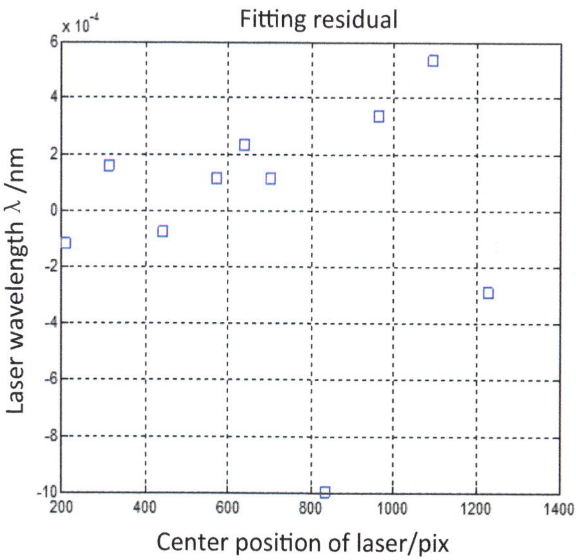


**Fig. 8** Residual of O2A band

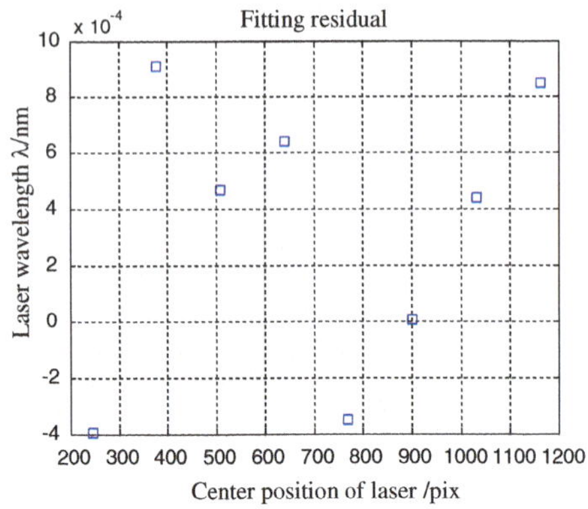


**Fig. 9** Validation of wavelength residuals in O2A band

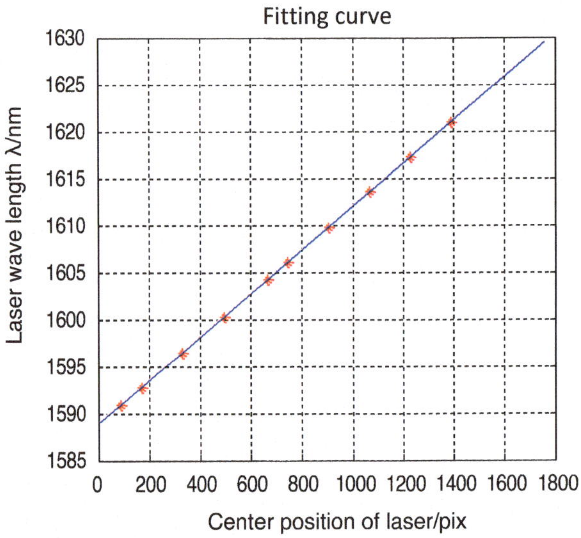


**Fig. 10** Spectral calibration fitting result of weak CO2 Absorption band

## *4.2 ILS Test*

In order to improve the measurement accuracy of ILS, multiframe spectral data are used to improve the sampling rate, monochromatic light scanning is carried out by tunable laser, and relative spectral response curve is fitted by Gauss function. Nearly 30 spectral response curves are merged to obtain the accurate distribution of ILS. Spectral resolution is obtained by calculating the full width of half peak value of ILS (FWHM). The wavelength of 756.3 nm, 759.5 nm, 763 nm, 766.5 nm, and 770 nm and 0 field of view, ±1 field of view are selected for the measurement ILS. The results show that the measurement accuracy of ILS is better than 1%, and the spectral resolution is better than 0.039 nm. Weak CO2 Absorption bands were selected at 1591.0 nm, 1598.5 nm, 1610.0 nm, 1613.5 nm, and 1621.0 nm and 0

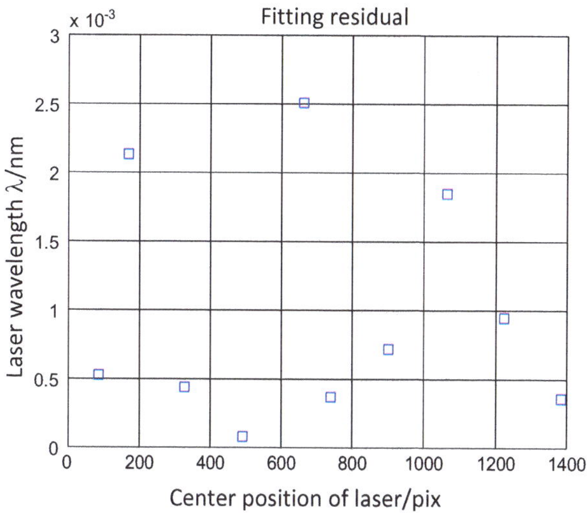


**Fig. 11** Residual of weak CO2 Absorption band

**Table 4** Original data for wavelength calibration of O2A band

| No. | 1 | 2 | 3 | 4 | 5 |
|---|---|---|---|---|---|
| Wavelength (nm) | 756.345 | 757.750 | 759.500 | 761.251 | 762.130 |
| Position (pixel) | 205.7100 | 311.0880 | 441.9029 | 572.4384 | 637.9143 |
| No. | 6 | 7 | 8 | 9 | 10 |
| Wavelength (nm) | 763.000 | 764.750 | 766.500 | 768.250 | 770.000 |
| Position (pixel) | 702.7397 | 833.3192 | 964.0595 | 1095.4570 | 1227.7420 |

**Table 5** O2A band validation wavelength data

| No. | 1 | 2 | 3 | 4 | 5 | 6 | 7 |
|---|---|---|---|---|---|---|---|
| Wavelength (nm) | 756.875 | 758.635 | 760.385 | 763.880 | 765.630 | 767.380 | 769.130 |
| Position (pixel) | 245.5389 | 377.2321 | 507.8717 | 768.3662 | 898.9809 | 1030.0470 | 1161.80500 |

**Table 6** Original data of weak CO2 Absorption band

| No. | 1 | 2 | 3 | 4 | 5 |
|---|---|---|---|---|---|
| Wavelength (nm) | 1591.010 | 1592.886 | 1596.523 | 1600.332 | 1604.261 |
| Residual (nm) | 0.0005 | −0.0021 | −0.0004 | 0.0001 | 0.0025 |
| No. | 6 | 7 | 8 | 9 | 10 |
| Wavelength (nm) | 1606.081 | 1609.787 | 1613.577 | 1617.287 | 1621.007 |
| Residual (nm) | −0.0004 | −0.0007 | −0.0018 | 0.0009 | 0.0004 |

field of view, ±1 field of view to test ILS. The results show that the measurement accuracy of ILS is better than 0.8% and the spectral resolution is better than 0.067 nm. It meets the spectral calibration requirements of greenhouse gas monitor (Figs. 12, 13, and 14; Tables 8, 9, 10, and 11).

**Table 7** Validation data of weak CO2 Absorption band

| No. | 1 | 2 | 3 | 4 | 5 | 6 | 7 |
|---|---|---|---|---|---|---|---|
| Wavelength (nm) | 1594.730 | 1598.438 | 1602.254 | 1607.925 | 1611.688 | 1615.461 | 1619.105 |
| Residual (nm) | −0.0023 | 0.0008 | 0.0017 | 0.0006 | −0.0022 | −0.001 | 0.0003 |

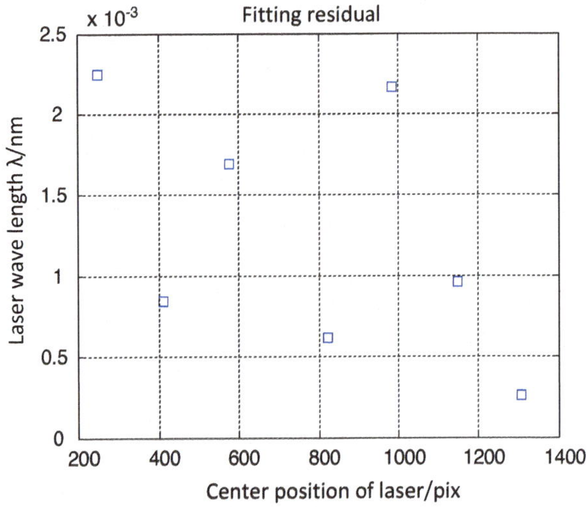


**Fig. 12** Validation of wavelength residuals in weak CO2 Absorption band

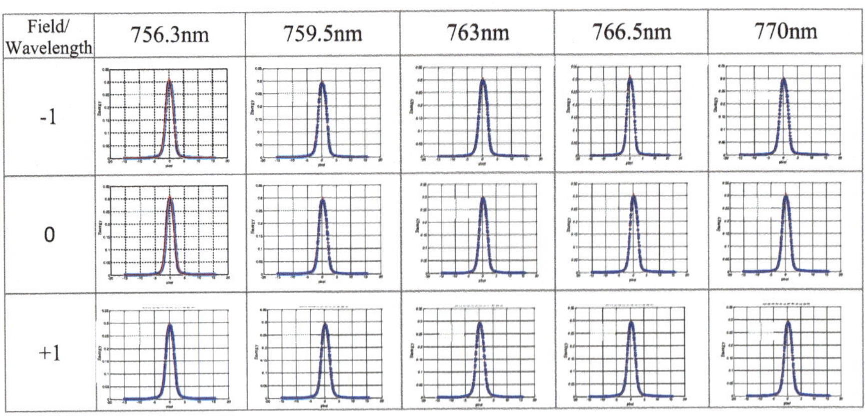


**Fig. 13** Measurement results of O2A band ILS

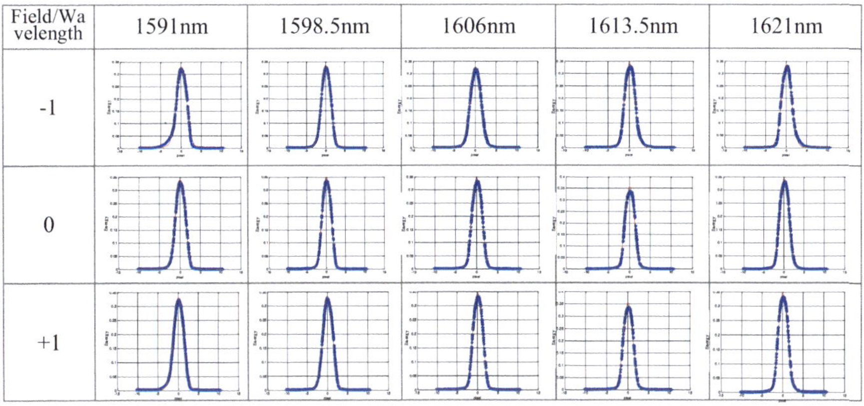


**Fig. 14** Measurement results of weak CO2 Absorption band ILS

**Table 8** Measurement accuracy of O2A band ILS

| Field/wavelength | 756.3 nm | 759.5 nm | 763.0 nm | 766.5 nm | 770.0 nm |
|---|---|---|---|---|---|
| −1 | 0.8% | 0.8% | 0.7% | 0.9% | 0.7% |
| 0 | 0.9% | 0.9% | 0.7% | 1.0% | 1.0% |
| 1 | 0.6% | 0.6% | 0.5% | 0.5% | 0.5% |

**Table 9** Measurement accuracy of weak CO2 Absorption band ILS

| Field/wavelength | 1591.0 nm | 1598.5 nm | 1606.0 nm | 1613.5 nm | 1621.0 nm |
|---|---|---|---|---|---|
| −1 | 0.7% | 0.7% | 0.6% | 0.7% | 0.6% |
| 0 | 0.8% | 0.8% | 0.5% | 0.8% | 0.8% |
| 1 | 0.7% | 0.7% | 0.6% | 0.6% | 0.7% |

**Table 10** Measurement result of O2A band spectral resolution

| Field/wavelength | 756.3 nm | 759.5 nm | 763.0 nm | 766.5 nm | 770.0 nm |
|---|---|---|---|---|---|
| −1 | 0.038 | 0.037 | 0.036 | 0.037 | 0.038 |
| 0 | 0.038 | 0.038 | 0.038 | 0.038 | 0.038 |
| 1 | 0.038 | 0.038 | 0.037 | 0.037 | 0.039 |

**Table 11** Measurement result of weak CO2 Absorption band spectral resolution

| Field/wavelength | 1591.0 nm | 1598.5 nm | 1606.0 nm | 1613.5 nm | 1621.0 nm |
|---|---|---|---|---|---|
| −1 | 0.067 | 0.065 | 0.068 | 0.064 | 0.065 |
| 0 | 0.064 | 0.063 | 0.063 | 0.061 | 0.063 |
| 1 | 0.066 | 0.066 | 0.063 | 0.062 | 0.063 |

## 5 Conclusion

The spectral calibration of greenhouse gas monitor is accomplished by using spectral calibration device. According to the requirement of monitor spectrum performance, the calibration method of tunable laser and wavelength meter is adopted. The composition and calibration principle of the calibration system are introduced, and the spectral calibration data of the monitor are given. The calibration results show that the spectral resolution of the O2A band is 0.036–0.039 nm, the ILS measurement accuracy is better than 1%, the spectral calibration accuracy is better than 0.9 pm, and the spectral resolution of the weak CO2 Absorption band is 0.061–0.068 nm, and the ILS measurement accuracy is better than 0.8%. The accuracy of spectral calibration is better than 2.5 pm. At present, the vacuum low-temperature adjustment of the strong CO2 Absorption band and CH4 absorption band of the greenhouse gas monitor is in progress. In order to obtain all the spectral calibration data of the whole greenhouse gas monitor, the next spectral calibration of the greenhouse gas monitor is carried out at the vacuum and low-temperature environment.

## References

1. Glumb, R., Davis, G., Lietzeke, C.: The TANSO-FTS-2 instrument for the GOSAT-2 greenhouse gas monitoring mission. Geoscience and Remote Sensing Symposium (IGARSS), Quebec City, Canada, pp. 13–18 (2014)
2. Day, J.O., O'Dell, C.W.: Preflight spectral calibration of the orbiting carbon observatory. IEEE Trans. Geosci. Remote Sens. **49**(6), 2438–2447 (2011)
3. Lin, C., Li, C.-l., Wang, L.: Preflight spectral calibration of hyperspectral carbon dioxide spectrometer of TanSat. Opt. Precis. Eng. Optics Precis. Eng. **25**, 2064–2075 (2017)
4. Du, G., Liao, Z., Jiao, W.: Spectral calibration of programmable imaging spectrometer. SPIE. **9676**, 96760C1–96760C7 (2015)
5. Kang, L., Pengmei, X., Weigang, W.: Simulation analysis of hyperspectral spectrometer grating profile on spectral performance. Spacecr. Recov. Remote Sens. **39**(2), 55–62 (2018)

# Index




H. P. Urbach, Q. Yu (eds.), *6th International Symposium of Space Optical Instruments and Applications*, Space Technology Proceedings 7,
https://doi.org/10.1007/978-3-030-56488-9

## O

Printed by Printforce, the Netherlands